Production

A Guide for
Authors and Publishers

Pete Masterson

Æonix Publishing Group
El Sobrante, California

ISBN-10: 0-9669819-0-1
ISBN-13: 978-0-9669819-0-2
LCCN: 2005901925

Published by
Æeonix Publishing Group
P.O. Box 20985
El Sobrante, CA 94820-0985

Visit *www.aeonix.com* for additional information, a list of printers specializing in books, and links to web sites of interest to publishers and authors.

Second printing, December 2007

Printed in the United States of America

This book is dedicated to all the self-publishers whose books instilled in me a desire to help them do a better job. And, to my dear wife, whose constant pressure and endless reminders kept me from losing faith in this project.

Contents

Foreword

This is an exciting time to be in the book business. Book writing, production, printing, and promoting are changing—for the better. Computers, desktop publishing software, digital printing equipment, and the Internet have reached a level of sophistication that allows an author (with a little training) to write, edit, and produce books that are fully competitive with the offerings of the major publishers.

In publishing books, such as *The Self Publishing Manual,* I have been an early adopter, preparing camera ready copy with a laser printer as early as 1981. Over the years, I have seen the professional tools of printing and publishing escape from the private sanctum of professional craftsmen and become readily available to anyone with the desire to purchase and use them. The current generation of page layout software is now able to achieve the quality once reserved to expensive, dedicated typesetting equipment. Digital printing technology is able to quickly produce books with such good quality that it takes an expert to tell them from those that were printed with the traditional offset press.

This book, *Book Design and Production: A Guide for Authors and Publishers,* covers the critical production aspects that are too easily overlooked when preparing to self-publish. While the tools are available to produce a professional-quality book, all too often, self-publishers will skimp on the production of their book with the mistaken assumption that "readers want to read what I have to say and aren't interested in what it looks like." Sadly, this attitude can result in a book failing to reach its audience.

Despite the old saying "don't judge a book by its cover," people who buy books do, indeed, judge them by their looks. All too often, the professional buyers with distributors and booksellers will reject a book based on poor production. Assuming a book does reach store shelves, a retail buyer, put off by the appearance, will never buy and read the book. Or, if they do buy the book, the message may fail to be communicated due to poor book design that limits comprehension or actually hurts the credibility of the author. Good book design supports the author's message by ensuring higher levels of comprehension and helps to communicate it effectively to the reader.

Humans have been communicating through graphic images and written words for thousands of years. As a result, there is a rich history in written communication, and experienced and talented craftspeople have learned the most effective ways to create readable books that have a high degree of comprehension. And that have, as well, a sensitivity to the style and message of the words within the book. These skills can be learned—and they are ignored at the peril of the author's message.

In this book, Pete Masterson, who has worked in the world of graphic arts and book production for nearly twenty years, has shared his research and knowledge of effective communication through typography and book design. The whole process of physically producing a book is covered from editing, through design, typesetting, and selecting a printer.

Whether you do it yourself or decide to hire a book designer, your understanding of the physical process will allow you to make more informed decisions, control your costs, and ultimately end up with a book you can be proud to sell.

Dan Poynter
Santa Barbara
February 2005

Preface

O NE DAY, SEVERAL YEARS AGO, while looking at books in a small bookstore in Stinson Beach (Marin County, California), I came across a little book with historical discussions, route descriptions and hiking maps of the local coastal area. As I looked through the book, I was horrified by the poor production values, the many basic design errors, and the truly ugly typesetting. It was obviously a self-published product. It did appear that it might have valuable and useful information (and helpful maps). It was probably in the store because the manager felt its value exceeded the many visible problems in the general design and layout of the book. (Or the author was a friend of the bookstore owner.)

I put it back on the shelf—I simply couldn't bring myself to buy it—as I felt the many obvious flaws overwhelmed the text. I simply didn't have confidence that the content would be well researched and accurate if the author-publisher couldn't bother to make the book look even close to being professional. While this book may have delivered far more than its humble appearance suggested, I didn't want to take the chance to trust that this book would serve me well as I navigated the various coastal trails. It also occurred to me that this author-publisher might have produced a much more professional and credible publication for not much more cost (but perhaps a bit more effort) if he had some means to understand how a book *ought* to look.

Then, I started looking around for a book on book design that was focused toward authors and publishers. While there are many books on typography (which includes many aspects of book design) and there are many books on graphic design (of a more general nature), these books all share a viewpoint in that they were written for experienced (or student) designers. To make matters more difficult, the portions specifically on book design were usually only a small part of the overall tome. You had to understand enough about design and typography to pick out the parts that applied to a book project and ignore those that applied to other purposes.

After reaching the realization that there were no current books on book design for authors and publishers (although I did find one book for self-publishers that is now both out of date and out of print), I decided that this was a topic that I might be able to beneficially address.

This book has been in development for over five years. It is not the book that I thought it would be when I began gathering information to put it together. Along the way, I have received input from many beginning self-publishers and, through my involvement with the (San Francisco) Bay Area Independent Publishers Association, I have learned the questions that are most frequently of concern to those just starting on the path of self publishing. I appreciate the questions, discussions, and ideas generated more than any who I have encountered may ever realize.

I only hope that the readers can enjoy using this book as much as I did in making it a reality.

Pete Masterson
El Sobrante
February, 2005

Introduction

IT ALL STARTS WITH A DREAM: I will write a great book. It will sell millions of copies. I will become rich and famous and then retire to Tahiti (or whatever your definition of paradise might be).

This dream is rather far fetched and is highly unlikely. The reality is more like a nightmare of having a garage full of unsold books. (Leave 'em to the kids—if they want an inheritance, *they* can sell the books!) However, if an author-publisher does a good job at publishing, a successful project may result in the dream becoming fulfilled at a more modest level.

Why do authors decide to self publish?

1. The author has been unable to find a satisfactory publisher; or

2. The author wants to retain greater control over the way the finished book looks.

3. Profit may play a role, but more often it does not. (When profit motive is involved, the author may feel that his "share" (royalty) is not enough or that a publisher won't do the right marketing job).

4. Usually there is a strong desire to tell a particular story or to communicate some particular ideas, no matter what the opinion of a publisher might be with respect to finding a market for the book.

Some industry statistics: Books in Print reports that about 70,000 publishers delivered 140,000 new titles in 2002 (the most recent figure I had available as this was written). About 80% or some 56,000 of those publishers are independents who publish fewer than 10 titles.

Large and medium sized publishers do not have their egos wrapped around a particular project, but an author-publisher can't help closely identifying with their "child"— the manuscript they've written. This ego involvement and desire to control every detail of the project may lead an author-publisher into serious problems. But ultimately, it's a major strength as a motivated author-publisher, who truly believes in the project, is more likely to make it a success.

Who will buy your book?

Well, everyone would want my book! (You say.) *Wrong answer*—even if it is true. The major publishers and other large consumer products corporations spend billions to reach the mass market. An author-publisher can not possibly obtain enough free publicity (initially) or to pay for enough advertising to be noticed on a large scale. (And it is a rare situation indeed, where advertising for a book will prove cost effective.)

The author-publisher must focus on market segments or small parts of a market segment (a niche market) to build awareness of the title. A small market allows a publisher of modest means to be noticed and make an impact.

One of my clients wrote a family history about his parents who immigrated to the United States at the beginning of the 20th century. Rather than writing a very personal family history, the author generalized and fictionalized it (slightly) to make a better story. The book, *Carved In Stone: The Greek Heritage* by Basil Douros, tells the story of how his family name was established by his great grandfather. Then it describes the conditions in Greece at the end of the 19th century that led many to desire to emigrate. Finally, he describes how his parents came to the U.S. and established themselves here. This story has broad parallels to the stories of many other Greek immigrant families. The author was able to focus his marketing on Greek-American organizations, Greek Orthodox churches, and even Greek restaurants. In about 18 months, he sold some 2,000 copies.

Subsequently, Mr. Douros wrote a similar treatment describing his wife's family's immigration from Ireland; *Roots of the Blackthorn Tree: The Irish Heritage.* This has been selling well through various Irish-American connections and through local museums in New Hampshire, where the family settled along with many other Irish. And the author has had a great time and has enjoyed the attention that the projects have created for him. So much the better that both projects have been modestly profitable as well.

You've decided to publish your manuscript, now what?

This book is designed to help you take your raw manuscript and turn in into an electronic file ready for a printer (either digital or offset). Or, to help you understand the book production process so you can hire the appropriate service providers to get the job done.

This book is not a general "self publishing" book. There are already a number of books on that topic. I can recommend both Dan Poynter's *The Self Publishing Manual* and Tom and Marilyn Ross' *The Complete Guide to Self Publishing.* (See the bibliography for the complete references.) Either of these books provides an excellent background on the overall questions you'll have about self publishing. Both books, in my opinion, are a bit vague when they describe the process of turning a manuscript into a ready-to-print book—and that is why I wrote *this* book.

Basic definitions

BOOK EDITING: The preparation of a manuscript necessary to make it ready for publication. This is discussed at length in chapter 2.

BOOK DESIGN: The plan and specifications for the physical structure and visual look of the book.

BOOK PRODUCTION: The execution of the design (typesetting) and arranging for the manufacturing and physical distribution of the book.

BOOKMAKING: Book editing, design, and production are all parts of a process: that of allowing an author to communicate a message to a reader in the best possible way. This includes creation of a book (or other media format) that can be marketed profitably in addition to satisfying the needs of both the author and the readers. The term *bookmaking* is sometimes used to refer to this process. Unfortunately, that term is also used to refer to illegal gambling practices, so it will not be further used here.

A book can be one of several different products. In general, when we refer to a book we will be discussing a tradebook. That is, a hard or soft cover book sold to the general public through bookstores and other retail outlets. Our discussion can be applied to many other kinds of books (ebooks, textbooks, reference, mass market, etc.) as well, but those will not be our primary focus.

A WELL-DESIGNED BOOK: One that is appropriate to its content and intended use; is practical and economical; and is satisfying to the senses. Its visual design enables the reader to achieve understanding of the author's message with the least effort (considering the material presented) and the physical properties are such that the credibility of the author is maintained.

A note about computers and software

There are a number of sections where we endeavor to describe using certain software programs on a computer. First, we assume that you have a solid general understanding of your computer system. You should know how to start the computer, operate various programs, know how to save, copy, and back up files, make disks or CDs, and otherwise operate the computer with a reasonable level of competence.

We will assume that you have a moderate understanding of the programs we discuss. If, for example, you have recently bought InDesign, you should work through all tutorials and sample projects that came with the product to gain a basic understanding of the operation, capabilities, and features available. You may wish to obtain books or videos intended to teach you the basics of the program or, if you can, either take a basic desktop publishing course through a community college/adult education system or work with someone who's fairly experienced while you 'get the hang' of the program.

This book is not a basic tutorial. Book design using page layout software uses the features of that software at a fairly high level. While we will do our best to guide you through the process, we can not provide beginner level instructions.

For those who are more advanced, please bear with us as we do give fairly detailed instructions for locating menus, etc. We do not describe any keyboard short cuts, although there are many useful shortcuts that we regularly use.

Finally, instructions are based on programs running on a Macintosh using Mac

OS X 10.2.8 or on a Macintosh running Mac OS 9.2.x. We understand that the menus and operation of most programs discussed are quite similar under most versions of Microsoft Windows, however we do not have a Windows computer available to us to check for variations. In that regard, there may be some differences in the locations of some menus or commands.

Beyond the user manual that comes with the software, we can suggest that you select one or more of the following InDesign books: *Adobe InDesign CS Classroom in a Book,* by Adobe Creative Team, Adobe Press, ISBN 0321193776; *Adobe InDesign CS One-on-One* by Deke McClelland, Deke Press (in association with O'Reilly), ISBN 0596007361; or *InDesign CS for Macintosh and Windows: Visual Quick Start Guide* by Sandee Cohen, Peachpit Press, ISBN 0321213483. For more advanced users, *Real World Adobe InDesign CS* by Olav Martin Kvern and David Blatner, Peachpit Press, ISBN 032121921x.

–1–

Some History

A brief history of the book

CAVE WALL PAINTING DATES BACK more than 30,000 years and is the earliest form of "written" human communication. Over the next 25,000 years, as writing developed past cave wall paintings, the Babylonians first used clay tablets at about 5,000 B.C. In Egypt, papyrus plants were processed by a surprisingly complex process to create a relatively smooth and durable writing surface. It's uncertain when papyrus first started being used, but evidence suggests that it has existed for thousands of years. "Books" were devised by making scrolls from the papyrus.

Animal skins were the primary competitor to papyrus. However, papyrus was cheap to produce and in Egypt, the dry climate made it an ideal substrate for use by scribes. Outside of Egypt, animal skins were more commonly used. There is evidence that shows processed animal skins were used as early as 2750 B.C. The Assyrians and Babylonians wrote on clay tablets for thousands of years, but they also used processed animal skins from the sixth century B.C.

In approximately 280 B.C., the ancient Greek city of Pergamon, now called Bergama, located in western Turkey, south of Istanbul, about 16 miles from the Aegean Sea, began to develop a great library in competition with the famous Library of Alexandria. In retaliation for the aggressive nature of the library acquisition process (through the force of arms) engaged in by the military establishment of Pergamon, the Egyptian government instituted trade sanctions against Pergamon, cutting off their supply of papyrus.

In response, Pergamon, a center of wool production, animal products, and tanning, substituted vellum, a processed calf skin that had been in use for some time. They called the material *charta pergamene*—which means, "paper of Pergamon." *Charta pergamene* evolved into our word parchment.

For convenience, the parchment was cut into rectangles instead of being left in the irregular shape of an animal skin. (Had the uncut animal skins been used, we might have had a totally different concept when someone says a book has "legs.") Parchment, when stitched end to end and rolled does not make a very good scroll as it is rather stiff and becomes somewhat brittle as it ages. At some point, a scribe (whose identity is lost in history) had the genius to stitch together the edge of a stack of parchment rectangles, creating the first book.

This new format also brought about the first data processing revolution. Once a book was paginated, this allowed the use of page numbers, footnotes, running headers, table of contents, and an index. (These elements were quite impractical for a papyrus scroll.)

What happened to the great library of Pergamon? By 40 B.C., both Pergamon and Egypt were under Roman control. In that year, Roman soldiers in Egypt accidently burned part of Alexandria's library. Anthony, in his obsessive love for Cleopatra, made her a gift of the Pergamon library to make up for the fire damaged volumes. Today, we remember Alexandria—and Pergamon is a virtually unknown footnote of history.

THE SECRET OF PAPER: The Chinese first discovered the secret of making paper from hemp fibers, mulberry tree bark, old fishnets, rags, etc. in approximately 100 A.D. While the Chinese managed to keep the process a secret for almost 600 years, by the seventh century, the Japanese had acquired knowledge of the process. In time, the paper making secret followed the caravan routes of central Asia and traveled across the Arab world.

Paper making reached Europe with the Moorish invasion of Spain. Paper mills were built in Italy in 1276, France in 1348, Germany in 1390, and England in 1494.

A brief history of printing

In about 400 A.D. the first printing ink was developed using soot from lamps mixed with linseed oil. (It is interesting to note that modern black printing ink is mostly a mixture of carbon black (soot) and oil, either from petroleum or vegetable sources.) Printing was used to reproduce pictures, playing cards, designs on cloth, and similar items. Pictures or designs were cut into wood, stone, or metal blocks, covered with ink, and pressed onto parchment, vellum,

WHY DO WESTERN COUNTRIES SHARE THE SAME BASIC ALPHABET AND LETTER SHAPES?

We can thank Charlemagne. After unifying France and much of western Europe, in 789 A.D. he decreed that the books of the church should be "revised" (recopied) using the most beautiful writing that had developed over the previous centuries. The writing is called *Caroline* (or *Carolingian) miniscule* after the Latin name for Charlemagne, *Carolus Magnus.* This script was rapidly adopted through Western Europe. The style had a foothold in England, but the Norman conquest in 1066 ensured its complete acceptance there. *Caroline miniscule* is the basis of the modern Western lowercase alphabet.

or cloth. Sometimes a few words were cut into the printing block, but that was the limit of text printing. Books and other documents were reproduced by monks or professional copyists hand-inking the words, letter by letter.

In 1450, Johannes Gutenberg developed a printing press that utilized movable, metal type. Various types of presses were being used for block printing, pressing grapes (for wine) or olives and seeds (for oil) for some time. Trained as a goldsmith, Gutenberg's real innovation was the invention of a simple mold that could be used to cast type with identical letters in the volume necessary to support the movable-type method.

Compositor (typesetter) standing at typecase. The upper case held the capital letters and the lowercase held the small letters.

Despite the criticism that the new movable-type printing press lowered the quality of the previously hand-copied books, it was an instant success. Even with the primitive presses of that time, the movable type printing process was more than 400 times more productive than the hand-ink method. The relatively short production time and reduced cost simply overwhelmed the old production methods. Professional copyists, scribes, and monks who previously performed these tasks were downsized from their positions and forced to seek retraining programs to enable them to re-enter the job market.

In the years following Gutenberg's movable-type press (called a "letter press"), printing technology was improved through many small innovations, but the basic process remained unchanged. Type was cast (usually from an alloy made principally of lead) into small individual pieces and sorted into bins (called a type case) by letter. Those in the upper part of the case were the capital letters *(magiscule)* and the small letters *(miniscule)* were in the lower portion of the case. Ever since, these have been called uppercase and lowercase letters. The individual letters were selected one by one and placed into a small hand-held metal tray, called a composing stick. When a line was complete, it was transferred to a metal frame.

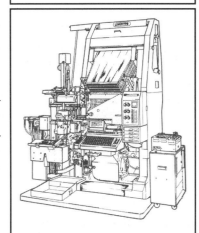

ETAOIN SHRDLU, WHERE ARE YOU NOW? The top row of the Linotype keyboard consisted of the foregoing letters, which are approximately in the order most frequently encountered in the English language. This positioning was created so that the letter molds most frequently called upon would have the shortest travel through the chutes of the machine.

When the frame was filled, it was tightened with wedged blocks, and placed in the press. This was slow and tedious work. Printers had to be good spellers and had to work with the alphabet in reverse, since all type was created in mirror image so it would be "right reading" when it was printed. After a sufficient number of copies of the page were printed, the frame would be removed from the press, disassembled, and the type would be dumped onto a counter in front of the type case, where an apprentice sorted the letters back into their proper bins. Printers' apprentices were often told to "mind your p's and q's," since those letters were simply the reverse of one another.

During the ensuing 400 years, printing became ever faster through improved presses, including steam and electric powered presses. Yet the basic process of setting type remained much the same until Otto Mergenthaler patented his Linotype machine in 1884. This Rube Goldberg-type contraption allowed the operator to set type by sitting at a keyboard and entering the text into the machine. The Linotype would gather matching letter-form molds as the text was entered and when a line was complete, would adjust the letter and wordspacing, then allow molten lead to enter the mold to be cast into a complete line of type. These line "slugs" were then collected into galleys to be placed later into frames for individual pages. These machines were large, heavy, noisy, and hot—with a caldron of molten lead waiting to be cast.

These new machines were greeted with the criticism that the quality was not up to the standards of "traditional" typesetting. Yet the vastly improved efficiency of these machines quickly displaced many movable-type setters, who were downsized from their positions and forced to seek retraining programs to enable them to re-enter the job market.

Linotype machines, along with similar competing machines by Monotype and Intertype, reached their peak during the 1960s. Then, more than 500 years after the basic printing process had first been developed, a truly revolutionary change swept the printing industry with the arrival of offset lithography.

Lithography, which literally means

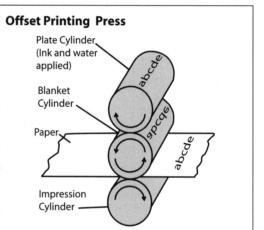

Offset Printing Press

Plate Cylinder (Ink and water applied)

Blanket Cylinder

Paper

Impression Cylinder

A "perfecting press" (one that prints on both sides of the paper at the same time) replaces the impression cylinder with another blanket cylinder and lower plate cylinder. Not shown are ink and fountain (water) rollers.

"stone printing," was first discovered more than a century ago when an artist was experimenting with transferring ink to paper from an image on a smooth rock. It was soon determined that certain properties of "lithographer's stone" would attract greasy ink to chemically treated portions while water easily washed away the ink from the untreated parts of the rock. In the early history of lithography, it was used primarily as a means of creating works of art.

Offset lithography was developed early in the twentieth century when a misfeed on a lithographic press resulted in the back side of a sheet being printed by the rub-

roller that normally pressed the paper to the stone. The innovative printer noticed that the accidental print was clearer than the print obtained directly from the lithographer's stone. He then developed a lithographic press that purposely offset the image to the rubber roller prior to the transfer to paper. Subsequently, lithographers' stones were replaced by chemically treated metal, plastic, polyester, and paper printing plates that worked just like the original stones (see diagram on opposite page).

However, offset lithography needed one other innovation to become the ascendant printing process: the arrival of phototypesetting (also known as "cold type"). Typesetting by photographic methods was proposed as early as 1866, but practical machines did not become commercially available until the 1950s. Photo composition coupled with offset lithography began a transition that relegated the hot metal Linotype machine to the museum by 1980. The early phototypesetting machines projected a beam of light through "slides" of the alphabet onto photographic film or paper. Later phototypesetting machines used cathode ray tubes (like a small TV screen) to expose the film instead of the slides. Eventually, lasers similar to those used in laser printers were used to expose the film. Of course, there was criticism that this new method of phototypesetting was not up the quality standards of "traditional" hot-metal typesetting. Yet the vastly improved efficiency of photo composition quickly displaced many Linotype operators, who were downsized from their positions and forced to seek retraining programs to enable them to re-enter the job market.

Finally, in 1985, the combination of PageMaker from Aldus Software and the Apple LaserWriter printer with the Adobe PostScript page description language set off the desktop publishing revolution. Suddenly, anyone with some modestly priced (as compared with the commercial typesetting products) computer equipment could typeset

Photo courtesy of Timsons

Timson T-48A "Zero Make Ready" lithographic offset press. This is a very high-efficiency press used by many book printers. It allows one signature to be made ready (prepared for printing) while the press prints a second signature. This is one of several brands of web press commonly used by short run book printers.

their own publications. Naturally, there was criticism that this new method of desktop publishing was not up the quality standards of "traditional" phototypesetting. Yet the vastly improved efficiency of DTP quickly displaced many photo composition operators, who were downsized from their positions and forced to seek retraining programs to enable them to re-enter the job market.

So what?

The point of this discourse about the history of books, paper, printing, and typesetting is simply to remind you that there is a lengthy tradition to the book design process. Humans have been looking at, reading, using, and criticizing books for centuries. The cautionary note is that expectations about books are deeply ingrained in the subconscious. It is very important to the success of a book project that the resulting product meets or exceeds the reader's expectations. The traditions of well over 2,000 years of book design are well established. Don't ever forget it.

– 2 –

The Process

Stages of publishing

THE FOLLOWING IS A DISCUSSION of the stages that a manuscript might pass through at a large publishing house. While the author-publisher certainly won't have all the specialists and personnel of a large publisher, it's important to understand that a manuscript will need to pass through equivalent stages in the small publishing house.

DEVELOPMENT

At this stage, an editor works with an author to develop an idea into a manuscript. The process usually includes writing a proposal and/or an outline as well as a series of drafts. The manuscript includes any specifications for artwork or photographs. Depending on inconsistencies in the material, other editors or subject experts may review a draft of the manuscript.

With the now almost universal use of digital tools, typically, authors submit their manuscript as a word-processing file on disk. Electronic files eliminate the need for a typesetter to re-key a manuscript, resulting in the introduction of fewer errors—assuming that all edits are input to the electronic file before the typesetting stage either by the author or the editor. In a more polished-looking manuscript, a copy editor may not detect all errors. (Don't worry, they will suddenly become all too apparent after the manuscript is set in type.)

The Process:
develop and
write manuscript
proof
edit
proof
edit
proof
edit
proof
edit
proof
design
typeset
proof
typeset corrections
proof
typeset more corrections
proof
make ready (for press)
proof "blueline"
print
bind
ship
distribute books

Participants at a larger publisher might include: an acquisitions editor who reviews the project from a marketing and financial forecasting standpoint; a developmental editor who may perform a "heavy" edit on the manuscript; a technical editor to review the manuscript if the subject involves technology or science; a permissions editor who identifies copyright permission issues and initiates requests for permissions from other copyright holders; and a project editor who will arrange scheduling and coordinate the work of the other people involved in the project.

Other participants at this stage may also include co-authors, a ghostwriter (or "book doctor"), subject experts, reviewers, and the art or design director.

MANUSCRIPT PREPARATION AND DESIGN

At this stage, a series of editors (overseen by a project editor) prepare the manuscript for production. Their goal is to minimize the number of changes necessary later, *because making changes once the text and visuals are set in type is more costly and increases the likelihood of errors in the final product.* Each editor checks the previous editor's work and/or inspects the manuscript for consistency, accuracy, and other issues determined by the nature of the project. A project editor also drives the design process by writing a design survey—an analysis of the manuscript's elements, on which the designer bases the design and specs—and codes the manuscript for formatting by identifying all the various elements, such as chapter titles, heads, subheads, etc. The designer bases design samples on representative portions of manuscript and art specs.

Digital preparation and design makes editing easier. Features such as spell checking and search-and-replace operations make it relatively simple to ensure accuracy and consistency in a manuscript. However, editors and proofreaders working with digital files may have difficulty reading text on screen; often, errors that are easily seen on paper are missed on screen. (It's wise to plan for checking to be done both on screen and from hard copy.) Also, many digital editing methods do not leave an editing trail to enable tracing the author of a particular change. This is less of a problem with a small publisher where fewer hands are actually touching the manuscript, but the cautionary note is that dated prints of interim versions of the manuscript should be made along with backup discs (or CD-ROMS) just in case some particularly wonderful phrase gets blipped out of existence.

The cast of participants at the manuscript preparation and design stage might include the following kinds of editorial functions: copy editing to ensure that spelling, syntax, and grammar are correct; heavy editing if more basic structural revision is required; technical editing to handle science and technology subjects; further work by a permissions editor to identify and secure additional permissions that may have become necessary during the editorial process; fact checking to insure that objects, dates, events, times, places, and people are correctly identified; and finally the project editor who is involved with keeping the project under control and on schedule.

Other participants at this stage may include a designer and a photo researcher.

PRODUCTION

At this stage, a project or production editor or a copy chief oversees implementation of the plans laid in the manuscript preparation and design stage. Once the manuscript is typeset or formatted according to the specs, a proofreader reads the proofs against the manuscript and marks text, art, and format corrections; the typesetter (and the illustrator, if applicable) or printer makes the corrections; and those corrections are again proofed. The cycle is repeated until every correction has been made and checked. The schedule and budget may limit the number and kinds of corrections that may be made; proofs are most commonly made with a laser printer although once the files are handed off to the book printer, a blueline proof made from negatives produced by an imagesetter might be used. Recently, printing companies are more likely to provide yet another laser proof as the underlying electronic file is used to directly image the printing plates. The later the stage, the more expensive the proofs and the more expensive it is to correct errors. It is standard practice to use a different proofreader—a fresh set of eyes—at each stage. Publishers generally expect a proofreader's work to result in ninety to ninety-five percent accuracy, although this expectation varies depending on the condition of the copy at the previous stage. Where complete accuracy is of paramount importance, such as on a cover or in the tables in an annual report, extra proofreaders may be called upon either to proofread or to read the proofs "cold"—that is, to read them without checking against the previous version. The ultimate in proofreading is to have one person read the material out loud while another person reads along silently marking problems or interrupting with queries as necessary.

Desktop typesetting programs, such as PageMaker, InDesign, or Quark XPress are used to go directly to individual pages rather than traditional galleys. Paginated laser proofs are easily produced as needed to review and correct. It has been my, all too common, experience that author-publishers "short circuit" the manuscript preparation phase and submit a less than fully polished manuscript to their designer/typesetter. Then, the inevitable corrections must be entered by the typesetter at considerable additional expense. It's truly better to take the additional time to properly prepare the manuscript before submitting the electronic file to the typesetter. This is less of a problem for those who are using the desktop publishing software themselves.

The larger publishers may use specialized software to help automate the creation of elements such as tables of contents, indexes, and bibliographies. There are several vendors who offer "plug-ins" to work with Adobe InDesign or Quark XPress that offer these and other advanced features, but smaller publishers are usually served well enough by the built in functions of the desktop publishing software.

The larger publishers often work with specialized indexing software—that automates alphabetizing, formatting, and other functions—to free the indexer to concentrate on providing readers with efficient subject access. Smaller publishers will probably prefer to use embedded indexing performed on electronic files in the word processor and imported into the page layout program, or entered directly in the page layout program, that embeds invisible index codes at the point in the text to which they refer. This allows indexing to begin before pages are fixed. Indexers using older dedicated indexing software or simply working semi-manually must work from page proofs and can not begin indexing before page breaks are fixed. This usually puts the index preparation off to the last possible moment before the book goes to the printer.

An author-publisher can do a reasonable job inserting the index codes for the index in the basic word processing file. Some review of books on indexing (see the suggestions in the bibliography) and time spent learning indexing techniques will help improve the index. Often, it makes good sense to hire a professional indexer to re-work the author's choices or to create an index from scratch. Once the indexer has finished his work, the author should review the index to ensure that important concepts weren't overlooked.

Printing technology now allows publishers to go straight from disk to film or disk to plate, eliminating a step in which errors can be introduced. When electronic files are imaged directly to plate, the quality loss of the intermediate steps is eliminated as the final printed copy is one generation closer to the original. This direct to plate process is called "CTP" for "Computer To Plate," but to confuse matters, so called "DI" ("Digital Imaging") presses that image plates on press use the initials "CTP" to mean "Computer To Press."

The people involved in this production stage may require the following kinds of editorial expertise: proofreading, indexing, and project editing.

Additional activities may involve the photo researcher, acquisitions editor, developmental editor, author, or others who may review the proofs as well.

Other participants at this stage may include illustrators, a book designer/typesetter, a subject expert or foreign language reader, and the art director.

Levels of editing

Every manuscript should be edited, preferably by an experienced, professional editor. Let me tell you this story before you object and say, "But my writing is so good I don't need an editor!"

A few years ago when I was general manager of a typesetting service in San Francisco we worked on a book by Pope John Paul (you know, the Pope—the one in Rome). The book was being published by HarperSan Francisco, the new age and religious book division of HarperCollins.

When the manuscript was received, it had obviously been edited (it was marked up in blue pencil). As the book moved through the typesetting process, each subsequent proof was reviewed and edited as well.

There are those who believe that the formal writing of the Pope is the inspired word of God. There are those who believe that the Pope is infallible.

If the Pope had an editor, could we mere mortals have a lesser need?

The fact is, no matter how good your writing, even if you are receiving direct inspiration from God, or even if you are the *highest* authority, an *editor* yourself, we *all* need an editor to check our work so we stay focused and don't embarrass ourselves.

LIGHT COPY EDIT

This is the least thorough edit and may be appropriate for a manuscript that has been through more serious levels of editing or for a book that's being prepared for a reprint. A light edit (or baseline edit) will:

- Check spelling, grammar, and punctuation.
- Correct incorrect usage.
- Verify specific cross-references.
- Ensure consistency in spelling, capitalization, etc.
- Check for proper sequencing of material.
- A light copy edit does not involve rewriting, changing heads or text or ensuring parallel structure or fact checking.

MEDIUM EDIT

In addition to all the tasks performed in a light copy edit, the medium copy edit includes:

- Changing text and headings for parallel structure.
- Style (using a standard reference, such as the *Chicago Manual of Style)* and word usage are verified.
- Consistency is checked to ensure proper handling of key terms and that specifications are met in vocabulary and index items.
- In fiction, the medium edit tracks continuity of plot, setting, and character traits. In a multi-author manuscript, the medium copy editor will enforce a consistent style and tone.
- The medium copy edit should identify ambiguous and/or incorrect statements for correction by the author.

SUBSTANTIVE EDIT (OR HEAVY COPY EDIT)

A substantive edit includes all the tasks performed in the light and medium copy edit levels, and will:

- Eliminate wordiness, triteness, and inappropriate jargon.
- Ensure smooth transitions and improve readability.
- Check heads for appropriateness—new or changed levels of heads will be applied to achieve a logical structure.
- The substantive editor will suggest additions and deletions to improve the flow and coverage of the manuscript.

The difference between the various levels of copy editing is the amount of judgment and rewriting involved. The substantive edit may involve rewriting rather than simply flagging problems and may involve enforcing a uniform level, tone, and focus as specified by the publisher or developmental editor.

PROOFREADING

After the editor is finished with the manuscript, an author-publisher should review the changes. As with any hired professional, it is appropriate to exercise your judgment about how much of the advice you'll accept. This last review (by the author) of the manuscript also insures that there are no errors inadvertently introduced by the editing cycle. The manuscript then goes to the designer and compositor (who can be the same person). In a large publishing house, the supervising editor would review the manuscript before it is forwarded to the designer and compositor for typesetting.

An observation here is that a publisher (depending on the publishing contract) may have the right to force certain editorial choices. An advantage of independent publishing is that the author-publisher has complete control over what eventually gets published.

The proofreader checks the typeset copy word for word against the final edited manuscript and identifies errors for correction or marks queries for editorial errors. Proofreaders may also review copy for conformity to type and design specifications, check against (or create) a style sheet, and check the typography for good letterspacing, margins, wordspacing, stacked hyphenations or words, bad wordspacing or bad breaks, etc.

Aren't we ever going to get into laying out the book?

Patience, dear reader. These preliminary steps are vital to the later success of the project and are essential to keeping your project on budget. The most common cause of cost overrun during the layout/typesetting phase is the failure to adequately complete the editing function. Editing is inherent in the book production process.

We'll cover a few other preliminary topics, as well, before we actually design and lay out a book.

- 3 -

Decisions, Decisions...

BOOK DESIGN INVOLVES A LARGE NUMBER of decisions to arrive at a final design. We need to select a trim size (which can also impact the cost of printing), margins, typeface and leading, additional typeface(s) for headings, and determine any special treatments required for elements such as side bars, illustrations, tables, and/ or photographs.

What is a book?

There are some commonly accepted definitions that apply. In the definitions used by the Library of Congress, the primary factor is the number of pages:

A book has 49 or more pages.

A booklet has from 17 to 48 pages.

A pamphlet has from 4 to 16 pages.

A brochure has from 1 to 4 pages.

A children's picture book (for ages 3 to 6) is usually 32 pages, but is sometimes 24 pages. When a children's picture book is published in hard cover, you can use an additional 2 pages by printing on the back sides of the end papers. These page counts are dictated by production issues. Books are usually printed in signatures of 8, 12, 16, 24, 32, 48, or 64 pages, depending on the specifics of the press being used. A children's picture book should be written, illustrated, and designed to be either 24 or 32 pages to ensure reasonable production cost and to meet market expectations.

A book for youths (ages 7 to 12) is generally between 50 and 150 pages. There are examples that ignore this guideline, most notably the *Harry Potter* series.

A board book, for children of 9 months to 3 years of age, is usually quite short (fewer than 20 pages) and is printed on very heavy, paper-covered boards (also made of paper-like material). Board books often have novelty features, such as shoelaces to tie. Since board books are more in the category of toys, rather than books, they are beyond the scope of our discussion here.

Gift books are simply a category of trade book, that used to require a UPC bar code in addition to the ISBN (EAN) bar code to facilitate sales through gift and specialty shops. [Note: As this book is being reprinted, UPC codes have been transformed into EAN compliant codes. Scanning equipment is now required to be able to read EAN-8, EAN-13 (used by ISBN), and UPC bar codes. To make ISBNs compliant with these new standards they have been converted to the new 13-digit format by adding the 978 "Bookland" prefix to the original 10 digit ISBN. As the last of the 10-digit ISBNs are used up, new 13-digit ISBNs will be issued. This also forces a change to the check digit of the 10-digit ISBN to accommodate the 13 digits. These changes have eliminated the need to obtain and use UPCs in addition to ISBNs on books intended for non-bookstore outlets.]

The Library of Congress (LoC) will provide Preassigned Control Numbers (PCN), commonly called the Library of Congress Control (formerly "card") Number (LCCN), only to books of 49 pages or more, unless they are primarily for children. Similar restrictions apply to the LoC cataloging in publication (CIP) program. (Visit the Library of Congress web site for information about LoC programs for publishers. Note that most self/small publishers categorically do not qualify for the Library of Congress Cataloging-In-Publication program.)

Trim size

Choose a reasonable size for your book. Look at other books in the same genre. Unless you have a specific and significant reason to choose another size, stick with the sizes used by similar books. Some common sizes are:

Trim size in inches	Common name
4 x 5	Pocket book
4¼ x 6¾	Mass Market Paperback
4 x 9	Rack Brochure
5⅜ x 8⅜ (or 5½ x 8½)	Digest (Half Letter)
6 x 9	Standard Novel
7 x 10	Text Book
7½ x 8, 7½ x 9	Software Manual (varies)
8 x 8	Juvenile
8½ x 9¼	Software Manual
8⅜ x 10⅞	Standard Catalog
8½ x 11	Standard Letter
9 x 12	Music Book
11 x 11	Sunday Insert

Digest and standard novel are the sizes most often used by independent publishers. These sizes are practical to produce on most printing presses with a reasonable level of efficiency. If you are considering a trim size that is larger than standard letter, you should verify its economic production with your printer. Many short run book printers use presses that do not economically produce some of the larger page sizes.

The trim size choice should be based on the market and intended uses for your book. Consider: what size are competitive books? Where is the book going to be read? How are readers using the book, e.g., as a reference, as a guide, as an educational resource? Will they read it from cover to cover or will they be referring to it again and again?

Another factor in choosing the trim size of the book is your budget. Some sizes are simply more expensive to produce.

Practical uses of the listed sizes

THE POCKET BOOK (4 x 5 inches) is a suitable size for books that might fit a pocket. Like the rack brochure size it will easily stow in an automobile glove box, suitcase, or tool box. A book of 200+ pages will easily fit in half-sheet size envelopes and mailers. Small quantity reproduction can be done on regular letter size paper with considerable efficiency, if printed four-up. There is a danger that a book of this size will be confused with a mass market book.

THE MASS MARKET PAPERBACK size (approximately 4¼ x 6¾ inches) is to be avoided by author-publishers. Mass market paperbacks are distributed and sold using practices and policies that generally prove toxic to the financial success of a beginning publisher. These paperbacks are distributed and sold through drug stores, supermarkets, and other mass merchants in addition to book stores. They are cheaply produced using "pulp" papers. When unsold books are left over, the merchants remove the covers (with the bar codes) and return them to the publisher for credit. The books themselves are disposed of in the trash. This is not a business model that will work for a publisher who only has one or two titles. Even if your book is produced with higher quality materials, use of the mass market size may result in the assumption that it *is* a mass market title to your considerable disadvantage.

THE RACK BROCHURE size (approximately 4 x 9 inches) is great for smaller booklets, travel guides, or other carry-along type books. It will fit purses, automobile glove boxes, and (depending on thickness) many pockets. A booklet up to 20 or even 32 pages will fit a standard #10 business envelope, facilitating mailing.

THE DIGEST SIZE (5⅜ x 8⅜ or 5½ x 8½ inches) is most often recommended for independent publishers. Only slightly smaller than the standard novel size, it makes very efficient use of letter size paper. For short runs, the smaller variation of this size can be printed two-up on a Xerox Docutech or other similar machine. Packing boxes, envelopes, and mailers are widely available that will easily accommodate this format. (While 8½ x 5½ is often the "recommended size" in self-publishing books, the slightly smaller 8⅜ x 5⅜ is truly the better choice. Many binding machines grind off approximately ⅛ inch to prepare the spine of the book block for gluing. This also leaves room for required trimming to even up the edges of a perfect bound book. If the interior design allows reasonably generous margins, this size may be easily set up for printing at either size with no other design changes as the smaller size is achieved by removing ¹⁄₁₆ inch from each side of the page.

THE STANDARD NOVEL (6 x 9 inches) is the most frequently used size for trade books. It fits the most common web press sizes and can be efficiently produced in large quantities by the major book printers. This size, however, is not especially efficient for the small publisher. It wastes a considerable amount of paper when produced in small quantities on digital printing equipment. It is also an awkward size that does not fit the most common packing boxes, envelopes, and mailers available through office supply stores. (Ordering shipping supplies from specialist catalog/Internet sources reduces the problem.)

THE SOFTWARE MANUAL size (approximately 8½ x 9¼ inches) is a common size for software manuals. It is also a good size for any publication that has many illustrations that need to accompany the text. It also allows generous margins that can be used for side-bars, photos, illustrations, or other supplemental text purposes.

THE STANDARD CATALOG (8⅜ x 10⅞ inches) is slightly smaller than the U.S. letter size sheet. It offers production advantages for very short production runs using standard copier equipment. The smaller dimensions allow for losses due to binding and allow the book to be trimmed to avoid any misalignment in the book block that might detract from the appearance. This is a handy size for cookbooks, training guides, and materials with substantial numbers of illustrations, tables, and photographs.

THE STANDARD LETTER (8½ x 11 inches) is the usual size for business reports and term papers. It is easy to produce in short runs on standard copy machines. It is useful for manuals, reports, training guides, workbooks and other materials that are likely to be placed in loose leaf binders, spiral, plastic comb, or other mechanical bindings. If you are planning a perfect bound book in this size, the standard catalog size is a better choice as it leaves room for edge trimming in the binding process.

The standard letter (particularly if trimmed to standard catalog, 8⅜ x 10⅞ inches) and digest (half letter) sizes offer a higher degree of efficiency than the other sizes when being produced on a sheetfed Xerox Docutech or other sheetfed digital printer, as the amount of trimming required and paper waste is reduced. The slightly smaller 5⅜ x 8⅜ measure allows printing two-up on letter size paper, leaving an allowance for trimming the books after binding. This is an especially efficient size if you are considering digital printing. It is also quite efficient with some of the presses in use for offset book printing.

THE SUNDAY INSERT (11 x 11 inches) size is rarely used for books. This might be a format used for promotional materials or it might be used for a "coffee table" book. However this may cause considerable difficulty in easily finding suitable packing materials. Small runs would require use of 11 x 17 sheets on digital printing equipment resulting in substantial waste.

THE TEXT BOOK SIZE (7 x 10 inches), like the software manual size, easily accommodates illustrations with text. It is suitable for nearly any publication that has a mix of such materials. Short run production is somewhat less efficient than other sizes, but the amount of trim from a letter-size sheet is not unreasonable.

THE JUVENILE SIZE (8 x 8 inches) is commonly used for children's books. Children's book publishing has a number of standards that are discussed elsewhere in this book. (See chapter 12.)

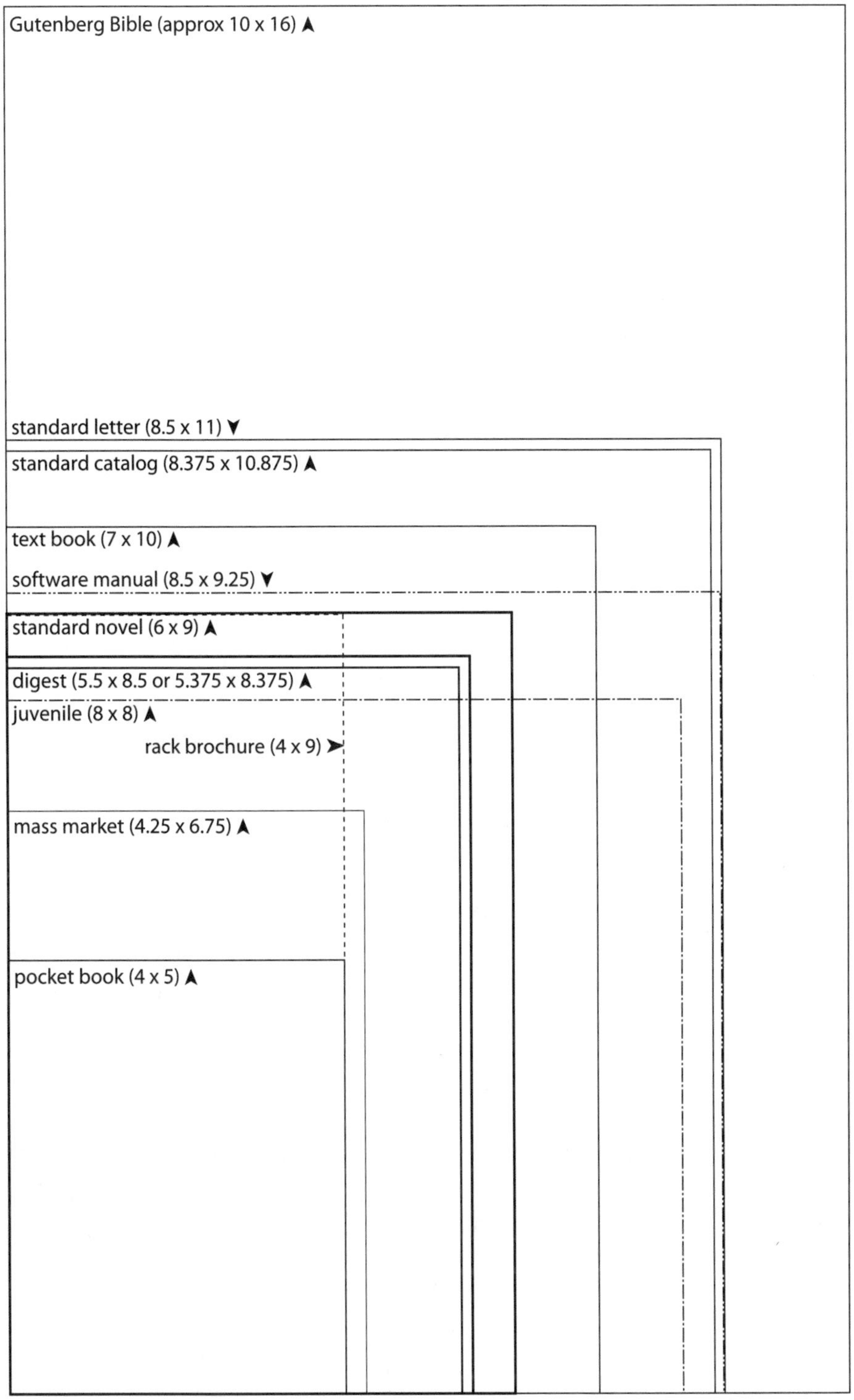

Gutenberg Bible (approx 10 x 16) ▲

standard letter (8.5 x 11) ▼

standard catalog (8.375 x 10.875) ▲

text book (7 x 10) ▲

software manual (8.5 x 9.25) ▼

standard novel (6 x 9) ▲

digest (5.5 x 8.5 or 5.375 x 8.375) ▲

juvenile (8 x 8) ▲

rack brochure (4 x 9) ➤

mass market (4.25 x 6.75) ▲

pocket book (4 x 5) ▲

Relative sizes of standard book trims (inches, decimal—sizes reduced for publication).

THE MUSIC BOOK SIZE (9 x 12 inches) should come as no surprise to be the one most often used for music books. It is also frequently used for high production value coffee table (art) books. Normally, this format may cause some additional expense for printing as it may not be a good match for the equipment of many printers. The web width of the smaller presses used by many short run book printers does not easily accommodate the extra size of the 9 x 12 format, causing considerable waste—hence extra printing expense. Sheetfed presses frequently are designed to accommodate paper sizes that are a multiple of the "standard" letter size. This oversize format then doesn't utilize the sheet as efficiently, also causing more waste and printing expense.

OTHER SIZES. There is no requirement that you use any of the trim sizes discussed here. You can design your book to any size that you can imagine. However, if you do select an odd size, you may be faced with increased printing costs, increased handling costs, and other potential marketing problems. For example, unless it is an "art book," a large 10 x 14 format book very likely won't fit the shelves at most book stores resulting in its not being displayed with other books in its category. Such a format also will not fit library shelves—possibly hurting sales to that market. Choose a "nonstandard" size only if you have *specific market-driven needs* to make such a choice. Realize, too, that the choice may result in increased costs that you will need to pass along to your customers—which may hurt sales. Don't be misled if you notice large-publisher books in odd sizes. Often, the large publishers print such huge quantities (20 to 50 thousand copies) that the extra costs involved are no longer significant when pro rated on a per unit basis. These extra costs are considerably more significant when printing the one to five thousand copies that's usually reasonable for a small publisher.

Extra-small books and other unusual trim sizes can cause difficulties or extra expense in production. One client desired to publish a gift book in a square 5⅜ x 5⅜ format. We discovered that many printers could not efficiently produce a book of that size. While it might seem that this would be a typical 8⅜ x 5⅜ press run, but cut smaller, we found that the binding machines used by many book printers could not be adjusted for such a small book. Fortunately, there were a few printers who had equipment or alternative methods that allowed the project to go forward at a reasonable cost.

The important lesson is to consult with several printers in the early stages of a project if it seems that you might need an unusual size or any other variations from "normal" books.

One final consideration is to check the availability of packing and shipping supplies to handle the book trim size. Books need to be firmly secured in their packaging with sufficient protection for the corners to avoid damage during transit. Unusual trim sizes may require extra expense to find correctly sized containers or to pay for additional packing material to fill voids. The marketing considerations may override these practical concerns, but then you will be able to adjust shipping charges to cover the extra expense before you have printed materials committing you to a lower shipping charge.

Margins

Once a trim size has been selected, we next need to consider how much text and other material to put on each page. You will note that *this* book seems to have rather large margins. (Indeed, it does.) The trim size was selected to have the same proportions as the Gutenberg Bible—the first (western) book printed with movable type in 1450 A.D. Since the Gutenberg Bible was printed on sheets approximately 10 x 16 inches, it would be impractical for this book to be that large. A trim size of 10⅞ x 7¾ was selected to remain proportional, yet fit within a standard trim size (standard catalog) to ensure reasonable reproduction costs.

It would be fair to ask, why did I select this particular size? The Gutenberg Bible (often called the 42-line Bible because it has 42 lines of type per page) is considered one of the most beautiful books ever printed. Of course, Gutenberg crafted his book according to the general book designs used by the scribes who provided book production services before printing became available. Gutenberg was a careful and talented artisan. He crafted some 300 pieces of type to include all the various ligatures (combined letters) commonly used by scribes. On the overleaf following, I have set a facsimile page using a typeface very similar to the one used in the Gutenberg Bible, however, to make it easier to appreciate, the text, part of Genesis Chapter 3, is in English from the King James Version (c. 1620). To view actual images of the Gutenberg Bible visit the Harry Ransom Humanities Research Center at the University of Texas, via the Internet at *www.hrc.utexas.edu.* (or do a search at *Google. com).*

> **Widows and Orphans.** An *orphan* is the first line of a paragraph at the bottom of a page or column. A *widow* is the last line of a paragraph at the top of a page or column. Both should be avoided, if possible. However, an orphan is less damaging to the page appearance. A widow may be accepted if at least two-thirds of the line is filled.
>
> Be especially careful to avoid widows and orphans between the bottom of a right-hand page and the top of the following left-hand page, as you must turn the page to finish the paragraph. (Also, do not break a word with a hyphen in the same circumstances.)
>
> Some writers also call a single, short word (or worse, a single syllable from a hyphenated word) on a line by itself an orphan. This situation should also be avoided.

These pages are set with margins in decimal inches of 1.0278 (top), 2.0555 (bottom), .9375 (inside or gutter), and 1.5625 (outside). These margins are rather generous for a modern book, particularly for the bottom. However, with this page size, these margins give a text block of 5¼ inches wide by (approximately) 7⁵¹⁄₆₄ high. (The height is established by the number of lines and leading. This book is set with 11 point type on 13.3 points of leading to make 42 lines of type per page. What struck me was that this type size and leading settings are very reasonable for a modern book and are also proportionate to the typesetting in the Gutenberg Bible, which is estimated (in modern terms) as 22 or 23 point type on 26.6 points of leading (baseline to baseline). This is a demonstration of how the design relationships of typography have carried down over the centuries.

Now the serpent was more subtle than any beast of the field which the Lord God had made. And he said unto the woman. Yea. hath God said. Ye shall not eat of every tree of the garden. And the woman said unto the serpent. We may eat of the fruit of the trees of the garden: But of the fruit of the tree which is in the midst of the garden. God hath said. Ye shall not eat of it. neither shall ye touch it. lest ye die. And the serpent said unto the woman. Ye shall not surely: ffor God doth know that in the day ye eat thereof. then your eyes shall be opened. and ye shall be as gods. knowing good and evil. And when the woman saw that the tree was good for food. and that it was pleasant to the eyes. and a tree to be desired to make one wise. she took of the fruit thereof. and did eat. and gave also unto her husband with her and he did eat. And the eyes of them both were opened. and they knew that they were naked and they sewed fig leaves together. and made themselves aprons. And they heard the voice of the lord God walking in the garden in the cool of the day: and Adam and his wife hid themselves from the presence of the lord God amongst the trees of the garden. And the lord God called unto Adam. and said unto him. Where art thou And he said. I heard thy voice in the garden. and I was afraid. because I was naked and I hid myself. And he said. Who told thee that thou wast naked Hast thou eaten of the tree. whereof I commanded thee that thou shouldest not eat And the man said. The woman whom thou gavest to be with me. she gave me of the tree. and I did eat. And the lord God said unto the woman. What is this that thou hast done And the woman said. The serpent beguiled me. and I did eat. And the lord God said unto the serpent. Because thou hast done this. thou art cursed above all cattle. and above every beast of the field upon thy belly shalt thou go. and dust shalt thou eat all the days of thy life: And I will put enmity between thee and the woman. and between thy seed and her seed it shall bruise thy head. and thou shalt bruise his heel. Unto the woman he said. I will greatly multiply thy sorrow and thy conception in sorrow thou shalt bring forth children and thy desire shall be to thy husband. and he shall rule over thee. And unto Adam he said. Because thou hast hearkened unto the voice of thy wife. and hast eaten of the tree. of which I commanded thee. saying. Thou shalt not eat of it: cursed is the ground for thy sake in sorrow shalt thou eat of it all the days of thy life Thorns also and thistles shall it bring forth to thee and thou shalt eat the herb of the field In the sweat of thy face shalt thou eat bread. till thou return unto the ground for out of it wast thou taken: for dust thou art. and unto dust shalt thou return. And Adam called his name Eve because she was the mother of all living. Unto Adam also and to his wife did the lord God make coats of skins. and clothed them. And the lord God said. Behold the man is become as one of us. to know good and evil: and now. lest he put forth his hand. and take also of the tree of life. and eat. and live for ever: Therefore the lord God sent him forth from the garden of Eden. to till the ground

Above is a facsimile of what the Gutenberg Bible might have looked like if set in English (KJV used) using the same typeface as the German version. (There are archaic forms of the letter "s" that look more like an "f" in this text that make reading more difficult.) Gutenberg left spaces at the beginnings of chapters so scribes could hand draw (illuminate) the book in the same style as hand-done books. (Clip art used here.) Often these were quite elaborate and decorated with colors and gold leaf. Indicating paragraphs, as we do today, was an innovation that came later.

MARGINS IN MODERN BOOKS: With modern books, the page sizes have been significantly reduced compared to the pioneering books, and to create a reasonable line length the margins have been similarly reduced. Researching the many books on typography, there are many opinions about what "good margins" might be. However, all the writers on the topic agree that margins are preferred to be wider rather than slender, that they should not be equal on all sides, and that they should follow a progression of widest at the bottom, next widest on the outside edge, followed by the top, then the narrowest on the inside (gutter) edge. (This may be counter-intuitive.)

Other considerations about margins suggest that the thumb should fit comfortably on the outside edge without significantly obscuring the type and that the gutter margin should not be so narrow that the type seems to curl around the edge of the page and disappear into the binding. From behavioral studies of eye movements while reading, it is shown that a "landing area" providing "eye relief" is necessary to give sufficient "rest" to prepare the eye to scan the next line. Here we must note that one of the most common errors in books produced by nondesigners is to have overly narrow margins of equal size. (To further confuse the beginner, most word processing program's default margin settings include a generous binding allowance on the inside [gutter] margin to accommodate a 3-hole binder. This misleads amateurs into giving the inside margin a larger size than is truly appropriate.) Visually, it's better that the facing pages "relate"

Image of a Gutenberg Bible. Some of the type was colored by hand-inking into the printed book. Gutenberg left space in the appropriate places. More elaborate illumninations were used at the begining of each chapter. This copy shows some comments and corrections added by hand. (Folios 128, 129, Volume 1, from I Samuel). Image courtesy of Harry Ransom Humanities Research Center, The University of Texas at Austin.

to one another by being closer together. The double margin, allowing for the take-up in the gutter, should *visually* be approximately equal to the outside margins.

On the next page we have a 5½ x 8½ page represented to show a suggested starting point with margins of (decimal inches) .985 (bottom), .875 (outside), .835 (top), and .75 for the inside margin. These margins give us a writing line of 3⅞ inches. The same margins can be used with a 6 x 9 book to yield a writing line of 4⅜ inches. Either of these writing lines is comfortable for most readers. This book would also have a normal page with 37 lines if the type is set at 11 points on 13 points of leading.

Note that the numbered lines extend below the main text block. Pages on a spread (facing pages) are most attractive if the bottoms end at the same line. Good page breaks are those that avoid widows or orphans. Pages may run long, even, or short with respect to the standard text block to eliminate bad page breaks. This gives a range of 3 lines to "play" with. Occasionally, a spread may need to be run 2 lines long (or short), but that should be rare.

While typefaces vary in average character width, a typical line is best between 60 and 70 characters (including spaces). That measure would allow a page to hold between 2200 and 2600 characters. If the 5 characters per word (used in my typing class of long ago) estimate holds true, that would suggest that a page would hold from 440 to 520 words. (This is a very rough estimate.)

This sample page also shows a baseline grid—horizontal lines where the baselines of the type will fall—with a leading of 13 points baseline to baseline. (In a later chapter, we will talk about type size and leading in more detail.) Other elements are also identified: the header line, which is not counted with the text block, is spaced about one blank line above the main text block. If running headers are used, the page numbers (also called folios) can be positioned on the same line. While there are no absolute rules, page numbers are most commonly placed at the top or bottom on the outer edge of the page or centered at the bottom of the page. These positions simply make good sense as those placed to the outer edge are positioned so a reader can thumb through a book to find a particular page without having to fully open the book to determine the location.

If headers with folios (page numbers) are used, they are omitted on blank pages, the opening page of a chapter, or other pages with an open space before the printing starts, such as a section title page. On the opening pages of the chapters (called the chapter opener) and on section title pages, a *drop folio* may be used (the term given to a page number at the bottom of a page).

Traditionally, books were designed with the header and folio (page number) running along the top. The physical act of setting type made it simple to set up headers before the body copy was ready but difficult to pre-position matter at the foot of the page. The body text would next be set in the frame and blocked off at the bottom and tightened with a wedge. Drop folios (numbers at the bottom of the page) required the body text to be in place before the page number (and any other footer material) could be inserted. Of course, with modern page layout programs, this is no longer an issue, so page numbers can be placed wherever convenient, keeping in mind that a reader will have an expectation that the page numbers will be positioned in a consistent and logical manner.

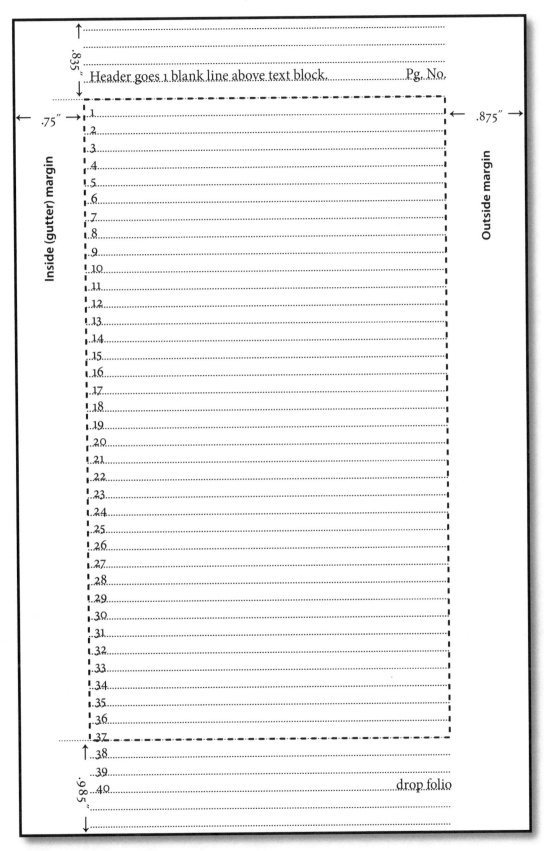

.835″

Header goes 1 blank line above text block. Pg. No.

.75″ .875″

Inside (gutter) margin

Outside margin

1
2
3
4
5
6
7
8
9
10
11
12
13
14
15
16
17
18
19
20
21
22
23
24
25
26
27
28
29
30
31
32
33
34
35
36
37
38
39
40

drop folio

.985″

A 5½ by 8½ inch page showing margins and baseline grid.

The traditional margin layout method

The margins arrived at in the prior example are *arbitrary margins*. Traditionally, various methods were arrived at to devise "natural" margins using mathematics, proportions, or geometry. Here we'll examine a geometric approach that is written about in many typography books and seems to date back to the earliest era of page layout. The Gutenburg Bible very nearly coincides with this method. (Gutenberg may have used this method, but the wear and tear and probable re-bindings over the centuries have made the current copies of Gutenberg's Bible not quite fit.)

(1) Begin by drawing a two page spread. Use squared paper and measure out the pages in proportion to the exact size. Or, if you have a large enough sheet of paper, draw out the actual size of the finished trim of a two page spread, e.g. you can fit a spread from a 6 x 9 trim on an 11 x 17 sheet by marking out a rectangle 9 x 12 inches and bisecting it into two 6 x 9 rectangles. (See following page.)

(2) Draw diagonals for both the full spread and for each of the pages.

(3) Draw the additional lines as shown. Then draw the text block so that the upper-inside and lower-outside points align with the individual page diagonal. Drawn this way, the inside and top margins are one-ninth of the width and the height of the page, respectively. The outside and bottom margins are double the size of the inside and top margins, respectively.

(4) As shown by the dashed lines, additional, smaller rectangles (only shown on one side here) can be drawn in to establish new page width fractions at one-eighth, one-seventh, etc.

(5) *Traditional arbitrary margins* ignore drawing the small box at the top of the spread, but simply establish the text block by ensuring that the inside top and outside bottom corners intersect the short (single page) diagonal and that the outside top corner point intersects the long (spread) diagonal. This will ensure that the outside margin will always be twice the inside margin and the bottom margin will always be twice the top margin.

The main failing of this method, even using step 5 to set arbitrary narrow margins, is that it results in rather large margins, especially to the outside and bottom of the page. Economics of modern book production rarely allow such extravagant use of material.

Yet another margin layout method

Finally, there are those who advocate setting margins arbitrarily using *numerical relationships*. Measuring the page in picas (6 picas = 1 inch), we calculate the margins as inside 4 picas, top 5 picas, outside 6 picas, and bottom 7 picas (or some other variation of incremental numbers). This example would result in an inside margin of .667 inch, top of .833 inch, outside 1 inch, and bottom 1.166 inches. While this is relatively easy to calculate (if you are used to working with picas), it makes for slightly wider margins than absolutely necessary, using a larger amount of material. Still, it will ensure a reasonably attractive page.

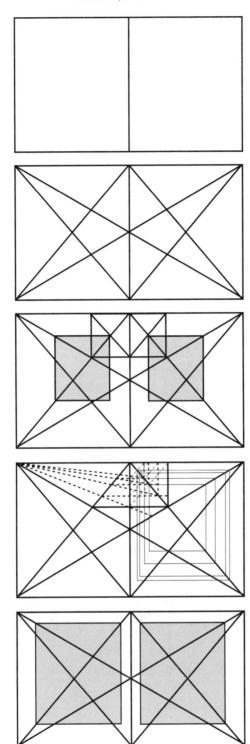

1

2

3

4

5

Parts of a book in brief
Front Matter
 Half Title Page
 Title Page
 Copyright Page
 Dedication
 Acknowledgments
 Epigraph
 Table of Contents
 List of Illustrations
 List of Tables
 Foreword
 Preface
 Introduction
 List of Abbreviations
 Editorial Method
 List of Contributors
 Chronology/List of Events
 List of Characters
Body
 Main Text (divided into)
  Parts
  Chapters
Back Matter
 Appendix
 Notes
 Glossary
 Bibliography
 Index
 About the Author
 Colophon
 Order Form

4

Parts of a Book

BOOKS ARE ORGANIZED IN A LOGICAL SEQUENCE. There are three basic parts: the front matter, the body, and the back matter. These major divisions are further broken down as described below. (Note: book pages are called recto [Latin for right], if it is a right hand page, which is normally given an odd page number; or verso [Latin for reverse or back side] if it is a left hand page, which is normally given an even page number.)

The traditional book numbers the pages in the front matter with lowercase Roman numerals, while the body and back matter are numbered with Arabic numerals. This custom arose as the typesetting of the body often preceded the preparation of the front matter, and the number of pages of front material might not be known. Modern production methods make it easy to add pages at any point in a book with automatic renumbering of all following pages. However, if an index is prepared before the front matter is ready, it might be best to use separately numbered front matter pages to avoid having to adjust the page numbers in the index. Absent this issue, I generally advise numbering a book sequentially from the first page through to the last. Most potential readers will, in part, evaluate a book by looking at the last numbered page to determine its length. If the numbering starts with the main body, the potential reader will ignore the front matter and simply assume the book is shorter than it actually is.

Front matter

HALF TITLE PAGE. This is normally the first page of the book and displays only the book's main title. The subtitle, if any, and author's name are omitted. The use of the half title page developed in the early days of book manufacturing to enable easy identification of an unbound book, as books were often not bound until sold. The half title page is now a traditional part of the book. It may be omitted to avoid an extra signature if page makeup requires. Originally, this first page was called the "bastard title" page as there may be a "true" half title page just before the body, if the front matter is extensive. The seemingly pejorative term has fallen into disuse as typesetting skills have moved outside the print shop environment. (Indeed, the 15th edition of the *Chicago Manual of Style* now simply calls the first page of a book, "the Book Half Title page."

The back side of the Half Title (second page of the book) is often left blank or is used as part of a double page spread for the title page, if the designer should so choose. If the book is in a series, this may be used to give the title of the series, the volume number (if any), and the name of the general series editor (if any). The series information may optionally be placed on the title page of the book.

Depending on the page make up for printing, a few blank leaves may be inserted before the Book Half Title page (and/or at the end of the book). These extra leaves are not normally counted as pages of the book. If there are extra pages due to the printing needs, it's handy to put one or two of them before the Book Half Title page as they can be used for additional blurbs (excerpts from reviews) in future printings of the book without affecting the page make up at that time.

TITLE PAGE (third page) displays the full title and subtitle of the book, name of the author, name(s) of editor(s) or translator(s) (if any), and the name of the publishing company with the publisher's city (and fully spelled out state name, if not obvious), e.g., Æonix Publishing Group, San Francisco (obvious) or Æonix Publishing Group, El Sobrante, California (not so obvious). This information is required on this page if it is to receive a Library of Congress Control Number (LCCN).

COPYRIGHT PAGE. The back of the title page, usually the fourth page. Since the 1976 and subsequent amendments to the copyright statutes, it is no longer *required* that the copyright notice appear here, but most publishers continue to place it on the back of the title page. (It has to go someplace, so they may as well follow the tradition.)

The copyright notice consists of three parts: The symbol © or the word "copyright," the year in which the book is published (or the year the copyright is registered), and the name of the copyright owner. In addition, most publishers add the phrase "All rights reserved." This expression establishes the maximum protection in the U. S. and most Latin American countries under the Buenos Aires Convention.

A typical copyright notice might read:

> © 2005 Pete Masterson. All rights reserved. (Often followed by additional legalese notifications indicating that it is not permitted to make copies or store the material in a computer without written permission and it may include blanket permissions granted to the purchaser, such as allowing making copies of maps in a hiking book for personal use.)

The name of the country where the book was printed must appear in the book (not just on the cover) and is generally placed on the copyright page. (All manufactured products in interstate commerce must indicate their country of origin.) The publisher may choose to place a "printing history" on the copyright page, although it is not required. The printing history normally contains the year of first publication; date of second or other edition; date (year only) of impression (printing) if other than the first. A typical notice might read:

Æonix Publishing Group, El Sobrante, CA 94803
© 2002, 2003, 2005 Pete Masterson. All rights reserved.
Published 2002.
Second printing 2003.
Third printing 2005.
Printed in the United States of America
09 08 07 06 05   10 9 8 7 6 5 4 3

The last line indicates the printing number and year. It is arranged so that the numbers can be easily deleted from negatives (once) used for making printing plates. Of course, if a full printing history is shown, then the page must be re-generated as a negative, so the easy deletion is moot. Indeed, in this era of fully electronic workflow with output direct to plate, there is no practical advantage to the "code numbers" used to show the printing number. You may as well use the full history, or simply state the latest print run year or month and year.

It is helpful to the publisher to be able to quickly identify in which printing an individual book was manufactured. Often, small typos and other minor errors are corrected between printings, and if a typo is brought to the publisher's attention, it's useful to know if it may have already been fixed.

The copyright page also traditionally has the cataloging-in-publication (CIP) data block. It might read as follows:

Library of Congress Cataloging-in-Publication Data
(or Publisher's Cataloging-in-Publication Data)
Masterson, Pete
   Book Design and Production : A Guide for Authors and Publishers

   Bibliography:
   Includes index and glossary
 1. Design   I. title
LCCN 2005901925   [Dewey Decimal Number]
ISBN 0-9669819-0-1

Cataloging-in-Publication data is provided by the Library of Congress or by commercial sources when the Library of Congress is unable or unwilling to create the data. Small publishers generally obtain cataloging data from Quality Books, Inc. or another vendor.

(See the Resources Appendix.) Some publishers may have cataloging data prepared by a willing librarian. It is important that only data provided by the Library of Congress be called *"Library of Congress* Cataloging-in-Publication Data." CIP information from any other source is simply called *"Publisher's* Cataloging-in-Publication Data."

There is some debate over the worth of having CIP data in a book. If a book qualifies for the Library of Congress (LoC) service, the information is periodically distributed to libraries and is a source of information for soon to be published books. Some libraries that wish to maintain significant depth to their collection in certain subject areas may use the pre-publication cataloging information database to identify potential acquisitions for their collection. Since most author-published books do not qualify for the LoC CIP service, they will not benefit from that information flow.

Some librarians have told me that the absence or presence of CIP data would not affect the buying decision for a book. Yet, I have heard others state that CIP data was essential. Yet another managing librarian told me that their system catalogs all new acquisitions without paying much attention to the CIP data, if present. In this case, the librarian indicated that the cataloging might be adjusted to reflect their local collection.

The Library of Congress Control Number (LCCN) is only obtained from the Library of Congress. It is a unique number for every title. Small publishers are able to join the Library of Congress' Preassigned Control Number (PCN) program which allows obtaining an LCCN from the Library of Congress via the Internet at no charge. (In exchange, publishers are required to send a book to the LoC Cataloging Directorate when it becomes available after printing. Go to *http://pcn.loc.gov* to sign up.) The LCCN used to be called the Library of Congress Catalog Number prior to the year 2000. With the new millennium, the numbering system was changed to accommodate the century designation. You should never change an LCCN, even though it identifies the year the book was registered and provides an indication of its age. While there may be marketing considerations that affect the publication date and/or copyright date, the LCCN is fixed at the time the LCCN is obtained. Any discrepancy in the date derived from an LCCN compared to the copyright date can be attributed to the potentially lengthy delay between PCN registration and actual publication.

The row of numbers below the "Printed in the United States of America" line are used as a code to show the printing history. The sample row indicates that the particular edition is in its second printing. Future printings will remove the "3," then the "4," etc. in sequence. The year of the printing will also be adjusted by removing the numbers of years past, e.g. the "05" will be deleted if the next printing occurs in 2006, and so on.

This method of displaying the printing history enables the publisher to tell at a glance in which print run a particular book was produced. (This can be useful in tracking typos and other errors in the book.) This method also minimizes adjustment to (printing) negatives and/or plates, as it only requires the printer to use a dab of correction fluid on either the plate or negative. This technique is not particularly relevant if you use all digital files and produce your work with Print On Demand, disposable plate, or direct to press technologies. However, it is helpful to have some means to identify what printing a particular copy may have come from.

In addition, the copyright page is an excellent place to put any disclaimers or other legalese notices to the buyer. This may include specific permissions wording required in connection with use of previously copyrighted materials. Sometimes various production credits will be placed at the top of the page above everything else. (See Colophon in back matter.)

Finally, note the spelling of "copyright." The word refers to the *right* to copy written material. It is not "copywrite" as I've sometimes seen it spelled.

DEDICATION PAGE (optional) is the next recto (right hand page) following the copyright page, usually the fifth page. If a dedication is made, it should be simple: To Mary or To My Wife, Mary. It isn't necessary to give last names or otherwise identify the person(s) named; nor do you need to give the life dates of a deceased person; although, you may do so if you wish. Extravagant dedications are long out of style. A humorous dedication should be avoided as it is likely to fall flat—particularly as time passes.

ACKNOWLEDGMENTS (optional) appear on the next recto, usually the fifth or seventh page. Sometimes an author would like to acknowledge a number of individuals for their emotional or financial support during the writing and publishing process. These acknowledgments should be brief. (If you have more extensive and specific acknowledgments, put them in the preface.) It is not necessary to give the reader any meaningful indication of the identity or role of those being acknowledged. Those named will know who they are. (Avoid thanking "all the little people who made this work possible.")

EPIGRAPH (optional) page is the next recto. An epigraph is a pertinent quotation at the beginning of a book. The quotation is listed, followed by the author's name on a separate line (last name only, if the author is well known). If there is no dedication, the epigraph appears on the fifth page. If there is a dedication, the epigraph usually will appear on a recto (right hand) page (seventh page), leaving the back of the dedication page blank. Since an epigraph is not part of the text, it should not be referenced with a number to a note to give its source. To save space, the epigraph may be placed on the verso (back) of the dedication page or on a verso (left) page at the beginning of the body of the book.

TABLE OF CONTENTS appears on the next recto. The table of contents usually follows the dedication and/or epigraph. However, some publishers prefer to place the table of contents last after all the other front matter, so it is the last thing before the start of the main text of the book. The table of contents page may start on either a right (recto) or left (verso) page. Sometimes, if the table of contents fits two pages, it starts on the left page and finishes on the right page. If this "double page spread" is used, the recto page just before the table of contents should not be left blank. The title should read "Contents" without "Table of."

LIST OF ILLUSTRATIONS (optional) follows the table of contents on the next recto. If the book has many illustrations and if it is appropriate to list them, then the list of illustrations (normally) follows the table of contents. The page is titled "Illustrations." (Dispense with the "list of.") A list of illustrations is not essential for a book with few illustrations or one where the illustrations are closely tied to the text.

LIST OF TABLES (optional) follows the list of illustrations on the next verso. Similar to the situation with the list of illustrations, a list of tables may be appropriate. The list is not essential for a book with few tables or where the tables are closely tied to the text. If page make up requires, the list(s) may follow the table of contents on the first verso page.

FOREWORD (optional) appears on the next recto (we're probably up to the thirteenth page or so). A foreword is a statement by someone *other than the author*. (An author *never* writes a foreword for his or her own book!) A prominent person may have his or her contribution shown on the cover and on the title page: "With a foreword by Dan Poynter."

The foreword is normally about two to four pages. The author's name (the one who wrote the foreword) is placed at the end of the foreword and may be followed by his or her title or affiliation. A particularly long foreword with a title of its own may show the author's name at the beginning instead of the end. A foreword usually starts on a recto page. Finally, please note the spelling of *foreword*. It is the words be*fore* the author's words.

PREFACE AND ACKNOWLEDGMENTS appears on the next recto page. A preface is written by the author and normally includes the reasons for undertaking the work, method of research (if applicable), acknowledgments (if more specific than the brief acknowledgments option listed earlier—avoid a double dose), and acknowledgement of permissions granted for use of previously published material. (This would be a more general acknowledgement rather than the legalese used on the copyright page.) If the acknowledgments are particularly lengthy, they may be separated and placed following the preface or may be placed in the back matter just before the index. If the preface only consists of acknowledgments, then the section should be entitled "acknowledgments" rather than "preface." In this latter case, it can appear before the table of contents.

Other front matter

INTRODUCTION. A relatively long, substantive introduction not part of the subject matter of the text itself should be paginated with the preliminaries (with Roman numerals, if used). An introduction written by an author to set the scene, however, such as the historical background of the subject, should be part of the text, paginated with Arabic folios.

LIST OF ABBREVIATIONS. In some heavily documented books, especially where there may be references to a few easily abbreviated sources, it may be a convenience to the reader to give a list of abbreviations for the sources before the text rather than in the back matter. If no more than one page long, such a list may be placed on the verso page facing the first page of the text. A long list of abbreviations used in notes and bibliography, and sometimes in the text, is generally best placed in the back matter, preceding the notes.

EDITORIAL METHOD. This is often a feature of a scholarly work. Such notes should be placed before the end of the preliminaries, just before the beginning of the main text. Short, uncomplicated remarks, perhaps that spelling and capitalization have been modernized, should be incorporated into the editor's preface and not put into a separate section.

LIST OF CONTRIBUTORS. In a multi-author book, such as the proceedings of a seminar, it is often desirable to list the contributors, leaving only the editor(s) appearing on the title page. The list may be headed "contributors," "participants," or whatever suits the particular work. The list is usually in alphabetical order with the names in their normal order (e.g. John B. Smith, not reversed as Smith, John B.). Other arrangements (geographical or by academic rank) may also be used, if appropriate. Sometimes short biographies may be included. However, it that case, such a list is more commonly placed in the back matter.

In some cases multiple authors are shown in the table of contents by each of their contributions (or group of contributions). Another treatment is to put a note at the beginning of each chapter indicating the author(s) who contributed that portion of the work.

CHRONOLOGY or a list of events important over a certain period of time may be useful in a volume such as a collection of letters or other documents where the sequence of events is not clear in the text itself. For easy reference, a chronology should be placed immediately before the main text or in the back matter—as is determined to be easiest for quick reference by the reader.

LIST OF CHARACTERS. In some works of fiction, memoirs, or biographical expositions, it is helpful to the reader to have a list of important characters named in the book with information showing their relationships. This may be a simple alphabetical list with prose comments or it might be depicted as an organizational or genealogical chart. A list of characters may optionally be placed in the back matter as determined for easiest quick reference by the reader.

The body

THE MAIN TEXT. The front matter is to serve as a guide to the contents and nature of the book. The back matter provides reference material. The main text should contain everything necessary for the reader to understand the author's message. The text should be organized to help the reader's comprehension. The writer should keep the potential reader in mind and present the material in a logical pattern, selecting the essential and omitting what is nonessential or repetitious.

CHAPTERS. Most books are divided into chapters, frequently of approximately the same length. Chapter titles should be similar in tone, if not in length. Each title should give a reasonable clue to what is in the chapter. Avoid whimsical titles in a serious book, as that can give the wrong impression. Many book buyers scan a table of contents to determine whether a book is worth their time and money. Shorter titles are preferable, both for their appearance on the chapter opening page and for the running heads.

Each chapter normally starts on a new page (verso or recto depending on the design choice). The chapter opening page (chapter opener) does not have a running head, but usually has a drop folio instead. The layout usually consists of the chapter number (without the word chapter), the chapter title and sometimes an epigraph as well. In titles of two or more lines, punctuation should be omitted at the end of a line unless it is essential for clarity.

Chapters are usually numbered sequentially with Arabic numbers. If a book has ten or fewer chapters, Roman numerals may be used, although this is uncommon. Use of Roman numerals for longer works is ill advised because most people have trouble deciphering the larger Roman numerals, e.g. "See chapter XLVIII."

Occasionally, chapter numbers are spelled out. Again, the advice is to restrain this affectation to shorter works to keep the chapter openers relatively clean.

Chapter titles are not required in fiction. If a work of fiction doesn't have chapter titles, the table of contents may be eliminated as well. However, careful selection of the chapter titles can establish a theme or scene or mark the passage of important events in a work of fiction. I usually suggest that fiction have chapter titles (and a table of contents) so that a potential reader can get more of a feel for the book while examining it in a bookstore.

In multi-author books (with each chapter by a different author) chapter numbers may be omitted. The author's name is always given in the display, but an affiliation or other identification is usually not considered part of the opening display but should be placed in an unnumbered footnote on its opening page. The source of previously published material may also be given in an unnumbered footnote on the opening page of the chapter, following the author's identity information, if any.

In some cases, the text may be logically divided into sections larger than chapters. Then, the chapters are grouped into parts. Each part is normally numbered (either Arabic or Roman) and given a part title. The part number and title appear on a recto page preceding the part. The back, (verso) of this page is usually left blank. Chapters within parts are normally numbered consecutively through the book, not beginning over with chapter one in each part.

Each part may have an introduction, usually short, and titled, for example, "Introduction to Part 3." A text introduction to an entire book that is divided into parts preceding part one does not need a part-title page to introduce it. Also, no part title is needed to precede the back matter of a book divided into parts, but part titles may be used for each section of the back matter, e.g. "Appendixes," "Notes," "Bibliography," "Index."

Other divisions

POETRY. In a book of previously unpublished poetry, each poem usually begins on a new page. Part titles need not be numbered but should appear on a separate page preceding the poems grouped under them.

LETTERS AND DIARIES. Correspondence and journals are usually presented in chronological order and are seldom reasonable to be divided into chapters or parts unless some event of historical importance makes such divisions reasonable, e.g. Teddy Roosevelt's journal before being elected to the presidency, journal after election, etc. Dates, used as guidelines rather than titles, are often inserted above relevant diary entries. The names of the sender and the recipient of a letter may serve the same function in published correspondence. However, in the case of letters written by (or to) a single person, the name of that person is not used each time. For example, in *Letters From Mom* "to my son, Jim" and "from my daughter, Mary" are sufficient. The date of the letter may be used in the guideline if it is not included in the letter itself. Such guidelines in diaries and correspondence do not begin a new page in the book unless page makeup demands it.

Subheads

In prose works where the chapters are long and the material complicated, the author (or the editor) may insert subheads in the text as guides to the reader. Subheads should be kept short, succinct, and meaningful. As with chapter titles, they should be similar in tone.

Many works only require one degree (level) of subhead throughout the text. Other works, particularly scientific or technical works, may require sub-subheads and even more subdivisions. Where multiple levels of subheads are used, the primary subheads are called "first level" subheads, the secondary subheads, the "second level" subheads, and so on. Only the most complicated works might require more than two or three subhead levels. (These are sometimes referred to as "A-level," "B-level," and "C-level" heads, using "A," "B," and "C" instead of "first," "second," and "third," etc.)

Subheads, except the lowest level, are each set on a line separate from the text, the levels differentiated by type and placement. The lowest level is often set at the beginning of a paragraph, perhaps in bold, italics, or small caps, followed by a period. This is called a run-in subhead.

Back matter

APPENDIX. An appendix is not an essential part of every book. Some of the material properly placed in an appendix include explanations and elaborations not essential parts of the text, yet helpful to a reader seeking further clarification; texts of (supporting) documents; laws; long lists; survey questionnaires; sometimes charts or tables. The appendix, however, should not be a repository for raw data that the author was unable to work into the main text.

When more than one appendix appears in a book, they should be numbered like chapters (Appendix 1, Appendix 2, etc.) or assigned letters (Appendix A, Appendix B, etc.). Each appendix should also be given a title. The first appendix usually begins on a recto page; subsequent appendixes may begin either recto or verso.

Sometimes, an appendix is more reasonably placed at the end of a chapter if the content is essential to understanding the chapter. In this case, the "chapter appendix" may start on a new page (either recto or verso) or it may run on at the end of the chapter after a suitable space, perhaps 4 to 6 lines. In a multi-author book, appendices should run following the chapter it supports. Otherwise, care should be taken that appendices following chapters do not interfere with the comprehension by breaking the flow of the main text.

NOTES. A section of notes or references follows any appendix material and precedes a bibliography. When notes are arranged by chapters, with references to them in the text, chapter numbers and titles should appear above each relevant group of notes. Each group of notes runs on after a short break of a line or two from the previous group.

As with appendices in multi-author books, notes would appear at the end of each author's contribution. Optionally, notes might appear at the ends of each chapter or may be placed as footnotes. For the sake of minimizing complexity of page composition, a section of notes in the back matter is preferred.

When importing text from a word processing program into a page layout program, footnotes are usually automatically moved to the end of the block of imported text. This transition can sometimes wreak havoc with automatically numbered notes. It is better to isolate notes in the word processing file and copy them to a separate word processing file before importing them to the page layout program, or to import each chapter separately into the page layout program so that the notes are automatically placed at the end of the relevant chapter.

GLOSSARY. If a work contains many foreign words, technical phrases, jargon, or slang, a glossary is often helpful, particularly if the work is intended for a general reader who may be puzzled by words and phrases not in the common vocabulary. Words or phrases to be defined should be arranged in alphabetical order followed by its definition. A glossary precedes a bibliography. For example, due to the amount of jargon used in the crafts associated with this book, an extensive glossary has been included.

BIBLIOGRAPHY. The form of a bibliography varies with the nature of the book, the author's approach, the advice of an editor, and the guidance, policies, or suggestions of the publisher. In some cases, the bibliography may represent suggested additional reading or potential resources for the reader to do further research on the primary topic. Consult a style guide for specifics in setting a bibliography or refer to the bibliography of this book for one example. The bibliography may be entitled "Resources" if that is a more accurate descriptor for the material.

INDEX. The index usually begins on a recto page. If there are additional indexes, they begin verso or recto. If there are name and subject indexes, the name index precedes a subject index. Consult a style guide for additional index types and forms. (*The Chicago Manual of Style* has an excellent and extensive discussion of indexes.)

ABOUT THE AUTHOR. A brief biography and photo are often placed in the back matter. The about the author page is usually a recto and may appear immediately following the main text or it may be the last page (a verso) of the book. Of course, both the about the author page and the photo are optional.

COLOPHON. A Colophon is a listing, usually on the last page of a book, providing design and production credits. The inclusion of a Colophon is not common, but it is a sign that the publisher particularly cares about the production values of the book. Some or all of the information usually put in a Colophon may appear at the top of the copyright page above any disclaimers, copyright notice and cataloging data. A Colophon might read as:

> Book Design and Production
> A Guide for Authors and Publishers
> Cover and interior designed by Pete Masterson, composed by
> Æonix Publishing Group in Adobe Minion with subheads in
> Myriad and display lines in Rubino Sans using Adobe InDesign
> CS. Printed by XYZ Printing Company on Hammermill Offset
> Opaque No. 5.

Some printers object to having a credit in the books they print. I find this hard to understand, but it is fair to ask your printer before putting their name in.

Sometimes, there isn't room for a full colophon, so a brief version may appear at the beginning of the copyright page. This brief "colophon equivalent" may simply credit the cover and interior book designer(s).

ORDER FORM. While the traditional book concluded with a Colophon, it is now appropriate to place an order form at the back of the book. This is a place to offer the reader the opportunity to order additional copies of the book for friends or business associates or to direct readers to a web site where current information or updates in support of the text might be obtained. Some books also lend themselves to selling associated items, such as difficult to find items mentioned in the book (for example, a testing kit from a book on particular health issue), or advertising specialties, such as T-shirts or coffee mugs. A small publisher may also offer other, similar books or related titles.

It is important to be mindful of the annoyance that some booksellers feel when they see an order form in a book. To soften such annoyance, I usually insert wording at the top that reads, "If you enjoyed this book and would like to give a copy to someone else, you may obtain one at your favorite bookstore, online bookseller, or use a copy of the form below."

Page numbering

Books are numbered from the first page traditionally (in the U.S.) with lowercase Roman numerals for the front matter and Arabic numerals (starting again at page 1) for the main text. In the alternative, many books are now numbered with Arabic numerals from the half title page as page 1 straight through to the end. A page number on the page is called a folio. (Numbers do not necessarily appear on every page—blank pages should have neither folios nor running heads.) Pages before the Table of Contents usually do not display folios.

Publishers with a shorter book may wish to use the straight through numbering pattern as readers often determine the book length by looking at the last page number in the book. If the front matter has Roman numerals, this technique will understate the number of pages in the book. (This is not deceitful, as all the pages are present. It is simply a means of causing the viewer's perception to coincide with the reality.)

Blurbs

Prepublication copies are often distributed to "opinion leaders" and other respected individuals in a field to obtain their comments. (These are referred to by the technical term "blurbs.") The cover letter with the prepublication copies usually requests the comments and clearly indicates that they are intended to be published on the cover or in the book. Sometimes, in a later printing, excerpts from reviews (particularly from reviewers of high standing) are added to the previously collected comments. While the "best" blurbs should be placed on the cover or dust jacket flaps (if published in a hard cover binding), overflow blurbs may be placed on the last pages or (better) before the half-title page. If the overflow blurbs are being added to a second or subsequent printing, often the half-title page is eliminated in favor of the overflow blurbs. (You don't want to force a repagination that changes the imposition from the earlier printing, as that can add considerable expense to a book produced with offset printing.)

If signature makeup requires extra blank pages, these might be distributed by placing one or two leaves at the front with the remainder at the back. Be sure to move an order blank, if present, to the last physical page. Such "bonus" pages might be used to acquaint readers with future projects planned by the writer or other promotional material. The blank leaves at the front can be used for blurbs in future printings. (These extra pages should not be included in the page numbering.)

Let's Talk About Type

TYPOGRAPHY, A CRAFT WITH MORE THAN 500 years of history, has developed certain terms, assumptions, and traditions. Before we can begin to actually design our document, we must first gain a basic understanding of type: How does it look? What different kinds of type are there? How do the various kinds of type relate to one another? And how do we use type to communicate?

Take a look at this diagram and then review the terms on the next pages:

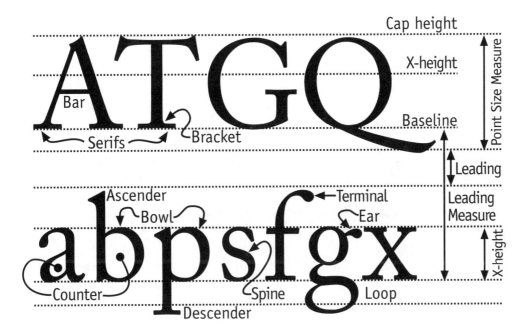

The basic type glossary

To talk about type, we must first understand the basic language of type. Look at the figure on the previous page to identify the various elements that are used to describe type and distinguish type into a particular typeface.

ARM: A horizontal stroke that is free on one end. The upper case E has three arms.

ASCENDER: The part of the lowercase letters b, d, f, h, k, l, and t that extends above the height of the lowercase x.

BAR: The horizontal stroke in the A, H, e, t, and similar letters.

BASELINE: An imaginary line on which the letters rest. Note that some curved letters (such as a capital "O") will extend slightly below the baseline.

BLACKLETTER: Type designs based on mediæval type forms; in England sometimes called gothic.

BOWL: A curved stroke which makes an enclosed space within a character. The bump on the letter P is a bowl. (The enclosed area is the counter, see below.)

BRACKET: A curved fill in the angle created by the intersection of the serif and main stroke of a letter.

CAP HEIGHT: The height of the capital letters above the baseline.

COUNTER: The full or partially enclosed space within a character. (e)

DESCENDER: The part of the letters g, j, p, q, y and sometimes J that extends below the baseline.

EAR: The small stroke projecting from the top of the lowercase g.

GOTHIC: Another name for type designed without serifs (sans serif). Sometimes confused with highly ornate early type designs known as blackletter.

HAIRLINE: A thin stroke usually common to (modern) serif typefaces.

LEADING: (pronounced *led-ing*) An additional space added between the bottom of the descenders and the top of the ascenders in subsequent lines of type. Usually it is expressed as a baseline to baseline measure, e.g. 10 point type with 12 point leading, therefore an additional 2 points of space is added between the rows of type. In the metal-type era, this was usually done with strips of lead.

LINK: The stroke connecting the top and the bottom of a lowercase g.

LOOP: The lower portion of the lowercase g (in most serif typefaces).

MONOWEIGHT: A typeface with no thick/thin transition, the width of the strokes making up the letter are all the same.

SERIF: A line crossing the ends of the main strokes making up a letter. There are many varieties of serifs, which will be discussed later. Type designed without serifs is called sans serif, gothic, or (from Germany) grotesk (grotesque).

The serif originated with ancient stone masons. When cutting letters into stone, the chisel was turned across the groove at the end of the letter stroke to make a clean, smooth finish to the line. These cross cuts also assisted the eye to line up the letters and see them as positioned evenly along the line of words.

SHOULDER: The curved stroke of the h, m, and n.

SPINE: The main curved stroke of a lowercase or uppercase S.

Spur: A small projection of a main stroke; found on many uppercase Gs.

Stem: A straight vertical stroke, or main straight diagonal stroke in a letter which has no vertical strokes.

Stress: The direction of thickening in a curved stroke. This is often a clue used to classify a typeface as old-style, transitional, or modern.

Swash: A type design with a flourish replacing a terminal or serif.

Tail: The part of a Q which makes it look different from an O, or the diagonal stroke of the letter R.

Terminal: The end of a stroke not terminated with a serif.

Thick/Thin Transition: The gradual change in the width of curved strokes.

X-Height: The height of lowercase characters excluding ascenders and descenders. The x-height is extremely important in determining the relative readability of different typefaces. A larger x-height generally makes a face more readable in a smaller point size.

Measurements

Agate: A measure equal to ¹⁄₁₄ of an inch. Agate type is a traditional name for 5½ point type (which is about ¹⁄₁₃ of an inch in height). An agate line is one column wide by ¹⁄₁₄ inch, traditionally used in publications for selling advertising space. Fortunately, Agate measurements are no longer frequently used in the United States.

Cicero: A typographic measurement system used in Europe. Approximately the same as our Pica.

Pica: A measure equal to 12 points or approximately ⅙ of an inch.

Point: Basic increment of typographical measurement, equal to .013837 inch. Computer programs usually round the point to .0138888 inch making exactly 72 points to the inch. This is a rounding loss of approximately ¼ point per inch. With the ascendancy of desktop computers, the 72 points (and 6 picas) per inch has become a de facto standard measurement. Care must be exercised as point/pica measuring rules (type gauges) can be found both with the traditional measure and the new rationalized measure.

Typeface names

Under current U.S. copyright law, the physical rendering of a type design is not protected by copyright. However, the names given to typefaces are able to be protected by trademark. The result is that competing type foundries have typefaces very similar in appearance but with different names. For example, Helvetica (Linotype-Hell) when produced by other sources is called things like Arial, Helvette, Geneva, or Swiss. These alternate typefaces are called analogs.

Some typeface names were never protected or the names came into use long before copyright and trademark law was devised. In these cases, you will find typefaces from different manufacturers with the same name, but the typeface itself may be quite different in design. More frequently, type designs with similar or identical names (but from different sources) may be very much the same in appearance, but with minor differences. It is, however, considered unethical to make exact copies of a designer's work and merely re-label it. Legal disputes in the past few years have established that *the soft-ware* that creates the display of type on a computer or controls the output in a computer-driven printer is protected by copyright, even though the letter forms themselves can not be copyrighted. Before the personal computer, each typesetting machine manufac-turer would re-draw the most popular faces to create a version to work on their propri-etary machines. With the arrival of the desktop computer, typefaces are now available in standardized formats (either TrueType, PostScript, or OpenType). The competition has caused some bitter rivalries and incidents of cheating have resulted in legal action against those who merely copied the underlying computer code. Despite the turmoil in the type industry, the result has been lower typeface prices for users.

Nonetheless, it is of extreme importance to know what brand of type you are using and how it may vary from other similar faces. While entirely similar in appearance, the spacing attributes of type (font metrics) from different sources may cause severe prob-lems with your layout if type from one source is substituted for type from another.

It is also important to know if you are using PostScript, TrueType, or OpenType in your project. Substitution of one for the other can cause utter chaos with a layout as the type widths and letterspacing of the fonts (the font metrics) are frequently differ-ent even if they are from the same manufacturer and have the same typeface name. It is also a poor idea to mix PostScript and TrueType faces in the same document. In the past, some desktop laser printers and many high resolution imagesetters would fail to generate acceptable output when both font formats are present in the document.

In recent years, this is less of a problem as equipment has been upgraded and/or replaced. Indeed, the new OpenType format can package either a TrueType or a Post-Script font within. The user has no simple way to determine the internal format of an OpenType file, unless a specialized font management program is used. Still, the bias against TrueType continues in the graphic arts field, due to the rather poor experiences encountered when TrueType was first introduced.

Another cause of a poor reputation for TrueType is the large number of "free" fonts available in TrueType format found on the Internet. Most of these were created by hob-byists and they may not have been carefully crafted or fully tested. They may work on home computer equipment well enough, but poor quality in the structure of the type-face may cause errors on high resolution output devices. (It's wise to avoid the "free fonts" that are widely available—or, if you must use one, send a test file to your book printer to check for problems before deadlines encroach.)

Some basic typographical considerations

Research has shown that serif faces are generally easier to read. Therefore it is recom-mended that body copy be set in a serif face with headlines set in a contrasting sans serif face. For example, a safe (if overused) combination is Times body copy with Helvetica heads. As you plan your design, try various combinations to see what looks best—look at examples shown here and in newspapers, magazines, or books—and copy what works. (The research shows a very strong advantage of serif typefaces for comprehension among

English speaking readers. Many non-English speaking European countries make extensive use of sans serif fonts in schoolbooks, etc. Research tends to suggest that readability is also strongly influenced by early childhood experiences with various typefaces.)

SET TYPE WITH ONLY ONE SPACE AFTER A PERIOD. Traditional typewriters had mono-spaced fonts (every letter takes up exactly the same space on the paper). It was therefore necessary to separate sentences with two spaces to clearly show that they were distinct from the separation of words. With the arrival of computers, word processing, and proportional typefaces, it is no longer

```
The quick brown fox jumped over
the dog.  Bad dog!  Now is the time
for all good men to come to the aid
of their party.  The knight rode away.
```

The quick brown fox jumped over the dog. Bad dog! Now is the time for all good men to come to the aid of their party. The knight rode away.

Sample contrasting monospaced (Courier) type with proportional (Minion) type. The Courier has two spaces after punctuation, while the Minion only has a single space following punctuation.

necessary (and is, indeed, an error) to separate sentences with two spaces. This is the one key "rule" that identifies the typography amateur from the typography professional.

THE NEXT BASIC RULE IS THAT "CURLY" QUOTES SHOULD BE USED instead of typewriter quote marks. Due to the limitations of the typewriter keyboard, universal opening and closing quotes were used, however in typography, true opening and closing single and double quote marks (and apostrophes) should be used. Take special care with the apostrophe—many word processing programs have "automatic quotes" that will put opening and closing quotes based on the relative position of the insertion point (cursor) to the between-wordspace. However, when used to indicate a date that is shortened (1997 = '97) the automatic quote feature normally assumes an opening single quote, rather than the apostrophe which always has the fat part at the top.

The exception to the curly-quote rule is in the case of marking feet and inches. When referring to measurements, the proper indicator is the prime mark. Since every computer program may not be able to access true prime marks (e.g. single prime ′ and double prime ″) it is most appropriate to substitute the single and double "typewriter" quote marks for the prime marks. (To accomplish this, it may require turning off the automatic quote feature of the software being used.)

AVOID SETTING TYPE IN ALL UPPER CASE LETTERS. In an experienced reader, the eye and brain processes words as a whole, not as individual letters. When type is set in ALL UPPER CASE the usual word-shapes are lost and the eye must slow down to allow the brain to spell out each word. If upper case letters are used, such as an acronym, the word should be set slightly smaller than the rest of the words (1 or 2 points), if possible.

Those of us who live in the USA (reduced 1 point) often take certain things for granted. I doubt if most people know what RSVP (small caps) *really* means.

IF YOU WISH TO SHOW EMPHASIS, YOU SHOULD USE ITALICS OR BOLD TYPE. While most computer typefaces come with a bold italic weight, this is typographical overkill and should be reserved for unusual situations or circumstances other than for showing emphasis. Underlines are also rarely used in typography. If the manuscript uses underlined words, normally such material is set in italics.

If you wish to show *emphasis,* (note that the comma is also set in italics) you should use *italics* or **bold** type. While most computer typefaces come with a ***bold italic*** weight, this is typographical overkill and should be reserved for unusual situations or other circumstances other than for showing emphasis. If the typeface offers several weights of bold type, use the next darkest weight from the body copy. For example, this paragraph uses Minion **semi-bold** for emphasis.

If the body copy uses Minion medium rather than Minion regular, then the **bold weight** would be used rather than the **semi-bold weight**. These differences can be fairly subtle. Bold, heavy, and black weights (if available) are reserved for special purposes. For example, in an ad where type is set reversed against a dark background, the body might be set with semi-bold with emphasis established with heavy or black weight type. This is done because reversed type always appears less heavy than when set normally as dark type against a light background.

Usually, the reading line should run about 60 to 70 characters for easiest eye movement. Longer lines should have more leading to keep the eye on course. Larger type (headlines) can be set with less leading.

Keep body copy in a range of 9-12 points. A frequently noticed "computer effect" is type set too large. Larger type is often easier to read (or just looks better) on a computer screen, but the result can be overwhelming on the sheet of paper!

Finally, watch out for closure of counters in small type. Low resolution printers (including desktop lasers) can not always render small type clearly. A typeface with a large X-height will hold up better in small point sizes.

Typeface classifications

There are literally thousands of typeface designs in use today. It may seem a daunting task to describe and classify each and every design. However, just like the Botanist, who classifies thousands of plant species, we can divide type into a number of classifications. Some typeface designs may be included in several classifications at the same time, while others may only fit a single classification. Sometimes the choice is obvious, other times the most subtle distinctions are used to make the determination—a few will cause students of typography to argue the classification without any final determination.

Humans classify things to create order from chaos. From foods (fat, carbohydrate, protein) to the game of charades (animal, vegetable, mineral), we have classified things to organize and analyze them. So it is with typefaces.

At the most basic level, all typefaces can be divided into four broad categories:
Serif
Sans Serif
Script
Decorative and other

SERIF TYPE has small points, lines, or other shapes at the end of each stroke. These evolved from scribes making an oblique hand movement to avoid formation of ink blobs at the ends of the letters—and from the work of ancient stone masons who discovered that striking their chisel at a 90° angle at the end of each letter stroke gave a more complete, even appearance to their work.

Adobe Garamond
LITHOS SANS
Zapf Chancery
Party by Letraset

Samples of the four basic typeface classifications.

SANS SERIF type are letters without any extra embellishment. Sans is the French word for "without." So, sans serif is literally "without serifs." While some think of sans serif typefaces as modern, they were frequently used in ancient Greek inscriptions. The typeface Lithos is modeled in the style of the ancient Greek inscriptions carved into monuments.

SCRIPT typefaces are those that appear to have been written by hand or with a brush. These come in styles ranging from very casual to extremely formal. Some appear as individual letters while others seem to connect together like true handwriting. With clever computer programming techniques, some script typefaces actually have random variations in the letters to make them appear as if handwritten.

DECORATIVE AND OTHER, naturally, is everything else. The range of this category is truly amazing. The main characteristic that puts a typeface in this basic category is that it (1) doesn't fit the other categories and (2) it's probably inappropriate for lengthy texts. This grouping also includes typefaces made up entirely of non-letter symbols, such as the symbols used to write music or for symbols on a map. These symbolic typefaces are often called "pi fonts" after the mathematical symbol pi (π).

More than four

This initial division of typefaces is sufficient to get started, but there are many subdivisions to consider to enable you to make reasonable and appropriate typeface choices for your project.

Serif typefaces can be further broken down into these additional categories:
Oldstyle
Transitional
Modern
Slab Serif
Clarendon
and Other Serif fonts (Those serif fonts that defy classification in any of the previous categories and don't otherwise belong in the general "decorative" category.)

OLDSTYLE TYPEFACES are characterized by serif formation, stress emphasis, and relative line weight. These faces have an appearance of having been drawn with a wedge shaped pen held in the hand. Oldstyle typefaces always have serifs and the serifs of the lowercase letters are always at an angle (the angle of the pen). Because of the pen, all the curved strokes in the letter forms have a transition from thick to thin (the thick/thin transition). The contrast between the thick part of the stroke and the thin part of the stroke is relatively moderate. If you draw a line through the thinnest parts of the curved strokes, the line is diagonal (oldstyle has a diagonal stress). This is the style that arises from the earliest Roman types and can generally trace ancestry to the *Carolingian miniscule* (script) imposed by Charlemagne.

The oldstyle typefaces are those most familiar to English speakers who learned to read in North America. These typefaces tend to look pretty much the same to anyone who hasn't studied typography. Their lack of unusual artifacts and distinguishing characteristics make them nearly "invisible." Think of oldstyle type as an old friend, since it is highly readable. **These are the prime candidates for most book projects.**

Bitstream Arrus	OHamburgetfiv
Berkeley Oldstyle	OHamburgetfiv
Bembo	OHamburgetfiv
Adobe Caslon	OHamburgetfiv
Centaur (Monotype)	OHamburgetfiv
Adobe Garamond	OHamburgetfiv
Goudy Oldstyle	OHamburgetfiv
Janson (Monotype)	OHamburgetfiv
Jenson	OHamburgetfiv
Minion	OHamburgetfiv
Palatino	**OHamburgetfiv**
Sabon	OHamburgetfiv

Samples of oldstyle typefaces, all set at 12 points.

As you look at the samples, note that typefaces with the same nominal point size result in significantly different apparent sizes. See also how the width varies as well. Your book length will change based on the actual typeface as well as the point size.

TRANSITIONAL typefaces share many characteristics with oldstyle faces, but also exhibit characteristics with modern typefaces. The stress of curved letters is often diagonal but the letter forms are less obviously based on handwriting. Serifs on ascenders are oblique and bracketed, and the bar on the lowercase "e" is (usually) horizontal. **Many transitional typefaces are reasonable candidates for a book project.**

Electra	OHamburgetfiv
Weidemann	OHamburgetfiv
Augustea	OHamburgetfiv
Baskerville	OHamburgetfiv

Samples of transitional typefaces, all set at 12 points.

If you choose a transitional typeface, be careful that it provides a comfortable reading experience to your intended audience. Books aimed at older and possibly vision impaired readers may have more difficulty with the generally lighter appearing transitional typefaces than they might with an oldstyle typeface.

MODERN typefaces were developed in the mid to late 1700s. At that time, they were "modern" and were so named. Of course, some 250 years later they are no longer modern in the full sense of the word, but the title remains.

Improvements in the technologies of printing, paper, and type manufacturing converged to support more "mechanical" type designs. Modern faces have serifs, but the serifs are horizontal instead of slanted, and they tend to be very thin. The thick/thin transition is very severe, with radical contrast between the strokes. Stress is perfectly vertical. Modern faces tend to have a cold, elegant look. Due to the extreme thin lines, modern faces need greater care in reproduction to avoid dropping out of the thin strokes. In extended amounts of copy, the thin lines tend to disappear with the thick lines being most prominent; the effect on the page is called "dazzling."

Of the modern typefaces, Bodoni (pronounced *bo-DOUGH-nee*) tends to be most popular with graphic designers. However, the high contrast between the thick and thin strokes can significantly hurt the comprehension of those reading this typeface. Bodoni

has been drawn (and redrawn) by many type foundries—and the various versions are not all equal. Some are considerably more readable than others. If you must choose a Bodoni, take care that you are using one of the better (more readable) variants of this typeface.

Acanthus	OHamburgetfiv
Bodoni	**OHamburgetfiv**
Didot	OHamburgetfiv
Walbaum	OHamburgetfiv

Samples of modern typefaces, all set at 12 points.

As a group, **the modern typefaces are generally a poor choice for most book projects.** This is especially the case if the project is to be produced with toner-based digital printing equipment due to the relatively low resolution of those devices. Learn to recognize modern typefaces so you can avoid them.

CLARENDON, SLAB SERIF, OR EGYPTIAN. With the industrial revolution came advertising. At first, modern typeface designs were merely thickened, but the result was difficult to read and from a distance almost looked like a picket fence. The obvious design solution was to also thicken the entire letter form. The design is called Claren-

don because Clarendon is the epitome of this style.

Slab serif is also descriptive, since the serifs are squared-off like a slab. The term Egyptian is used because many of the slab serif faces were designed during one of the periods of "Egyptomania" that swept our culture in the early part of the nineteenth century and again in the early twentieth century with the discovery of King Tutankhamen's tomb. Many typefaces in this category have Egyptian

Clarendon	**OHamburgetfivA**
New Century Schoolbook	OHamburgetfivA
Glypha	OHamburgetfivA
Memphis	OHamburgetfivA
Alexandria	OHamburgetfivA
Officina Serif	OHamburgetfivA

Slab serif typefaces come in two distinct styles. The Clarendon style has stroke variations more like old-style faces. The Egyptian slab serif style has an even stroke more like sans serif faces.

names for marketing purposes, e.g. Alexandria, Memphis, Cairo. Many of the slab serifs that have a slight thick/thin contrast (such as Clarendon or New Century Schoolbook) are highly readable, solid typefaces suitable for books or other extensive text. Due to their relatively low thick/thin contrast, slab serif faces tend to give the page a darker look ("color") than oldstyle faces. With their clean, regular letter forms, slab serif typefaces are frequently used in children's books. Slab serif faces with very little or no thick/thin contrast (monoweight) are somewhat more difficult to read and should be reserved for shorter works or as headings and titles.

Typefaces to avoid

Times Roman and its variant Times New Roman is the default for many word processing programs and it is available on almost every computer made during the last 20 years. If
for no other reason, this common use of Times (and its variants) should be a good reason to avoid it for producing books.

| Helvetica | OHamburgetfiv |
| Times Roman | OHamburgetfiv |

However, to make matters worse, Times was originally designed for the narrow columns of a newspaper (The London Times). It is a fairly narrow typeface (which may seem attractive since you can set a lot of text in a short amount of space) but the narrowness makes long lines of text uncomfortable to read. Times also shares the fine stroke lines of the modern typefaces. The original was intended to be printed with letterpresses on soft newsprint—both the paper and printing method cause the type to print with darker and heavier strokes than the type design would otherwise suggest. Modern printing methods don't have this darkening effect to the extent as the prior methods—with the result that Times is even less desirable than ever. While it might be inconvenient, use something other than Times for the layout of your book.

Helvetica and its variant Arial are also overused. It is also the primary typeface used by the IRS in its forms and instructions. Need we say more?

Sans serif (gothic, grotesk)

Sans serif (French for "without serif"), gothic, or grotesk (the German term) typefaces were not extensively used until the early part of the twentieth century. A primary proponent of sans serif typefaces was the highly influential typographer, Jan Tschichold (1902-1974). His writings on the philosophy of typography revolutionized the view of good typographical design. His seminal work with strong influence of the Bauhaus art movement, *The New Typography: a Handbook for Modern Designers,* (published in Germany in 1928) created an awareness of using modern æsthetics in typographical design. (While a number of English-speaking writers have produced many articles and books discussing the design philosophy advanced by Tschichold, it is interesting to note that an English translation of the complete 1928 book was not produced until 1995, published by the University of California Press, Berkeley, California.)

In his later years Tschichold backed away from many of the hard and fast "rules" he established in *The New Typography,* declaring that **good book design had, above all, to be perfectly legible as the result of intelligent planning.** The basic design concepts he taught have had strong influence on typographic design through subsequent generations of design professionals.

Sans serif typefaces are almost always monoweight, that is, there is virtually no visible thick/thin transition in the strokes; the letter form has the same thickness all the way around.

In the post-World War II era, sans serif type forms dominated the marketplace. While their use in advertising is well established, many books were also printed using sans serif faces. Recent scientific testing has determined that serif faces offer far superior reading and comprehension characteristics for long documents and books. However, good design uses contrast in color, weight, and style to add interest to the page. To achieve this

Franklin Gothic	Hamburgetfiv
Gill Sans	Hamburgetfiv
Optima	Hamburgetfiv
Serif Gothic	Hamburgetfiv

Samples of sans serif "gothic" typefaces, all set at 12 points with 14 points of leading. Observe that Serif Gothic has vestigal serif points at the ends of the strokes, while Optima has a definite thick/thin transition.

contrast, it is often recommended to use appropriate sans serif faces for titles, headers, and other elements.

Comprehension is also what we're used to. In non-English speaking Europe, sans serif typefaces are quite frequently used, so readers with that cultural background are more comfortable reading sans serif type. In the English-speaking world, serif typefaces remain dominant as the chosen form for publishing books. As a result, tests of comprehension show a very significant loss when material is set in sans serif type. These same tests, if done in France or Germany might have different results.

Sans serif typefaces do not have as many elements to manipulate as do the serif faces, however there are still subcategories:

Gothic, Grotesque (Grotesk)
Geometric
Humanist
Other

Gothic, Grotesque (Grotesk). The term gothic was applied to early sans serif type after the Boston Type and Stereotype Foundry issued a new series of types without serifs under the name Gothic. It may have been the bold weight of this type that prompted the designation. The Boston firm was the first American foundry to produce a serif-less design, then achieving great popularity in Europe, particularly Germany and England. In Britain, the type was described with the term "grotesque" which is echoed by the appellation "grotesk" used as part of the name of some sans serif designs originating from German foundries.

The best example of a traditional early-nineteenth-century sans serif face is Franklin Gothic. It retains certain features common to Roman serifed faces. For example the lowercase "a" and "g" are normal Roman characters, and in all of the letters there occurs a thinning of stroke at the junction of rounds to stems. Some of the contrast of

Roman letters also persists, although the overall appearance is monotone.

These type characteristics are sometimes used to describe the "humanist" typefaces. Humanist faces use the proportions of Roman inscriptions with more contrast between thin and thick strokes and sometimes with slightly fluted stems. The humanist term was used to contrast with the mechanical sameness of the geometric sans faces.

It was with the traditional sans faces that type foundries (1900–1910) started offering typefaces in a number of weights (light, bold, etc.) and widths (extended, condensed, etc.) suggesting to printers that they needed to have these variations to meet customer needs.

Metro	OHamburgetfiv
Akzidenz Grot.	**OHamburgetfiv**
Univers	OHamburgetfiv
Neuzeit	OHamburgetfiv

Samples of grotesque typefaces, all set at 12 points.

Grotesque sans typefaces, while still somewhat mechanical in appearance, but having somewhat more proportionality, are intermediate between the geometric sans and the humanist sans faces. Some of the grotesque sans faces may be used effectively in books for headings and other short passages.

GEOMETRIC SANS. During the period from 1926 to 1950, the geometric sans faces became popular. Indeed, the geometric unserifed faces, as represented by Futura, dominated commercial printing. But by the end of World War II a reaction to Geometric Sans faces set in with critics stating that Futura (and the other geometric sans faces) were too cold for modern tastes. Swiss typographers, who were in the vanguard of a search for a new typography that fitted the needs of a postwar generation, described the geometric sans as being "mechanical." However, the younger designers did not return to the Roman letter, but reaffirmed the sans serif ideal by a return to the traditional turn of the century gothics.

Avant Garde	OHamburgetfiv
Cable	OHamburgetfiv
Metro	OHamburgetfiv
Futura	OHamburgetfiv

Samples of Geometric Sans, all set at 12 points.

Geometric sans serif typefaces are characterized by having an even stroke weight and often appear to have been created with drawing implements—perfect round curves and straight lines.

In general, geometric sans serif faces are not the best choice to pair with oldstyle serif faces, but there are some exceptions.

HUMANIST sans typefaces are proportioned more like the oldstyle serif typefaces. Stroke weights vary and the shapes are more organic. They have a more human scale, hence the term "humanist" is used to describe them. The humanist typefaces are the work-horses of the sans serif form. They are generally easy to read and

Dax	OHamburgetfiv
Frutiger	OHamburgetfiv
Myriad	OHamburgetfiv
Syntax	OHamburgetfiv

Samples of humanist typefaces, all set at 12 points.

pleasant to look at. While not the best choice, if a book were to be set in a sans serif face, one from the humanist category would be the most reasonable choice.

Script categories

Script typefaces can be categorized into several groups. While most typographers are familiar with the groupings of serif and sans serif fonts—and can usually agree on the classifications, there is less standardization of classification for script typefaces. Script typefaces might be used as chapter titles or for the book title on the cover. The following are categories that I find useful:

Formal
Calligraphic
Brush
Marker pen
Handwriting
Child's handwriting

FORMAL SCRIPTS are highly ornate and very calligraphic. This category is the work-horse of social announcements, wedding invitations, and the like. They are difficult to read. They should *never* be set all caps, and are best if kept to very short passages. A formal script might be suitable for a book title or other embellishment on a cover, but they generally have no application in book production. The sample is Regency Script.

CALLIGRAPHIC SCRIPTS are strongly based on lettering done by scribes and have a strong pen and ink appearance. Some have been more regularized into type and have less of the pen and ink look—but still have the basic characteristics. Zapf Chancery was inspired by the "chancery" scripts used in the middle ages. Again, this should be limited to short passages and *never* set all caps. Some might be suitable for chapter titles or even subheads if the text can support that style.

BRUSH SCRIPTS appear to have been written with a very wide pen or an artist's brush. There are a number with the name "brush script." The one shown is from Bitstream. Limit application to book interiors with possible use on a cover for a title or other element. You may have heard this before, but I'll repeat myself: *Never* set a script in all caps.

MARKER PEN SCRIPTS appear to have been written with a felt tip pen. Often blobs are shown at the ends of letters as if the pen was allowed to sit for a moment and more ink was absorbed by the paper at that point. These tend to be newer designs (marker pens and felt tip pens don't have a long history). Some of these script styles might be used as chapter titles or even subheads. If the project is right, they could be used for a book title or other cover elements. Many are quite readable and have a "feeling" similar to a very casual variant of a humanist sans. Since the caps are relatively unadorned, if you really must, these may be set all caps.

Samples of script typefaces, all set at 28 points with 36 points of leading. Note that script typefaces often "set small" and a much larger than normal point size is usually required to get an expected size.

HANDWRITING SCRIPTS are those that mimic "normal" handwriting. While formal scripts may appear to have been written by hand, they are clearly mechanically designed and carefully rendered. Handwriting scripts may be smooth and even, like sample cursive writing in a text book, or rather idiosyncratic as most people write. The sample is based on the handwriting of the artist, Cezanne. Obviously, these should not be set all caps. (At least, I hope it's obvious!)

CHILD'S HANDWRITING SCRIPTS might be classified as a subcategory of "decorative" typefaces, but are included with scripts due to the appearance of having been written by hand. Obviously, these mimic a child's efforts to master the alphabet. Some may appear to have been written with a crayon while others might mimic other writing implements. Frankly, some feel more realistic than others, but most have a child-like quality. While setting lengthy passages is a poor idea, these faces might make nice touches in a children's book or in a book for adults on a child-related topic. Note: Both Cezanne and Toy Box are from the font foundry "P22," a small company that makes a wide variety of very creative and attractive typefaces.

Decorative and other

The last broad category is "decorative and other." This is by far the largest category of typefaces. There are few, if any, candidates for setting lengthy blocks of text here, but there are many possibilities for chapter titles and for cover typography. As with scripts, there is little standardization of subcategories. As my typeface library grew, I found these categories helpful:

Symbol (pi) fonts
Blackletter/Uncial
Antique Look
Victorian
Circus
Old West
Art Deco
Art Nouveau
Moderne
Grunge
(Other) Ornamental

SYMBOL (also called "pi") fonts are those made up of various non-alphabetic glyphs. They may be specific to a field, such as Mathematic Pi, or more general, such as Zapf Dingbats, Wingdings, or Universal News Pi. These special typefaces may have just the right symbol to establish a tone or create a theme for blocks of text. Pi fonts are available in an amazing number and broad range of styles and subjects, ranging from American Sign Language hand-signs to a menagerie of zoological figures.

Many of the oldstyle typefaces that come with "expert sets" (which contain additional glyphs such as small caps, swash capitals, or alternate number sets) will include a symbol font with printers' ornaments designed for the particular typeface. For example, the typeface used here, Minion, has these symbols in its extended character set:

The last three, in particular, make an excellent choice as a "break" symbol as used in a novel where there is a break in action or a change of scene. While a manuscript often uses three asterisks (* * *) for this indication, to use them in a book shows a lack of imagination and poor typographic style. If you are using a typeface that does not have any printer's ornaments (or does not have an extended character set), then it is often possible to find a compatible symbol in another typeface or an otherwise unrelated pi font.

BLACKLETTER AND UNCIAL fonts are those that most closely resemble the ancient texts. Uncials, used from the fourth through the eighth century, were developed as a faster means of writing. Eventually miniscule writing, the basis of our lowercase letters, replaced the uncial style. The blackletter typefaces began to develop in the eleventh century, reaching their most formal style by the fourteenth century. The type of the Gutenburg Bible (mid-fifteenth century) was modeled on this *textura* writing. The main title in the masthead of many newspapers uses blackletter type. Sometimes a particular example might appear to be either an uncial or a calligraphic typeface. These typefaces are best reserved for titles and as more decorative elements. Some of the uncials might work as subheads or on a book cover.

Certain blackletter faces are still used in formal German typography, but are rarely used in English language works. Blackletter is sometimes called Gothic in reference to the Gothic period of architecture in vogue at the time that printing was coming into general use. Since the term Gothic was (perhaps mistakenly) applied to sans serif faces, the preferred term applied to this style of type is blackletter.

As with the script typefaces, most of these are practically unreadable when set all caps. Of course, to our modern viewpoint, many of them are practically unreadable—period.

ANTIQUE LOOK typefaces are modern creations that appear to have been printed on presses from the 17th or 18th century. Some are modeled on common typefaces of the era, such as Caslon. Others, perversely, are made from Art Deco or other modern designs. Again, there are many possibilities for chapter titles or for cover typography.

VICTORIAN, CIRCUS, AND OLD WEST cover typefaces associated with the period from roughly 1850 to 1900. Selecting between these categories is probably somewhat arbitrary, but it's useful in sorting through a large font library. A smaller collection might combine these categories as Victorian or Victorian/Old West. As with most of the other decorative typefaces, these are suitable for setting a tone that supports the content of the book. Many are useful for chapter titles, subheads, and other ancillary items.

ART DECO, ART NOUVEAU, AND MODERNE are also groupings that are useful in sorting a large font collection but somewhat less useful when there are fewer examples in each category. As with the previous category, distinguishing between Art Deco and Moderne is rather arbitrary. (Perhaps less so had I studied more art history in college.) Do not confuse Moderne with Modern. While the Modern typefaces were created around the 1820s, the Moderne typefaces were designed about 100 years later.

GRUNGE typefaces are the contribution of the last bit of the twentieth century. These typefaces are often blurred and hard to read. They could be described as "deconstructionist" rather than grunge—but grunge is easier to spell. When used appropriately in a title or on a book jacket, a grunge typeface can have great emotional impact. Most, however, appear to have been designed by people who can't draw very well and/or don't have the patience to render a careful design. *Wired* magazine attracted considerable attention by extensively using grunge type and having generally obnoxious layouts which many designers thought were "cool." Eventually, under new owners who expected people to actually read the articles, the over-the-top grunge style was significantly moderated.

OTHER ORNAMENTAL. When every conceivable type design is categorized, a fairly large number of faces end up in the catch-all Decorative/Eccentric grouping. Type designs can be created that convey the meanings of letters with an extremely wide divergence of actual letter forms. Advertising and other artistic pursuits (such as movies and music) have demanded a wide variety of faces designed to convey particular feelings. Often Decorative faces are incomplete without having lowercase letters (or upper case) or not having a completely drawn letter set. Faces falling into this category include those drawn strictly for a logo or title. Sometimes a particularly popular logo will generate the interest so that a designer completes the alphabet. For example, at least one typeface is available based on the NASA "worm" logo.

These typefaces fall into the "other ornamental" category. They are unique. But, there seem to be endless ways to draw shapes that suggest our alphabet. The first line is SpaceBT (Bitstream), next is Fatty Patty from Cassady & Greene, and last is Dynamoe, sold through The Font Shop.

Again, these typefaces are not appropriate for long blocks of text, but are most useful in cover designs and for chapter titles or other book elements if they support the subject matter of the book. I was delighted, recently, to be able to use SpaceBT for the title and chapter titles in a novel (*The Payload* by Aaron Thiel) with a "space law" theme. It was also used on the cover.

Putting it all together

The general goal of a good book design is to bring together the various typographical elements with the structure of the book to create an environment that allows positive communication of the thoughts expressed in the work. To achieve this goal, it is necessary to understand the emotional impact of type and how the multitude of possible combinations of typefaces can be grouped and understood. Keep in mind, don't go overboard. Use one basic type family for the body text supplemented with a contrasting but compatible typeface for the heads and subheads, etc. Avoid the "ransom note" look!

In addition to the various classifications described, type may be separated into various emotional groups. This separation is more of a continuum rather than a yes-no determination. Consider if the typeface is formal or informal. Is the typeface light or heavy? Is the typeface smooth or rough? As you explore these questions, you can begin to match possible faces to the nature of your project.

For example, I recall hearing an art director refer to Electra as "intellectual." Perhaps this is true because I've seen a number of philosophy texts set in Electra and philosophy is certainly intellectual. Or, perhaps the philosophy texts were set in Electra because it "feels" intellectual. This illustrates an emotional response to a particular typeface and illustrates the importance of matching your typeface to your project.

If you are trying to select typefaces for your project consider the emotional response that each face represents. Keep in mind, typefaces appropriate for book production do not usually generate the strong responses that typefaces used in advertising frequently can create. Still, they can be of great value in establishing an emotional tone.

Alternate type classification systems

The foregoing classification system is my own, based on traditional classifications and on my struggles to bring some order to the chaos of my type library with more than 10,000 typefaces.

The organization I suggest is loosely based on this scheme:

1. Blackletter
2. Oldstyle
   a. Venetian
   b. Aldine French
   c. Dutch-English
3. Transitional
4. Modern
5. Square Serif
6. Sans Serif
7. Script-Cursive
8. Display-Decorative

As you can see, I followed this scheme, except I ignored the technical breakdowns in the oldstyle group (even I have trouble separating the nuances of Venetian from Aldine) and, I added many sub-categories under the last three groups due to the many script and decorative faces in my library.

The German national standards organization, DIN, has created the typically thorough list of type classifications that follows:

1.   Romans
1.1   Renaissance styles (Venetians): Only minor differences in stroke thickness inclines as is for curved portions of characters.
1.11 Early types and styles (Jenson, lowercase *e* with diagonal crossbar)
1.12 Late styles (Palatino, Weiss, Vendome; types designed after 1890)
1.2   Baroque Styles: Strong contrast in strokes, more angular serif formation, almost vertical axis for curved portions of characters.
1.21 Dutch styles (Van Dijk, Janson, Fleischman)
1.22 English styles (Caslon, Baskerville)
1.23 French styles (Fournier)
1.24 Modern styles (Horizon, Diethelm, Times Roman)
1.3   Classical Styles: Horizontal initial strokes for lowercase characters, strong contrast in stroke thickness, right-angle serifs, vertical axis to curved portion of characters.

1.31 Early styles (Bodoni, Didot, Walbaum)

1.32 Late styles (Bulmer)

1.33 Modern styles (Corvinus, Eden)

1.4 Free Romans: Romans with calligraphic modifications, etc.

1.41 Victorian styles (Auriol, Nicholas Cochin)

1.42 Non-serif Romans (Steel, Lydian, Optima)

1.43 Individual styles (Hammer Uncial, Verona, Matura)

1.5 Linear Romans: Grotesques and sans serif types of optically uniform stroke thickness.

1.51 Early styles (Announce Grotesque, Grotesque 9)

1.52 Modern styles (Futura, Spartan, Gill)

1.6 Block Styles: Egyptians and antiques. Types with slab serifs.

1.61 Early styles (nineteenth-century styles)

1.62 Late styles (Clarendon)

1.63 Modern styles, no brackets (Beton, Memphis, Rockwell, Stymie, Landi)

1.64 Typewriter faces (Courier, Pica)

1.7 Scripts

1.71 Stress variation (Legend)

1.72 Expanding strokes (Bernhard Cursive, Copperplate Bold, Invitation Script)

1.73 Uniform stroke (Signal, Monoline, Swing Bold)

1.74 Brush stroke (Mistral, Catalina)

2   Blackletter

2.1 Textura (Blackletter Gothic)

2.2 Rotunda (Wallau)

2.3 Schwabacher (Alt-Schwabacher, heavy-looking types). [Gutenburg Bible]

2.4 Fraktur (Unger, Fette)

2.5 Kurrent (Chancery)

3   Non-Roman Characters

3.1 Greek

3.2 Cyrillic

3.3 Hebrew

3.4 Arabic

3.5 Others

Had I been aware of the DIN classification system when I started seriously collecting typefaces (more than ten years ago), I might have used it as the basis of my organization. Still, the DIN classification system suffers from a weakness in classifying the multitude of decorative typestyles (i.e. 3.5 Others). If you have a small library with only a few dingbat and decorative faces, then it's not unreasonable to just lump them together. However, it is quite easy to obtain a sizable collection of decorative faces that then need further classification if you are ever to find the special one you're seeking for a particular purpose.

Type management software

There are a few utility programs that can help you manage your type library. They include Adobe Type Manager Deluxe, Font Reserve, Extensis Suitcase, Suitcase Fusion (that combines elements of Font Reserve and Suitcase), Font Agent Pro, Master Juggler and Apple's FontBook (with Mac OS X 10.3.x and later). (Not all are available for both Macintosh and Windows.)

While each individual may have preferences for one or another of these programs, I have been generally happy with Font Reserve for the past several years. It is the only font management utility that allows classification, searching, and viewing typefaces using the classification terms discussed in this chapter. All the font managers allow grouping fonts as "sets" (helpful when you have a number of projects). Of those products I've used, all are reasonable for a font library as large as 3,000 to 4,000 typefaces. When font libraries become significantly larger (as mine with over 10,000 typefaces), then there are notable performance issues. If you have a large type library, you might want to experiment with "demo" versions available from many vendors at their web sites.

Shortly before the first printing of this book, Extensis bought out the Font Reserve developers and subsequently issued Suitcase Fusion that has elements from both Font Reserve and Suitcase. I eventually started using Fusion and have been well satisfied with it. It seems to have fewer bugs that plagued the OS X version of Font Reserve. Some versions of Font Reserve are still available. See the Extensis web site for details.

Windows users should look carefully at Adobe Type Manager Deluxe or Font Reserve. Macintosh users have Apple's FontBook included with Mac OS X 10.3.x (or later). FontBook is certainly acceptable with a smaller type library, but it does not have the sophisticated features of the more robust third-party type management utilities. Those with a larger type library (anything above 400-500 typefaces) would likely find FontBook limiting.

6

Design With Type

Contrast, Alignment, Repetition, and Proximity (CARP)

THE ACRONYM DERIVED FROM THESE WORDS actually enumerates the basic principles behind all design. If one is not careful in the use of these principles, however, the letters can be reordered into an expletive declaration about the particular design.

CONTRAST is an avoidance of too much similarity. Design elements should reflect strong contrast to really show the differences. Avoid a bland sameness where everything looks almost the same. When trying to make elements have contrast, boldness is the best approach. Otherwise, the viewer will assume you tried to "match" the elements, but failed. Contrast is often the most distinctive visual attraction on a page.

ALIGNMENT is the avoidance of random placement of elements on the page. Page elements need to have a plan that gives a visual connection to every other element on the page. No element should be placed arbitrarily on the page that will break up the planned alignments. The result is a clean, sophisticated look.

Alignment is often used to guide the reader in a book. The eye becomes accustomed to returning to a similar place to pick up the next element. Use this to impose order to the layout.

REPETITION is where elements are placed throughout the document. You can repeat color (lightness or darkness), lines, illustrations, shapes, textures, and spatial relationships to develop the visual organization and to create a sense of unity. The familiarity of seeing similar elements used in the same manner directs the eye to the important parts and keeps the reader moving through the document in an orderly fashion.

PROXIMITY is the way that elements relate to one another. When several elements are placed close together, they become a visual unit rather than separate units. This is another way to provide organization and eliminate clutter.

The uninitiated tend to place elements helter-skelter around the page. There seems to be a fear of leaving empty space. When the elements are scattered all over, the page seems disorganized and communication is hampered by the reader being unable to quickly access the information.

Group related items together. Place them physically close to each other so that the group is discerned as a single visual unit, rather than as unrelated pieces scattered about. Likewise, elements that are not related should not be placed together (in close proximity). The reader then has an instant visual clue as to the organization of the page.

The page

When you begin your design, you can approach the page either as a single item or you can approach the layout as a spread—that is, the two pages of an open book side by side. The single page layout is appropriate for your "sell sheet," brochures, or other materials used to support the sale of our book. However, books are viewed as a sequence of spreads, so the left and right pages play an important role in the reader's appreciation for the work.

The paragraph

The width of the text block, or the length of the reading line is of critical importance. If the reader must move his or her eyes across too much material, there is a danger that he or she will lose his or her place or that he or she will become fatigued.

One "rule of thumb" is that a line of type should be about two or three lowercase alphabets in length of the size of type selected.

Let's see how we are in this book:

abcdefghijklmnopqrstuvwxyzabcdefghijklmnopqrstuvwxyzabcdefghijklmnopqrstuvwxyz

Two alphabets is 52 letters, three alphabets is 78 letters—and our line length here is about 74 letters. So, we're at the long end of the range, but still within. Experiments have shown that 60 to 70 letters are a comfortable line length for most people. Had we set this book in two columns, the lines would be too short and the reader would find the text broken up in "choppy" bits by the frequent line breaks. (Another way to look at it is to shoot for a minimum of 5 or 6 words per line.)

The maximum ideal line length would be no more than about 78 letters or 12 or 14 words per line. Had we used the suggested margins in chapter 3, the lines would have been too long (about 87 letters).

What if production requirements forced a wider line length? There are several options: 1. Use a larger point size (but not more than 12 points in most cases); 2. Use a wider-setting typeface (that's why it's important to note the relative widths of the typefaces at each nominal point size); or 3. Add more leading between the lines.

This last idea is often the solution for many seemingly difficult typesetting problems. More space between lines will give the eye a "track" to follow back to the beginning of the next line. Indeed, another rule of thumb (at text sizes, add 2 points to the point size of the type for the leading measure) would suggest that this book be set 11 on 13. But, to give me the 42 lines per page I wanted, the leading came out to 13.3 points. While the line length is within the rule of thumb, the small bit of extra leading also opens the lines up and helps the reader more comfortably read the page.

Ragged right or justified

A substantial majority of books are set with full justification. (That should tell you something.) But the decision isn't all that cut and dried. The answer depends, in part, on the tools you choose. If you use a professional page layout program, then the choice should be for justified text unless you have a really good reason for not using justification.

Keep in mind that medieval manuscripts are full of justified calligraphy, a style that was continued when Gutenburg produced his Bible—and it's a style that's continued in the majority of books for the past 500 years. That's a long tradition to ignore.

However, if you are using a word processor to typeset your book, then it may be better to use a ragged right style, since word processors generally do a poor job justifying text to professional standards.

Still, ragged right doesn't mean "ignore hyphenation." Well set rag type has a "gentle" rag. That is, the line lengths don't vary substantially from one to the next. (You want to avoid a "gap toothed" effect.) So, properly set ragged right type must be carefully monitored to hyphenate words as appropriate and thus to maintain a pleasant alignment. It may even be necessary to re-write (with the author's permission) some of the text to insure good spacing in the lines.

Rules of typesetting

Yes, it is said that rules are made to be broken. But first you need to know the rules, so that when you break them you do it with purpose and a logical intent. Here we will discuss the traditional rules used to create good typesetting. When you're on your own, you may

> ### WHY ARE WORD PROCESSORS A POOR CHOICE FOR TYPESETTING BOOKS?
>
> Word processors are great for writing and processing words. You can move stuff around with ease. You can spell check and grammar check (not that grammar checkers are much use). You can sort and index. Great stuff.
>
> Word processors were designed for writing letters and preparing business reports—and they excel at those tasks. They were not designed to layout pages or carefully control the setting of type.
>
> Typically, word processors will tightly set a line followed by a loose line followed by yet another tight line. This causes dark and light bands to appear across the page.
>
> Next, word processors are generally more willing to add space between words rather than hyphenate. (If you set ragged right, you will frequently have to force the hyphenation.) The result is that you will have rivers of distracting white flowing through your text rather than the even color that is desired.

break these rules if you have good reason and make reasoned judgments to reach the design decisions that support those decisions.

One of the messages that I've tried to pass along to those I've helped is that the computer allows us great liberty to experiment. But just because you *can* do something doesn't mean you *should*. So, please keep in mind the rules. If you need to break a rule, do it with knowledge in reasonable intent—not as a blunder. Honor the centuries of craftsmanship and tradition behind the setting of type.

There is a tight relationship between what you say and how you say it. The "how you say it" part of the relationship includes the typographic display placed before the reader. Unlike conversation, where facial expression, body posture(s), intonation, and gestures act to enhance (or detract from) communication, when you put words on paper (or as electronic bits to be displayed on a screen), you must make up for the unconscious communication through design, layout, and type.

Many of these issues are also covered in style guides like *The Chicago Manual of Style, Fowler's Modern English Usage,* or the *AP Manual of Style.* While these style guides are authoritative and of great use, it's important to know and understand the situations you're likely to frequently encounter. When the style guides are silent (or ambiguous), make notes and start your own *house style* guide. Indeed, every publisher has a house style, even if it's simply based on following certain conventions on an ad hoc basis.

EMPHASIS. It's better to use italics for emphasis rather than bold. Italics set off the word (or phrase) and give extra *oomph* to the message without affecting the overall color of the page. Bold is better reserved for titles. Of course, you already know that typesetting *never* uses underline as the effect on descenders is <u>ugly</u>. Good writers recognize that emphasis is best accomplished by the structure of the sentence or paragraph.

Other uses of italics include italicizing a word when it is used as a word (e.g. the term *font* is often misused); for foreign words not yet assimilated into English; as well as titles of books, movies, plays, operas, motion pictures, newspapers, magazines, names of parties to a lawsuit (e.g. *Marbury vs. Madison),* scientific names of genera and species (e.g. *homo sapiens),* and works of art. (Remember, titles of shorter works or portions of a larger work are enclosed in quotations marks.)

Since we mention italics, please note that the italics *include* any following punctuation. While most typefaces don't distinguish between an italic and non-italic period, most do have italic and non-italic variants of commas, parenthesis, and quotation, question, and exclamation marks.

SMALL CAPITALS. Small caps are useful for lower level subheadings, but are also used for A.M., P.M., B.C., and A.D. (or BCE and CE). They may also be used for acronyms. Some publications make a distinction between acronyms that are commonly pronounced as words (NASA, NATO, and RAM) and those that are pronounced letter by letter (CIA, FBI, and CPU) by using small caps for the former and full caps for the latter; although this practice isn't so widespread as to indicate acceptance as a standard. If setting acronyms as full-sized caps, it's generally more visually appealing to set them one point smaller than the type being used (although some editors object to this practice).

DASHES. There are three little horizontal lines in common use. The -, –, and —; normally referred to as the hyphen, en-dash, and em-dash. The last two are sometimes called nuts and mutts. The em-dash gets its name from being the approximate width of the capital "M," typically the widest letter of the alphabet. This also suggests that the em- and en-dashes vary in width by typeface as do the letters themselves.

Most people know to use the hyphen to break a too-long word at the end of a line, for certain compound words *(forget-me-not),* and to connect words that jointly modify another *(two-year-old child).*

The em-dash is also fairly well known as it is what most people will call "a dash." Its most common use is to indicate a break in thought or in the structure of a sentence—or to set off an aside or a parenthetical statement. In this second case, the em-dash is used much like commas or parenthesis. Watch out for that holdover from the days of the typewriter—don't use two hyphens instead of an em-dash. Even MS-Word will automatically convert two hyphens into an em-dash for you if you have the auto correct feature turned on.

The en-dash is much less well understood. Many people don't know that it even exists. It's mostly used to separate numbers *(Giants win 8–4, Home Open 2–5, see pages 24–29).* But it is also used to connect items made up of multi-word elements: *The train is on the Chicago–New York route;* or when one of the words is already hyphenated *(anti–blood-clotting agent).*

INDENTS. Beginning a paragraph with an indent may seem quite arbitrary, but the traditional practice was to have a one or two em-space. (The em-space is related to the em-dash in that both come from the relative measurement of the em-width or the width of the letter *M.)* The era of the typewriter gave us the half-inch indent (5 spaces on a Pica typewriter). While the large, fixed-width type and open feeling of a page produced on a standard typewriter made these half-inch indents seem normal, they are clearly excessive when placed into otherwise well-set type.

Sometime, somewhere, a graphic designer indented a paragraph by half a line width. Although this affectation can be forgiven in a single magazine article or other short piece, the "cool" look was picked up by many others. However, if the goal of your book is to communicate, you should not get in the way of that communication with faddish experimentation. (Some extra flair is allowed for the first line of a chapter.)

A new paragraph must be indicated in some manner. If your style is to set all text flush left, then separate the paragraphs with a blank line (or half-line). However, do not both indent paragraphs and separate them with blank lines. While the ultimate worry-wort may wear both a belt and suspenders, it isn't nearly so charming when you're trying to read and understand a message.

Fortunately, most page layout and word processing programs make indenting or spacing after a paragraph quite easy through the style sheet. You can set your "body text" to automatically indent or to automatically add space after a hard return. When you make this adjustment, set the automatic indent to ¼ inch as that corresponds reasonably closely to two em-spaces. It also facilitates setting up tabs for bullet statements or other needs. (The disadvantage of actually using em-spaces is that they are a relative measurement and are different in every typeface and at every type size.)

HEADINGS. Typography and meaning are especially closely entwined when it comes to headings. The typography of headings is used to communicate the relative importance of each subject and its relationship to other subjects and the book as a whole. The headings should follow the basic hierarchical outline of the work. In publishing, headings are described as having a "level" as in first level head, second level head, third level head where the first level is the most important heading below a chapter title, the second level is the next most important topic and so on. Most books only have one or two levels while others (such as this book) have four or more. You may need to rely on a professional editor to help you sort things out. Note that consistency must be an important consideration. If an element is a third level head in the first through third chapters, it shouldn't become a second level head in the fourth chapter (even if the fourth chapter doesn't have any second level heads).

Some people refer to the hierarchy of heads as A-heads, B-heads, C-heads, and so on. This is the same arrangement based on level of importance as first level, second level, etc. Neither of these descriptions is more correct than the other and seems to be a variation based on where or from whom you learned the topic.

Heads have special rules on capitalization. Words are capitalized based on their parts of speech—not based on æsthetic considerations. Normally, the first and last words are always capitalized along with all others except articles, conjunctions and prepositions. Some publishers capitalize the longer prepositions—those with more than four or five letters (e.g. "because"). This same convention also applies to titles of works cited with the text. If you don't know the parts of speech, then an acceptable alternative is to only capitalize the first word as a sentence. Generally, punctuation is not used, although a comma within the phrase may be used, if necessary for clarity.

The typeface used in headings gives very important clues to the organization of the text. The manuscript should be reviewed to determine what levels of heads are going to be needed. Then a reasonable arrangement of typeface, size, and position should be worked out to make the positional ranking obvious at a glance.

Sometimes book designers are more acquainted with the nuances of type than the reader. This can lead to problems if the subdivisions are too subtle. Most readers can discern that bold is more important than light, all caps more important than upper and lowercase, and larger is more important than smaller.

Other variations used to establish hierarchy include using centered type to show more importance over type set aligned at the left margin. The next level is then shown as words set at the beginning of the paragraph set in small caps. This last arrangement is called a run-in head.

Last, try to avoid stacking heads together without intervening text. If it can't be helped, certainly don't stack any more than two heads. (e.g. it's OK to have a chapter title followed by a first level head. It might look OK to have a centered first level head and a left aligned second level head, but to then have a third level run-in head would be too much.)

Punctuation Inside or Outside Quotation Marks. This is one of those issues where the experts and authoritative style guides differ. This leaves me free to suggest what I think is best: punctuation marks should always go inside quotation marks unless clarity requires otherwise. Usually, the punctuation mark involved is a period. The spacing of xxx". looks terrible. Of course, when a question mark is used, the gap below the quotation mark looks even worse. (xxxx"?)

For those who read a lot of material from our British cousins, you may notice that on the other side of the Atlantic the usual practice is to put punctuation outside of quotation marks. But, you'll also notice that they use single quote marks where we use double quote marks and double marks where we use single. The result is that the gap caused by the closing quotation mark is smaller and somewhat less distracting. I say, let the British do as they wish, and we'll do what looks best with standard U.S. practice.

We've already discussed the importance of using curly (typographer's) marks instead of the typewriter marks, so I won't repeat that here.

SMALLER POINT SIZE FOR ALL CAPS TEXT. When text is set in all caps, it can overwhelm the rest of the text. It almost seems like shouting. I generally reduce all caps text by one point (use 9 points when normally setting at 10 points or 10 when normally setting at 11 points). Here is an example using the acronym for "what you see is what you get" full size: WYSIWYG and reduced by one point: WYSIWYG. With some typefaces that have a smaller X-height, a reduction of 1½ or even 2 points might be more appropriate.

Another option is to use small caps if they're available with the typeface you're using. The computer generated small caps usually end up being too light and spoil the even color of the typeset page, so it's generally better not to use them. With some page layout programs you can adjust the small caps setting (usually about 60–70% of normal size) to be 85–90% of normal size. Then the computer generated small caps aren't quite so bad. Sometimes adjusting the character width to 105–110% of normal for computer generated small caps will make their lightness less apparent—but that may cause as many æsthetic problems as it solves.

USE OLDSTYLE FIGURES. As with small caps, some typefaces come with extended character sets that include oldstyle figures. They are designed to harmonize with lowercase text and are appropriate to use when numbers appear in prose passages. The use of computers and desktop publishing for the past decade have made use of old style figures much less common than they were in the era of traditional typesetting equipment. Few typefaces have oldstyle figures, and they can be annoying to use with many programs if they are located in an alternate font.

You may have noticed the use of oldstyle figures in this book. It's being typeset using Adobe Minion (which has expert set fonts available) in the OpenType version encoded in unicode (unicode typefaces may have up to 64,767 characters). InDesign allows automatic insertion of oldstyle figures through implementing various OpenType features. The figures appear based on the selections made in the style sheet. This paragraph uses the "body text" style and I have selected "use proportional oldstyle figures" in the style definition. Another paragraph or character style might specify the more common "tabular lining numbers" for use in a table of numbers where columnar alignment is desirable.

SMALLER POINT SIZE FOR LINING NUMBERS. When oldstyle numbers are not available, reduce the point size of numbers by one point (or so) when they occur in text. The reasoning is the same as for reducing the point size of all caps words or acronyms.

USE LIGATURES. Ligatures are combined letters. Gutenburg created over 200 ligatures when he printed his Bible. Fortunately, typefaces today don't have anywhere near so many. The most common are fi and fl, others include ffi and ffl and there could be even more depending on the typeface. Some decorative typefaces have a set of very distinctive ligatures that make the type more elegant. Those are intended for situations where maximum readability isn't necessary.

While Minion is a rather business-like typeface, it also has a more elegant side with a variety of swash characters and ligatures that you would not normally use in day-to-day documents. However, these can create an attractive tone to a formal announcement or award. (Here in this paragraph I've turned on the "discretionary ligatures and swash characters OpenType feature.) *When swash characters are available, the italics versions often have dramatic capital letters: A B C D E F G H I J K L M N O P Q R S T U V W X Y and Z.* Do keep in mind that the use of swash characters and unusual ligatures should be limited only to those situations where it is appropriate. Their use is not appropriate in everyday typography. In some cases, when it supports the material, swash characters might be used for chapter titles, and heads.

> **IN ORDER TO MIMIC** the 'look' of hand calligraphy, Gutenburg crafted many ligatures, abbreviations, and variants in his typeface for the first printed Bible. As a result, the Gutenburg Bibleschrift (the name of the typeface) had a 267 piece character set.

The OpenType version of Minion has a surprising number of ligatures in the normal text. In addition to fi, ffi, fl, and ffl, it has ff, Th, ffb, ffh, ffk, fft, fh, fj, ffj, fk, and ft ligatures. It is important when using ligatures that you do not allow significant letterspacing, as the ligatures will then appear very obvious.

LETTERSPACING. The story is told that Frederick Goudy, a famous type designer, went to a prestigious event in New York City where he was given an award for excellence in type design. Upon accepting the certificate, he is reported to have said, "Anyone who would letterspace blackletter would steal sheep." Of course, this was a very uncomfortable moment for the man in the audience who had prepared the certificate. Later, it is reported, Mr. Goudy apologized, claiming that he said that about everything. Somewhere along the way, "blackletter" got changed to "lowercase" and the typesetting urban legend that lowercase letters should never be letterspaced became common knowledge, even if it was incorrect.

The fact is that letterspacing is an appropriate means of helping make adjustments to line length to facilitate full justification. However, to maintain readability and minimize negative impacts on page color, letterspacing should be done with a very light touch. Most page layout programs default with rather broad ranges for the line adjustment tools. Often letterspacing defaults are overly generous. PageMaker, as one example, defaults to allow letterspacing to vary from -5% (minimum reduction from normal) to a maximum increase from normal of 25%. This is usually excessive, although the appropriate range may vary from one typeface to another. I generally set letterspacing to have a minimum of 0% (moving letters together not allowed) and a maximum of 15% (allow letters to be moved apart up to 15% of their designed spacing). Some typefaces will manage a minimum of -5%, so I sometimes choose a range of -5% to 15%.

Ideally, letters should be spaced so that each letter stands by itself without touching adjacent letters, but is attractively positioned by the adjacent letters so that they appear to be evenly spaced. The shapes of letters (some narrow and tall, some wide, and others round and/or curved) make a mechanical, even spacing inappropriate. The type designer has created "side bearing" measurements for each letter. Then (in a carefully constructed typeface) the designer has adjusted kerning (letterspace between individual pairs) for all letter pairs that need special adjustment, e.g. To, Yo, VA, etc. One of the more amazing features of InDesign is its ability to space letters using optical spacing. Sophisticated programming determines the optical mass of each letter and ensures that the apparent (not measured) space between each letter is the same. (This reduces our dependence on the skill of the typeface designer and/or the type foundry distributing a particular typeface.)

Since InDesign handles optical spacing for you (when the right options are selected) in most cases, it's not necessary to worry too much about any special letterspacing. However, with other programs you may want to add a small bit of letterspacing in words using small caps or when full caps are used in text. Caps are usually designed with an expectation that they will accompany lowercase letters, so they often appear too close when used together.

The usual letterspace control is the tracking function. (Quark XPress and some other programs call this same feature "range kerning"—it's the same thing.) A slight increase in tracking for caps and small caps will do the trick. Some typefaces (if by design or by accident) just seem too widely set for my taste. When I work with Adobe Garamond using InDesign's optical letterspacing, I'll set the tracking to -10% for all normal text. Some other typeface might need a +5 or +10% setting.

WORDSPACING. Wordspace is used as a primary means to allow type to be justified. Because the space between words is relatively large, adjustments spread along the line of type are not usually noticeable. Again, the default settings are much too broad (especially with Quark XPress). PageMaker defaults to allow a minimum of 75% and a maximum of 150% of the normal inter-wordspace. Quark XPress defaults with a range of 50% to 200%. (Why bother having limits?) In both cases, a more reasonable setting is a minimum of 85% and a maximum of 125%—although this range might need to be adjusted for some typefaces. (In all cases, the "desired" spacing, both word- and letter-is 100%.)

Sometimes, the software—in attempting to perform justification—makes the wordspacing too tight with some letter combinations. In those cases you may need to insert a fixed "thin space" to force the words apart. This should be done as a last step after other edits to the typeset material are completed, since such changes may impact such combinations, making the fix unnecessary.

Don't be afraid to experiment or to make adjustments when it seems reasonable. On one project, I discovered that a recently acquired version of Souvenir that I was using was returning excessive wordspacing. Ultimately, I adjusted the wordspace to range from 75% minimum, 85% desired, and maximum of 105%. The result was that the wordspacing was much more within my expectations. While these settings would make a mess of most fonts, in this case it fixed an otherwise serious problem.

GLYPH SCALING. All the page layout programs and some word processors allow you to stretch or compress letters. This usually has horrible consequences to the design of the letters and is not recommended. If you need wider letters, use a font that's naturally wider or has a designed "extended" variant. Myriad, for example, has a "semi-extended" variant. And Minion has a designed condensed variant. As you should be able to see, the word "condensed," set here with the condensed variant is much more even in appearance and doesn't look weird like the computer generated compression/extension above. Indeed, in the interest of the page budget, I set a 300 page book with the condensed version of Minion. Over 7,000 copies have sold, and no one has ever mentioned that the type was condensed.

Now that I've just told you not to extend or compress type, I'm going to suggest that there are circumstances where it is safe to do so. InDesign allows "glyph scaling" as one of the variables in the justification dialog. While the default is a conservative zero, I feel that a compression of 97–99% and an extension of 101–103% is not visible with most typefaces. (You will need to experiment with each typeface you use. Mostly, I set glyph scaling to 98/102%.) With the programs that don't provide this feature automatically, you may find situations where a glyph scaling of 97 or 98% will allow you to fit a paragraph and eliminate an otherwise bad break, without the adjustment being at all obvious. This is one situation where it's safe to say, if you can't see it, it didn't happen.

Buying typefaces

By now, you may have realized that the typefaces that came with your computer are rather limiting. You'll probably want to purchase typefaces for your project. The question is often asked: "What typefaces should I buy and were can I get them?

Actually we first need to talk about the file formats used to distribute type used on computers. (Please, don't blank out on this—it's important!) Type is distributed as TrueType, Postscript Type 1, and OpenType. Both the Windows and Macintosh platforms claim to be "neutral" to the type file format, but you may need Adobe Type Manager (a "lite" version is free) on some Windows computers to use PostScript type files.

For typesetting books, you should stick with PostScript Type 1 or OpenType files. TrueType is not trusted by most printers (although it isn't really a problem any more). Frankly, OpenType seems to be the way of the future, so I've focused my purchases in that direction. OpenType uses an internal coding called *unicode* that allows a typeface file to have over 65,000 characters. The older typefaces are limited to 256 characters. (While 256 may seem like a lot, once you account for upper and lowercase, numbers, punctuation marks, non-English characters and accents, you quickly run out of space.)

Some sources of type are listed in the resource directory at the back of the book. While type can be expensive, we have identified some reasonable collections of moderate price that can be satisfactorily used for book projects.

- 7 -

Professional Tools

THERE ARE TWO SCHOOLS OF THOUGHT among fledgling independent publishers, particularly those who are interested in keeping the cost of their project as low as possible. One school holds that doing quick and dirty (and cheap) layout in a word processing program will produce a book that looks "good enough." The other holds that only a layout program produces professional results.

The issue boils down to the perception of quality. For a book to look like it was produced with professional care, the typography must meet or exceed accepted trade book standards. Book buyers (those who are buying for their stores) won't buy a book that doesn't look like it was professionally typeset. Indeed, *retail* buyers often won't buy a book that doesn't follow typesetting standards. The wholesale buyer will know why he or she didn't buy the book. The retail buyer may not be able to pin down why he or she didn't buy, but poor typography, subtle though it is, may be the difference between a sale and a book that's put back on the shelf. That not-so-subtle choice may very well occur if you use a word processing program instead of a layout program.

Word processing programs and page layout programs are designed with entirely different basic concepts. Word processors look at a whole document; page layout programs focus on a page at a time. They also use very different algorithms to actually set the type. Word processors have many features that make them ideal for a creative writer or for business writing tasks. Many of the features of word processors make writing and revising text easier—and these features are superior to anything offered by a page layout program. Most writers wouldn't want to compose their work in a layout program. It's better to write your book or to have it key-entered using a word processing program.

A page layout program is designed to give the user maximum control over the appearance of the page and the position of type on that page. Many of the controls that may be hard or impossible to apply in a word processor are easy to select and apply in a page layout program. Indeed the global view aspect of word processing programs often cause simple changes to ripple through a substantial portion of a document while such minor changes usually only affect only one or a few pages in a page layout program. Page layout programs have superior algorithms for the setting of type.

Look at the following sample prepared with *Microsoft Word 98* for Macintosh (Later and Windows versions of Word produce similar output):

> The Mississippi does not alter its locality by cut-offs alone: it is always changing its habitat *bodily*—is always moving bodily *sidewise*. At Hard Times, La., the river is two miles west of the region it used to occupy. As a result, the original site of that settlement is not now in Louisiana at all, but on the other side of the river, in the State of Mississippi. *Nearly the whole of that one thousand three hundred miles of old Mississippi river which La Salle floated down in his canoes, two hundred years ago, is good solid dry ground now.* The river lies to the right of it, in places, and to the left of it in other places.
>
> Although the Mississippi's mud builds land but slowly, down at the mouth, where the Gulf's billows interfere with its work, it builds fast enough in better protected regions higher up: for instance, Prophet's Island contained one thousand five hundred acres of land thirty years ago; since then the river has added seven hundred acres to it.
>
> *—From Life on the Mississippi by Mark Twain*

This sample was set with the default typographic settings in Microsoft Word on a 4 inch line (comparable to many trade books) in 11 point Palatino. Notice the "rivers" (large spaces between words that align from line to line) that run through this text, particularly in the second paragraph. If you set this text as ragged right (right margin uneven), then apply (full) justification, you will see that MS-Word merely adds space between the words to make an even margin. The result is awkward (some would say ugly) typesetting.

This typographic quality may be adequate if you are preparing a simple book for a very small audience—perhaps a family history which will be produced with a press run of fewer than 50 copies. More serious work—that which you wish to sell to the general public—is better served by being prepared in a page layout program—or, at least, fiddling with the numerous internal settings in Word as described later in this chapter.

Now, look at the following sample prepared with *PageMaker 7.0:*

> The Mississippi does not alter its locality by cut-offs alone: it is always changing its habitat *bodily*—is always moving bodily *sidewise*. At Hard Times, La., the river is two miles west of the region it used to occupy. As a result, the original site of that settlement is not now in Louisiana at all,

but on the other side of the river, in the State of Mississippi. *Nearly the whole of that one thousand three hundred miles of old Mississippi river which La Salle floated down in his canoes, two hundred years ago, is good solid dry ground now.* The river lies to the right of it, in places, and to the left of it in other places.

Although the Mississippi's mud builds land but slowly, down at the mouth, where the Gulf's billows interfere with its work, it builds fast enough in better protected regions higher up: for instance, Prophet's Island contained one thousand five hundred acres of land thirty years ago; since then the river has added seven hundred acres to it.

—From Life on the Mississippi by Mark Twain

The PageMaker sample is set with the same 11 point type, but line spacing was adjusted to 13 points from the default 13.2 that Word used. Notice that PageMaker has closed up the text and generates a more even look. PageMaker also has adjusted the spacing of the letters and the words to make the whole paragraph appear smooth and consistent. The first paragraph only takes ten lines instead of eleven and the second paragraph takes only six lines instead of seven. This subtle space reduction can amount to a significant number of pages in a long document. Simply using PageMaker instead of Word could reduce the cost of printing some books by a considerable sum. (The PageMaker's typesetting engine is unchanged from version 6.5.2.) [Note: Adobe, who publishes the PageMaker software, has withdrawn the product from the market. We continue to discuss it here because there are still many copies in use and a reader may still wish to use it for their project.]

Here are the same paragraphs prepared with *InDesign CS3* (version 5.0):

The Mississippi does not alter its locality by cut-offs alone: it is always changing its habitat *bodily*—is always moving bodily *sidewise*. At Hard Times, La., the river is two miles west of the region it used to occupy. As a result, the original site of that settlement is not now in Louisiana at all, but on the other side of the river, in the State of Mississippi. *Nearly the whole of that one thousand three hundred miles of old Mississippi river which La Salle floated down in his canoes, two hundred years ago, is good solid dry ground now.* The river lies to the right of it, in places, and to the left of it in other places.

Although the Mississippi's mud builds land but slowly, down at the mouth, where the Gulf's billows interfere with its work, it builds fast enough in better protected regions higher up: for instance, Prophet's Island contained one thousand five hundred acres of land thirty years ago; since then the river has added seven hundred acres to it.

—From Life on the Mississippi by Mark Twain

With the InDesign version, we see that the type color (spacing) is even more improved. Unlike all other page layout programs, InDesign's type engine looks at complete paragraphs to select the best compromise of letter and wordspacing and line endings. InDesign also offers automatic hanging punctuation—if you look at the comma at the end of the first line of the second paragraph, you'll see that it extends slightly into the margin. Likewise, capital letters such as "A" and "V" will also have a similar extension into the margin should they appear at the beginning or end of a line. InDesign also offers "optical" letterspacing (used in the sample). This takes into account the physical shapes of the letter outlines and adjusts spacing accordingly (ignoring the built-in font metrics). A particular advantage of the optical spacing is that it sometimes can improve the appearance of a font that might otherwise be unusable due to poor quality or absent kerning tables. All other page layout programs consider one line at a time. InDesign is the first page layout program that approaches (and in some cases exceeds) the quality and control available in the previous generation of dedicated, computerized typesetting systems—this is a first for a Desktop Publishing program.

Yes, that's all there is to it... but the difference is obvious to any professional book buyer. The ordinary reader may not immediately notice, but tests have proven that poorly spaced type is difficult to read and quickly causes readers to become fatigued. Since the whole point of publishing your project is to have it read, you should make the effort to insure that it is readable!

What page layout program should you use

There are five professional quality page layout programs: Adobe InDesign, Adobe PageMaker; Adobe FrameMaker; Corel Ventura Publisher, and Quark XPress.

ADOBE PAGEMAKER. In January 2004, Adobe announced that they will not develop PageMaker beyond its current version. We have long expected that PageMaker was reaching an end of life situation due to changes in the operating systems it depends on. (PageMaker does not run as a native Macintosh OS X program and may not work with future upgrades to Windows.) PageMaker is no longer available from Adobe, but may be found on eBay or at used computer/software shops. In all honesty, I can't recommend PageMaker any longer, despite that it was the easiest to learn of the professional quality page layout programs—and has long been popular with professional book designers. PageMaker is credited with starting the desktop publishing revolution when it was combined with the Macintosh computer and the PostScript page description language in the mid-1980s.

Concurrent with the suspension of further PageMaker development, Adobe Systems announced Adobe InDesign CS* PageMaker Edition, an extension to InDesign designed to make it possible for PageMaker users to make a quick and successful transition. Adobe is offering an upgrade to Adobe InDesign CS PageMaker Edition for Mac OS X and Windows with an estimated price of $349 for registered users of *any* version of Adobe PageMaker. The PageMaker Plug-in Pack adds some enhanced features to InDesign that were previously unique to PageMaker or only available as extra-cost plug-ins

---

*As we go to press for a second printing, Adobe InDesign CS3 (version 5.x) is the current release. It includes the PageMaker Edition features mentioned as well as a number of other improvements. InDesign CS3 will open PageMaker 6 and later files as well as all earlier InDesign files. It also opens Quark XPress files saved in version 3.3 or 4.1 formats.

from third party software developers. In addition to various templates similar to those available with the last few versions of PageMaker, the Plug-in Pack adds the ability to open PageMaker 6.0 documents (PageMaker 6.5–7.0 documents are already supported in InDesign along with the ability to open Quark XPress 3.3–4.x files). Included in the PageMaker Plug-in Pack is the *ALAP InBooklet Special Edition* plug-in that automatically rearranges a document's pages at print time with complete control over margins, gaps, bleeds, creep, and crossover traps for making booklets (replacing the similar functionality of PageMaker's "booklet maker" feature). Other unique functions include automated bullets and numbering, data merge (similar to the mail merge feature in many word processors), and the ability to easily switch the keyboard shortcuts in InDesign CS to match those of PageMaker—an absent function for which I criticized Adobe when InDesign 1.0 was first announced with a function that switched InDesign's keyboard shortcuts to those of Quark XPress. Those who already own InDesign CS (Version 3.x) may purchase the PageMaker Plug-in Pack from the Adobe online store for $49. I understand that new copies of InDesign CS include the PageMaker Plug-in Pack on the release disc.

ADOBE FRAMEMAKER* offers a number of features that are particularly attractive to academic publishers (e.g. automated footnoting features and table-making features), computer documentation publishers (support for multiple formats, languages, and numerous versions), and large publishing operations (a $7995 "server" version is available and some UNIX versions support multiple users). FrameMaker was specifically created for long document creation (and is somewhat difficult to use for shorter documents). It is also available in versions for some UNIX computers as well—an advantage if UNIX is an issue. Adobe FrameMaker 7.1 sells for $799 ($1329 to $1989 for various UNIX versions). Upgrades from previous versions is $199 (Windows) to $379 (UNIX). Apparently FrameMaker + SGML 7.0 ($1449 Windows/UNIX only) has been withdrawn since the 7.1 version features extended XML support. (If you don't know what SGML (Structured Generalized Markup Language) or XML (eXtensible Markup Language) is, then you don't want to know.) FrameMaker for Macintosh remains at the 7.0 version level and runs under Mac OS 9.x and above with operation in Classic Mode under Mac OS X. (Adobe recently announced that FrameMaker will not be further developed on the Macintosh platform.) It appears that Adobe is continuing to develop FrameMaker toward large organizations creating highly complex documents. Frankly, FrameMaker is probably not the most appropriate program for small, independent publishers.

QUARK XPRESS is particularly popular with designers/graphic artists and magazine publishers. It also provides a number of features and add-on programs that make it particularly desirable in large, networked publishing organizations where many people work on a project at the same time (magazines). Quark has tried to improve its long document features, but it still falls short of PageMaker and InDesign in that respect without purchasing special utilities (Quark Xtensions) from third party software developers that add to the already considerable cost. The typographic spacing defaults in XPress result in rather poor letter and wordspacing. You will need to learn how to make adjustments to the default values to get acceptable typography. XPress has a reputation

---

*Subsequent developments would indicate that Adobe may be planning to eventually phase out FrameMaker at some time in the future. The CS3 version of InDesign contains some of the previously unique FrameMaker features (e.g. automated footnote handling). While Adobe hasn't made any announcements in this regard, it is a speculation on my part that this may occur at some future time when InDesign reaches a much higher level of development. After all, it makes little sense for Adobe to maintain several programs that perform, essentially, the same functions.

of being expensive, and the Quark company has a reputation of sometimes cantankerous customer support. XPress is available for both Mac and Windows platforms, but has a relatively miniscule share of the Windows platform due primarily to the heavy "tilt" toward Quark XPress by the professional graphic design market (Macintosh has an 80 to 90% share of the professional graphics market depending on how it is measured). If you are experienced with XPress, there's little advantage to learning another product, but if you are just starting out, there are better choices to be made. Quark XPress 6.0* was released in the Fall of 2003. Its list price is $1045, although it is generally available for $945. List prices of upgrades from earlier versions range from $199 to $499, depending on the version you're upgrading from. XPress 6.0 has been received with mixed reviews. It has a couple of new features, particularly the ability to combine several projects in one package—for example, business cards, letterhead, envelopes, and a promotional brochure can share a common project designation and share various style sheets, etc. While a nice feature for general graphics production, this offers no particular advantage to book publishers. There have been many reports of dissatisfaction with registration and "authorization" procedures. To minimize piracy, Quark requires that each copy of the program receive an authorization code from Quark. Initially, Quark wanted to force someone who used XPress on a desktop computer to buy a second license to run the program on a laptop. Quark has apparently backed away from that stance under the onslaught of complaints from customers who wanted to work with the program on an office desktop machine, but then use (the same licensed copy) on a laptop while at a client location or at home. Another complaint is that installation of version 6 "wiped out" version 5 on the same computer—this makes it difficult to deal with legacy documents—and is a further complication for sharing files with those who might still be using version 4.x as Ver. 6 can only save down to Ver. 5. Version 5 is required to save to Ver. 4. (So, how can you save to Version 4 if installation of Version 6 wiped out your copy of Version 5?) This wouldn't be much of an issue, except that Version 5 was not well received by the XPress user community and many designers never upgraded to it. Quark has announced that they will make a 'plug-in' available soon to make it possible to save directly from Version 6 to Version 4.

Late in 2004, Quark announced version 6.5. Under competitive pressure from Adobe's InDesign, Quark has shaken up its management, claims to have "reformed" its support system, and has addressed many of the criticisms of Version 6.0.

So, Should You Buy Quark XPress? If you are an experienced Quark user and you aren't disturbed by the restrictive and annoying policies of Quark—then stick with the "devil you know." XPress 6 (6.5 and 7.0) supports Mac OS X and the latest versions of Windows. (My opinion is "who needs these headaches?" when there are excellent [and possibly better and cheaper] alternatives.)

Adobe InDesign originally began as a complete rewrite of PageMaker. Ultimately, as the development process progressed, the project surpassed PageMaker and it was decided to introduce it as a completely new product. (This also left PageMaker as a viable product while InDesign went through the inevitable growing pains.) InDesign is the

---

*As we return to press, version 7.x of XPress has been released. Significant changes in management at Quark have improved customer service. I'm still of the opinion that InDesign is the "Quark Killer" that it was described as upon release of Version 1. (But that was an over-hyped assertion when it was originally made.)

page layout program with the best typography of any of the competing products. It is the first desktop program that rivals the dedicated computerized typesetting equipment used in the previous generation in typographic quality. Adobe InDesign CS (version 3.0) has excellent support for long documents with table of contents and index creation features similar to those in PageMaker. InDesign, in its original form, was criticized for leisurely operation, but version 2.0 was much improved. Now, InDesign CS (version 3.0 and later) is most pleasantly responsive. (New computers are faster, too.) The excellent typography of the first version—which used a "multi-line" composer—was improved with the software now checking the appearance of the type in each complete paragraph at one time. Adobe, keeping the ease of use lessons of PageMaker in mind, has created a very robust program that is only slightly more difficult than PageMaker to learn. (The additional difficulty is mostly due to the large number of additional features.) You can download a demo copy of InDesign from the Adobe web site. **InDesign is probably the best choice for book layout and design.** Most of the professional book designers using PageMaker have converted to InDesign.

Adobe has radically changed the way they are marketing their software*. While you can still purchase the individual programs separately, InDesign is now sold as part of Adobe Creative Suite which comes in several editions. The standard Creative Suite includes InDesign, Photoshop (photo editing and manipulation), Illustrator (vector graphics/drawing program) and Version Cue (file manager) at a list price of $999.99. The premium version adds Dreamweaver (web site development software) and the full Adobe Acrobat Professional (create, view, and limited editing of PDF files) software. The premium version carries a list price of $1229.99. The programs within the Creative Suite aren't just packaged together, they have also been modified to work together more closely than ever before. Either of these editions represent a substantial savings when compared to the list price of individual components. Most small book publishers will find either package of the Creative Suite useful. Those who plan to prepare their own web site will probably be most attracted to the premium version. InDesign, as a stand-alone program, is available for $699.99 and may be upgraded from an earlier version of InDesign from $169.99. Adobe chose an interesting route for upgrades to the full Creative Suite: from *any* version of Photoshop, you may upgrade to the standard Creative Suite for $549.99 or to the premium Creative Suite for $749.99.

> **AS WE GO TO PRESS,** Adobe has released InDesign CS2*. The new version has a number of improved features that make InDesign even more suitable for book production. An updated review of InDesign CS2 will appear on the Aeonix website.

COREL VENTURA has gone through several owners over the years. Once a strong PC-based program, it suffered from lack of attention during its "middle age." Now owned by Corel (of Corel Draw fame), Ventura Version 8 has been released for Windows computers. Not widely supported by book printers, you might consider this program if you plan to deliver camera ready hard copy output or if you can create an Acrobat PDF file. If you're a Mac user, Ventura's Mac version was seriously bug-ridden and should be

---

*See the Adobe web site *(http://www.adobe.com)* for details on the latest iterations of the Creative Suite packages. They now offer suites tailored to various design interests, including print, web, and video. The basic suite will cover the needs of most independent publishers, but those with other interests beyond publishing my find one of the more advanced suites more attractive.

avoided—if you can even find it. (It appears to have been withdrawn.) Ventura 8.0 for the Mac has not been (and probably won't be) released. Corel states that Ventura is being targeted at "corporate users." This is a reflection of its low penetration of the design market. Ventura has not been widely supported by book printers in recent years, making it a less desirable product for publishers expecting to pass their project to a book printer. Corel Designer Suite 11, priced at $529, claims "page layout features" but that is far short of a real page-layout product. You may be able to effectively use Ventura if you use a suitable utility to translate the output into Acrobat PDF files. While there is some merit to Ventura—and a strong contingent of satisfied users—it is hard to suggest that someone new to page layout should select the Corel product when there are such superior, if more expensive, alternatives.

Other page layout programs

Other low-cost page layout programs are also available. Most aren't quite ready for prime time and others present problems that may seriously delay your project.

MICROSOFT PUBLISHER is inexpensive, heavily promoted, and very popular, but it shares some of the flaws that word processors do. It's not really suitable for book production as it is not supported by most book printers. Its streamlined feature set may cause severe limitations in creating book-length projects. Its poor support for CMYK colors make it inappropriate for creating book covers. I haven't directly observed any output from Publisher, so it is hard to critique its typesetting algorithms. However, I suspect that they are closer to the simple algorithms used in word processors. (I've had conversations with a couple of folks who used MS-Publisher for their books. The usual assessment is that they wish they'd used almost anything else...)

STONE STUDIO is a $299 package that includes "Create" which features page layout, web page creation, and illustration functions, "GIF Fun" that makes animated GIF images for web sites, and "PStill," an advanced PostScript distillery to generate PDF files from EPS and PS files. Stone Studio runs on Mac OS X only. It is unlikely to be supported by any book printers, but you may be able to deliver acceptable files in the PDF format. The few reviews I've seen aren't particularly encouraging and did not address book-length projects. There are better choices. The Mac OS X only aspect may limit interest by the many small publishers who are operating with Windows computers.

If you're working on a Mac, the PStill component is available separately, and might be a viable alternative to the more expensive Adobe Acrobat for creating PDFs with greater control than the Mac OS X built-in PDF creation feature.

READY, SET, GO! was a pioneering page layout program that almost disappeared. It's been kept alive by overseas developers and is available in version 7.6.2 for Mac OS X for $175; in version 7.2.8 for Mac OS 9.1 or 9.2 for $150; and in version 2.3 for Windows for $99. (The Mac and Windows versions can share files.) While this program, that even predated PageMaker, is a reasonably solid page layout program, it is unlikely that you will find any printers able to accept native RSG files, however RSG claims to save in PDF format. I've only seen one magazine review of the program—and it wasn't particularly encouraging. It is likely the small market penetration of this product is well deserved.

T_EX (pronounced "tek") is, in my opinion, *a really bad choice* unless you have an engineering mind set. (Over the years, I've received much harassment from the T_EX *true believers* for my opinion.) The cost conscientious may be drawn to T_EX, as it is, essentially, free. T_EX is not an integrated program but is a series of batch processing utilities and a coding scheme. Developed a few years before WYSIWYG (what you see is what you get) desktop publishing became available, it was devised by a mathematics and computer programming professor to enable him to "easily" publish math text books with complex equations. T_EX has a powerful typography algorithm (which is the basis for the algorithm used by InDesign) and is capable of producing excellent typography. Unfortunately, it's difficult to work with, requires using macros to encode your material and can be frustrating for the non-technically gifted to use effectively. There are some commercial applications available that support T_EX and there is an extensive network of "true believers" who will support newcomers. From my point of view, T_EX should be avoided. I am not without significant, relevant experience with T_EX: when I supervised the publishing contractor staff at NASA Ames Research Center, my staff received many T_EX documents—which (1) never could be processed through the system on schedule, (2) often had macros that failed and required difficult troubleshooting, and (3) required us to retain an expensive outside consultant to troubleshoot the worst problems. I tried to discourage use of T_EX, but I found it rather difficult to explain anything of this nature to "rocket scientists." (However, even "rocket scientists" couldn't always make T_EX work right. That says it all...)

Distilling Acrobat PDF files

Most of the discussed programs can export or save as PDF files. Some program generated PDFs, including those from Adobe programs, aren't fully compatible with the plate making or other high resolution output devices used by book printers. (As one example, InDesign 2.x* generated PDFs that handled unicode fonts in a manner that caused difficulties with many high resolution output devices, but this handling was an advantage in supporting non-English languages.) The best way to generate an Acrobat PDF, from most programs, is to print the document "to disk," saving it as a plain PostScript file, then process the PostScript file into an Acrobat PDF using Adobe Acrobat Distiller with the appropriate job options set. A copy of Acrobat Distiller was included with PageMaker 6.5.x and 7.0. Normally to get Acrobat Distiller you must buy the full Adobe Acrobat program, (now version 7.x), standard version, $299.00, or the professional version, $399.99. Alternatives include Stone Studio PStill (Mac OS X only) available for $69.00 (as a separate application from the Stone Studio Suite), pdfFactory (Windows only) or Jaws PDF Creator (available for both Mac and Windows) for $79.00. These alternates (and possibly others) will generate PDF files, but one can never be certain that the alternates will have the functionality to distill the most complex documents properly. However, they may serve book publishers quite well, as books aren't usually extremely "complex" even if they are long. Macintosh users, who have upgraded to Mac OS X, can use a built-in feature to create PDFs in the print dialog, although the OS X feature has limited capability for setting job options.

---

*Later versions of InDesign CS do not share this problem and produce acceptable PDFs directly from the internal generator suitable for print production.

Which one?

I recommend that a new user learn InDesign unless they have particular and strong needs for the special features of FrameMaker. Those who prefer Quark XPress should stick with that program. Ventura, *if it is already available,* is practical for Windows users; although it will require processing into an Acrobat PDF file for submission to a printer. Finally, PageMaker might be a reasonable choice if ease of use and low cost outweigh the considerable disadvantages of working with an already discontinued program. Among the alternative page layout software, those to directly avoid are TEX and Microsoft Publisher. Ready, Set, Go! may have the functionality and capability to do page layout on a budget, and it's cheap enough that those with an adventurous heart and more time than money may find it worth the effort (but help with problems may be hard to find).

- 8 -

Using Word for Layout

First, I must again caution that Microsoft Word is *not* an appropriate program to lay out a professional quality book. Word, and other word processing programs, do not have the correct features or programming to produce good quality typesetting with good letterspacing or wordspacing. Word processors tend to favor adding wordspacing rather than hyphenating, the result being that books set with a word processor have unattractive and distracting rivers of white running down through the type, especially when full justification is used (as it generally should be for a book). Aside from the typographic flaws, a word processor is simply more difficult to use for book layout. Word processors view a document as a complete whole, not as individual pages, so establishing "good" page endings is difficult. Other controls (for leading, letter- and wordspacing) are considerably less convenient to manipulate.

If a document has graphics (photos, charts, or tables), the word processor has various tools that accommodate these elements. However, as the number of elements (especially imported images) increase, the word processing software becomes overwhelmed and takes longer to work with. (The work-around is to break up really long documents into chapters to limit the ultimate file size.) I have had the experience where a very long document with many imported images took more than 45 minutes to open or save. My employee (who created this disaster) would have been well advised to have broken that file into several sections. Also, such long, complex files run a much greater risk of becoming corrupted during editing.

Word processors have been getting more sophisticated over the years as "improved" versions are released. Still, the software design choices to make word processors efficient for business letters and reports cause the appearance of the final output to be rather short of the quality of typography expected for a book.

Word processors are great for processing words. I highly recommend using one to prepare a manuscript before importing the file into a true professional quality page layout program for book production.

With the cautions out of the way, there are cases where the practicality and cost containment make using a word processor **a reasonable choice for books with limited distribution or a short expected lifespan.** (There are page layout alternatives that can be quite economical. See Chapter 7 for a discussion.) Examples are books written to share a family history or personal memoir with a fairly small audience. Likewise, business reports or other long documents of very limited distribution may not justify the effort and expense required to use a page layout program. So, if you must use a word processor, here's how to do it.

File clean-up

There are a few things that cause problems and make the typography less attractive. Number one is extra spaces between sentences. (I know, you were taught, "period–space–space" when you learned to type.) There are also instances of other flaws. Before you begin preparing your document as a book, you'll need to execute a series of find/change procedures:

First, go to the menu: **Edit>Replace…** and enter **<space><space>** (hit the space bar twice) in the "**Find what?**" box. Click into the "**Replace with:**" box and hit the space bar once, i.e., search for two successive spaces and replace them with one space. When the first instance is found, select the "**replace all**" choice, unless you have specific places where multiple spaces are intended (if that's the case, you probably should have used tabs instead). Re-run the Replace procedure as many times as necessary until you get a "not found" response.

Next, select the contents of the "**Find what?**" box and enter **<space>^p** and then select the contents of the "**Replace with**" box and enter **^p**. This removes any single spaces that may remain before the paragraph ending character (non-printing). While this situation is usually benign, there are cases where mysterious problems might be related to the presence of these extra spaces. Usually, one "**replace all**" is all that's necessary to remove these extra spaces.

Repeat the find/change procedure, but this time, make the first entry **^p<space>** and replace with **^p**. This eliminates spaces at the beginning of paragraphs. These sometimes appear during the writing/editing phase of a project. They become very noticeable when the book is printed, but are practically invisible on screen or even on printed proofs.

One more issue is the removal of extra blank lines. Often writers prefer to put two returns at the end of each paragraph. Correct typographic practice is to either have a line (or half line) space between paragraphs or paragraphs may be indented. Never should there be both an indent and extra space between paragraphs. Assuming that you want to use indents, you can do a find/change by entering **^p^p** in the "**Find what?**" box and **^p** in the "**Replace with**" box. You may want to go through the document with the **change/find next** button rather than the **change all** choice if you have extra returns used to create the spacing on your chapter opening pages. (These larger spaces are better created by using the "space after" function in the style sheet.)

Sometimes, you'll be faced with a document where there is a "hard return" at the end of each line. This presents a dilemma as all paragraphing may be lost if you simply search for all instances of **^p**. Hopefully, any such document has two returns separating each paragraph (it's usually the case—if not, you'll need to go through and ensure that each paragraph is separated from the next with two returns). In this situation, select a character that does not otherwise appear in your text. (Usually the tilde (~) works well for this.) First step, do a find/change to remove all extra spaces (as described above). Next step, search for ^p and replace it with the tilde (~) character (**replace all**). Now, search for each instance of ~ ~ (two tildes in a row) and replace with **^p**. Last, search for ~ (single tilde) and replace with **<space>**. Repeat the find doubled spaces search to eliminate any extra spaces that may have slipped into the document. (If all went well, there shouldn't be any.)

Options in the preferences dialog

Now, spend some time setting your preferences. You'll want to ensure that you're using "typographer's quotes." Go to the menu **Tools...>AutoCorrect** and select **AutoFormat** tab and make sure that the "**smart quotes**" option is selected. Take a moment and review the other options and select or deselect as you wish. (Hint: you should *deselect* "**Internet paths with hyperlinks**" if that option is selected.)

Now, visit the **Tools...>Preferences** dialog and select the **Compatibility** tab. Depending on earlier setup when you installed the program or when working on other projects, various options may have been selected. The safest course would be to simply clear all options—that is accomplished by selecting your current version of Word from the drop down menu.

Sometimes, you might want to experiment with some of these settings in the **Tools>Options... (tab) Compatibility** dialog. Those that most directly affect type are:

"Do full justification like WordPerfect 6.x for Windows" (usually desirable)
"Don't add extra space for raised/lowered characters"
"Don't add leading (extra space) between rows of text"
"Don't add space for underlines" – deselect
"Don't center 'exact line height' lines"
"Don't draw underline on trailing spaces" – deselect
"Suppress extra line spacing at bottom of page"
"Suppress extra line spacing at top of page"
"Swap left and right borders on odd facing pages" (usually desirable)
"Use printer metrics to lay out document" (experimental – may be good or bad)

As you select/deselect these options, you'll note that the drop down menu will show "**custom**."

The first option (**...like WordPerfect...**) lets MS-Word *contract* wordspace as well as add wordspace. This should markedly improve the normally terrible wordspacing generated by Word to justify type. The last option (**...printer metrics...**) may not work well on a Mac unless you also select "**use fractional widths**" from the **print tab** (but we assume you're probably using Windows). These settings may cause the type to look odd or mis-spaced on screen. Try setting the zoom view to 125% or so and see if it improves. (When publishing a book, the screen rendering isn't all that important. It's what the book looks like in print that counts.)

For descriptions of what the other options do, check your manual, look them up in the help dialog, or visit the Microsoft web site (*support.microsoft.com*) and search on "Word compatibility tab."

Print a couple of pages and see if any of these options improve the output. Should things seem worse, start by turning off the "**...printer metrics...**" option and print another few sample pages. Continue the process, selecting and deselecting options as you print sample pages until you find the most satisfactory results.

Make written notes listing your final selections. These compatibility options "stick" to the document, but do not become "default" selections. So, you'll need to re-set your selections for each document or template that you use.

Hyphenation. For justified type to have good wordspacing, hyphenation is essential. Return to the menu and select **Tools...>Language>Hyphenation**. (The menu may read **Tools...>Hyphenation**.) If the command isn't there, then you may not have installed Word's hyphenation tool. (The very fact that it is possible that hyphenation might not be installed is a shocking disregard of decent typography!)

Set the hyphenation zone to .5 or .75 inch. Set the limit for consecutive hyphens to 2 or 3. (University presses tend to limit consecutive hyphens to 2 while many trade publishers allow 3. Personally, I prefer to opt for no more than 2 consecutive hyphens.)

The dictionary in Word is pretty decent, so the hyphenation is usually pretty good. However, it is still wise to watch for poor hyphenation choices, as you wouldn't want *ther-apist* to become *the-rapist*.

There may be times when you want to turn off the hyphenation for a paragraph or two. Select the paragraph(s). Go to the menu: **Format>Paragraph...**, when the dialog appears, select the Line and Page Breaks tab and select "Don't hyphenate."

To turn off hyphenation for particular words within a paragraph, select them and go to the menu: **Tools>Language...** and choose "**no proofing**" (top of the list) or "**Do not check spelling or grammar**" [depends on your version of Word]. This stops the words from being hyphenated by excluding them from being checked by the dictionary. *Be certain that the words so excluded are spelled correctly, as Word will not check them.*

The reverse of the problem is that Word will not hyphenate words not in its dictionary. When you do a spell check, be sure to add the words to the dictionary and insert the appropriate hyphenation instructions (see the Word manual). In addition, be vigilant and manually hyphenate any words that slip by. Generally, in such cases, use the discretionary hyphen feature.

Automatic Kerning. Kerning is the adjustment of spaces between individual characters. In theory, this is supposed to make the letterspacing look more natural based on the optical shapes of the letter combinations. Most fonts have kerning tables built into their coding—but cheaper fonts or those from less than reliable sources may not be well kerned, one of the reasons that we recommend that "name brand" fonts be used. In some cases, the kerning tables may be flawed or have become corrupted.

The automatic kerning available from Word is rather primitive. It usually makes body-sized text look worse. However it often does improve the look of larger type used in headlines or chapter titles, etc. To kern a word or phrase, select the text, then go to the menu: **Format>Font…**, select the tab Character Spacing and click the checkbox "**Kerning for font:**" and enter 14 in the "**Points and above**" box. If the automatic kerning doesn't work well, you can manually kern each pair of letters in the word or phrase. To accomplish that task, place the cursor between two letters you wish to kern. Go to **Format>Font…** and select the Character Spacing tab. Then select "**Expanded**" or "**Condensed**" from the drop down menu on the line starting with "**Spacing**"—then enter an amount (Word designates the amount in tenths of a point). Be sure the "**kerning for fonts**" checkbox is deselected. Some experimentation will be required.

MANUAL SPACING. There are two tools used to resolve spacing issues: the manual line-break (shift-return) and the optional hyphen (Mac: Command-hyphen; Windows: Control-hyphen). The manual line break forces a new line but does not start a new paragraph (with indent). The optional hyphen tells the program where to split a word only if it is needed. (You may want to go to the **Tools>Preferences…** menu, select the "view" tab and, in the middle section, click all the checkboxes on to show the nonprinting characters. You will then be able to see the items you're placing into the file as you work.)

Now, working the book from front to back, look for lines where the wordspacing seems too loose or too crowded (tight). For a line that's too tight, put the cursor into a potential breaking point (either before a short word or between syllables of a longer word) and enter a line break to shift the letters to the next line. (If you broke a word, insert a hyphen). This may clean up the whole paragraph or it may cause a new problem further down.

An optional hyphen may shift letters up or down (depending on placement). The optional hyphen locally overrides the hyphenation settings established for the whole document.

You may want to insert hyphens or force line breaks when:

- There are too many hyphens in a paragraph. While the consecutive hyphen setting was established at two or three, if there is a line without a hyphen, then the counter is reset to zero. Sometimes a paragraph simply ends up with too many hyphens. Often manually "fixing" one hyphen will cause the paragraph to reflow and eliminate several other hyphens. (Or, it could make things worse—you'll need to pay attention!)
- There is a bad break where a word is hyphenated across a page break (especially if it's from a recto to a verso) or column break.
- An "orphan" word or syllable occurs. A very short word or syllable on a line by itself will look bad, especially if the number of letters is less than the paragraph indent for the following paragraph. Try to move the orphan back up or bring down additional letters/words to eliminate the problem.
- Spacing between words creates a river through the text. This is the most common problem with typesetting justified text in MS-Word. Try to tighten the lines by inserting manual hyphens and/or optional hyphens.

- There are stacked words. This one can be tricky. If the same word or phrase aligns at the beginning or end on two lines it can cause confusion to the reader. Try to break this up with inserted hyphens or line breaks. In some cases, you may need to re-word the sentence or phrase.
- Word fails to break a line after a dash. (This tends to be more of a problem with Word on the Mac than the Windows version.) Insert a space or line break after the dash to force the break.
- The "WordPerfect" justification causes lines that are too tight. Force line breaks or use optional hyphens to open up the tight lines.

If there is a paragraph that is too tight that won't adequately respond to your adjustments, remove the various forced breaks, etc. then select the paragraph and go to the **Format>Font...** menu and set the "**Spacing**" to "**Expanded**" with the drop down menu and click the arrows to set a value of one or two tenths of a point. (Any more is likely to be very obvious and probably unattractive.) Likewise, if there is a loose paragraph, you can try the "**Condensed**" choice. Again, a light touch is best.

Selecting the basic document format

The first step is to set the page size and margins. Using Microsoft Word, open the **File>Page Setup... (tab) Paper Size** (Windows) dialog. Use the pull-down menu (Macintosh) (that displays **Page Attributes** when the dialog is opened) to select **Microsoft Word** then click on the **Custom...** button and enter 5.5 (inches) for the page width and

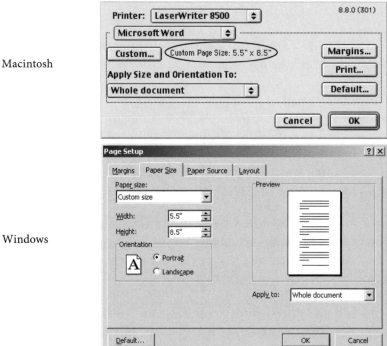

8.5 (inches) for the page height. Make sure that **Apply size and Orientation To:** reads **Whole Document.**

Next, click the **Margins** tab/button to move to the *Margins* dialog. (This can also be reached through the **Format>Margins...** menu.) Enter values of .9 inches for the top, 1 inch for the bottom, .75 inch for the left, and .88 inch for the right. Gutter is left at zero. Check the box to **Mirror margins.** (Note: Macintosh and Windows dialogs are the same for the rest of the settings.)

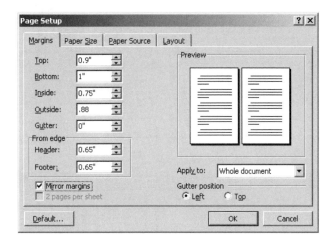

Then select the **Layout** tab and check the boxes **Different odd and even** and **Different first page** in the **Headers and Footers** section. Be sure that **Apply to: Whole document** is displayed. (We will set up different running headers for left and right pages and will have no header on the beginning page for each chapter.) Finally, click the **OK** button.

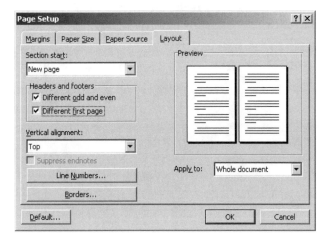

We have now completed the basic page size and margin arrangement. In the next section we will establish the basic design characteristics for the main text.

Selecting body text criteria

Open the **Format>Paragraph...** dialog. With the **Indents and Spacing** tab open, set the **Alignment:** to *Left* and **Outline level:** to *Body text*. In the **Indentation** section, **Left:** and **Right:** should be set to *zero* and **Special:** is set to *First line* and **By:** to *0.25".* The **Spacing** section should be set to *zero* for **Before:** and **After:** and **Line spacing:** should be set to **Exactly:** *13 points* for type set at 11 points. (You should set leading [line spacing] to a value approximately 2 points higher than the point size of the type. See Chapter 4 for a discussion of typefaces [fonts].)

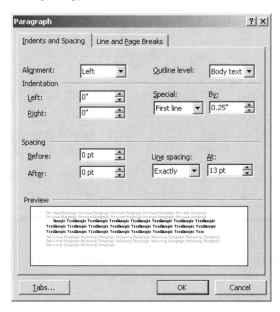

Next, select the **Line and Page Breaks** tab. Under Pagination, you may select the **Widow/Orphan control** checkbox (unless it proves to be a problem later). Make sure the rest of the checkboxes are *not* selected.

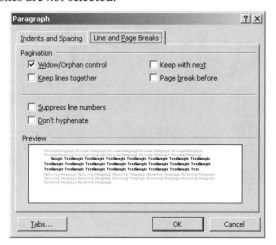

After completing the selections in the Paragraph dialog, click the **Tabs** button at the lower left. (This can also be reached from the menu item **Format>Tabs...**)

Set the default tab stops to *0.25″*, select the **Alignment** *Left* button and the **Leader** *1 none* button. Finally, click the **OK** button. We have now completed the basic paragraph formatting for the body text.

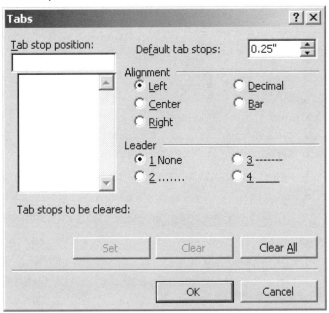

Selecting the body text typeface

Note that the program uses the word *font* instead of typeface. See the glossary for a more complete explanation, but this usage of the word is now common, but incorrect.

Open the **Format>Font...** dialog. On the **Font** tab, select the body text font you wish to use. (In the example, we have selected **Font:** *AGaramond Regular,* **Font style:** *Regular,* **Size:** *11,* **Color:** *Automatic [black],* **Underline:** *(none).*)

On the **Character Spacing** tab set **Spacing:** and **Position:** to *normal* and check the box **Kerning for fonts:** and enter *9* **Points and above.** Under the Text Effects tab, (not shown) the **Animation** setting should have "*(none)*" selected. With these settings complete, you may now click the **OK** button.

Creating a style sheet

The style sheet feature of Microsoft Word (and other word processing programs) allows you to be consistent throughout your document. You should use the style sheet feature and apply the appropriate styles to all text elements. This will ensure that the document is consistent throughout.

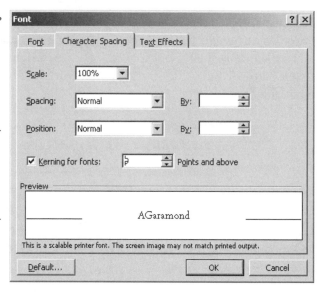

The easiest way to create each style is to go through the steps to manually format the particular text as you desire, then select that text and use the style sheet dialog to create the new style based on the selected text. Give the body text an obvious name like "BodyText 11/13"or "BT 11/13" so you will remember what the style is when you see it on the list. Set **Style type:** to *Paragraph*; **Based on:** to "*(no style);*" and leave **Style for following paragraph:** (the same as the style you just established). For other styles you

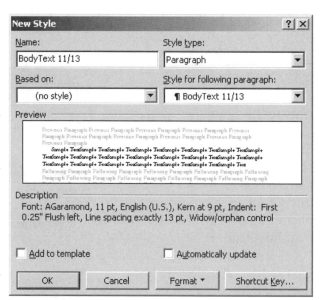

may wish to select the most likely style to be used in the next paragraph. For example, with the style for A-level heads, the following style is most likely to be the main body text style you established.

The default styles that most programs offer are inappropriate for laying out a book. You may wish to delete the unused (default) styles to lower the chance of confusion; however, in Word, the styles "**default paragraph style**" and "**normal**" cannot be deleted.

After you create the basic body text style, you will then want to develop styles for Chapter Number, Chapter Title, Head1 (for the "A-level head"), Head2 (for the "B-level head") and so on through the various levels of headings in your book. See the discussion in chapter 6 to help you select appropriate settings for these elements.

Page endings

A book normally has the bottoms of the pages on each spread line up. (e.g. if the left page has 39 lines of type, the right page will also have 39 lines—not 38 or 40.) This is one of the aspects of book typography that's somewhat difficult to accomplish with Word. You can turn on "widow and orphan" control, but that often creates more difficulties than it solves. It's better to set your bottom margin one line longer than you normally might prefer, i.e. if the normal page has 39 lines, set the bottom margin so all pages will have 40 lines.) Then, as you work through the book, you can put a forced page break at the *end of each page* (on the 38th or 39th line) as necessary to balance the spreads—or you can let the two pages run one line long (don't enter the forced page break) when that is necessary to make the pages balance. Remember when you're balancing the page endings to always work from the front to the back, as changes made in the earlier pages may flow through many following pages as well. It's also a good practice to save the page balancing to the end of the project as interim edits may force you to rework all your page endings. Keep in mind, this is a tedious, iterative process. You may need to go back and forth over several pages to find the best combination of normal, long, or short pages to solve all bad page breaks

If your book has subheadings (as this book), then you can use the "keep with next" feature in the **Format>Paragraph...>Line and Page Breaks** dialog. Turning on that check box (set this feature in the associated style) will cause subheads to move to the next page if it is the last line of the page. Be sure to check the page balance when this happens.

Chapter breaks

Use "**insert Section Break**" at the end of each chapter. When other options are set correctly, the Section Break should delete the running header from the following page (the chapter opener). The Section Break will also isolate each chapter from those that go before. Without a Section Break, changes earlier in the book may force text to flow onto the chapter opening page from the previous chapter. With the section break marked, an overflow will add a new page before the section break.

Most books up to 300 pages or so may be accommodated in a single file, although Word is notorious for slowing down as files become larger—especially if there are graphics or images in the file. Depending on the specifics of your equipment, you may want to break the book into several files. Be sure to adjust the page numbering to start each file with the appropriate page number. (Odd numbers are always on right hand [recto] pages.)

Since word processing files are frequently subject to becoming corrupted as their length increases, be sure to make frequent, successive backups several times each day or just before you do a major operation on a large file. Be sure to save before you print, as a common time for a computer to lock-up (crash) is when a print command is given. At least once each work session, copy your book onto a floppy or CD-R, note the time and date on the label and set it aside as a "safety" copy. From time to time (as per your judgment and depending on how complete the project is) you may wish to make a CD-R and store it in a safety deposit box or with a friend or relative as "insurance" against total loss in the case of a major event (which could be anything from a stolen lap-top to a fire or natural disaster).

Other Word features

If you have a book with many photos or other graphics, Word is even less desirable for making the layout. Word stores all graphics inside the document file, which causes the files to become truly enormous. The larger the file (1) the slower that Word becomes and (2) the more likely that the Word file will become corrupted. If you *must* make a Word document with many graphic images, be sure to break it up into individual chapters.

If your book has a few (very few) photographic images, be sure that they are saved as grayscale TIFF (tagged image format file) with a resolution of 300 dots per inch at the size of reproduction. (It's best to resize images in a program other than Word.) If you have vector graphic images (EPS) Word may not handle them very well. You may use the drawing features in Word or other, related Microsoft programs. Keep them simple for best results.

Turn off "Allow fast saves" or "Allow background saves." These features are an invitation for file corruption with longer files. Make regular backups of your files. (I like to make rotating backups as I work. For example, I start the day working on a file with the name "Book.doc"; at mid-day, I'll save my file, then perform a "save as" and give the file the name "BookX.doc." At the end of the day, I'll again perform another "save as" and call the file "BookX1.doc." As the project progresses, I'll make successive versions—but at some point, to avoid running out of disk space, I'll go back and reuse the "Book.doc" name and repeat stepping through the series of file names. In this manner I'll have 3 or 4 previous versions. If (when) the dread file corruption occurs, then I will usually only lose a few hours of work by going back to the immediately previous version (or an earlier version, if the corruption managed to affect more than one file).

The Word spell checker is good, but if you have a correctly spelled but wrong word, then you'll still have an error, e.g. I have a problem where I type "form" instead of "from." I'm sure that you may encounter similar words as you type. The Word grammar checker is, frankly, not much use. It may catch a few minor glitches, but it will miss important errors and will nag you about stylistic "errors" such as passive voice. (Some topics are best treated in the passive voice!) In both cases (spelling and grammar) you will still need the help of a skilled proofreader. You may be able to find a student through a community college or university—call the head of the English department for referrals as the typical school "job center" doesn't always list such positions. At the very least, get a friend or relative to look over your work. I suggest paying for this service, as friends or relatives are notorious for taking a quick look and telling you everything is "OK." If they are paid, they are much more likely to actually make a reasonable effort at providing the feedback you truly need.

Creating a PDF

If you call a book printer and tell them you have a book "ready to print" as a MS-Word file, you're likely to have them hang up on you or you'll hear them run screaming from the room. You'll need to make your Word file into an Adobe Acrobat PDF (Portable Document Format) file. This requires using a conversion program (Acrobat Distiller) to process the Word file and generate a PDF.

Depending on your version of Word, you may have "PDF Writer" installed on your system. While the natural expectation might be that this will create the PDF for you, it is unlikely that this will prove satisfactory. PDF Writer is a very simple conversion program designed for business documents. It's generally incapable of doing a good job with anything as complex as the average book.

Adobe Acrobat is a fairly expensive program, but will certainly be able to process your document. Ask around, you may have a friend or coworker who has access to a copy of Acrobat Distiller. Another alternative is the $79 program, Jaws PDF Creator. See the Æonix web site for more information.

When you create your PDF, be sure that you select the "press" quality output choice. That will satisfy most printers. However, once you select your printer, you should verify with them if there is any need for special "job options" when you make your PDF.

If you're using a Macintosh with Mac OS X, the creation of PDFs is built into the operating system. You simply select "Save as PDF" from the print dialog. PDFs made in this manner do not give you all the options available in the full Acrobat program, but a "press ready" file can usually be generated from a simple Word file, assuming that the printer doesn't have any special job option requirements. (Some printers supply a "job options" file that works with Acrobat Distiller. Others will want you to use a particular "PPD" (printer description file) when creating an intermediate PostScript file.

What do you get?

When done with care and with considerable attention to details, you can get reasonably decent typesetting out of Word. I still don't recommend it for regular production of books as it requires much more work than the professional page layout programs to get acceptable results. And some tasks, such as page balancing, are truly tedious with Word while they are fairly easy with a page layout program. On the next page is a paragraph produced in Microsoft word similar to the samples in Chapter 6. Compare this sample with the one on page 86 that shows the output from Word with default settings.

Other word processing programs may give better results. I'm told that WordPerfect produces "much better" typography than does MS-Word. (That's hinted at by the use of "... Justification like WordPerfect..." as one of the selected preferences.) Still, even the best word processing programs require more work to achieve the same result—work that a true expert with a word processing program may be able to manage but is difficult for the individual with moderate word processing skill. The same moderate skill level can result in excellent typesetting when a professional page layout program is used.

The Mississippi does not alter its locality by cut-offs alone: it is always changing its habitat *bodily*—is always moving bodily *sidewise*. At Hard Times, La., the river is two miles west of the region it used to occupy. As a result, the original site of that settlement is not now in Louisiana at all, but on the other side of the river, in the State of Mississippi. *Nearly the whole of that one thousand three hundred miles of old Mississippi river which La Salle floated down in his canoes, two hundred years ago, is good solid dry ground now.* The river lies to the right of it, in places, and to the left of it in other places.

Although the Mississippi's mud builds land but slowly, down at the mouth, where the Gulf's billows interfere with its work, it builds fast enough in better protected regions higher up: for instance, Prophet's Island contained one thousand five hundred acres of land thirty years ago; since then the river has added seven hundred acres to it.

—*From Life on the Mississippi by Mark Twain*

Here is a sample of the the same text used in chapter 7 as an example of typesetting quality. This was set in Adobe Garamond on a Windows computer using the Word options described in this chapter. This sample is not directly comparable with the chapter 7 samples as Palatino (used there) was not installed on the Windows computer that was available to me.

In hindsight, I see that I should have forced a hyphen in "settlement" (fifth line) as the three lines above are rather loose. (This is a perfect example showing how Word prefers adding wordspace rather than using hyphens.)

- 9 -

Page Layout with InDesign

HERE WE WILL DESCRIBE HOW TO lay out a book with InDesign. (InDesign is sold as a stand-alone program and as part of the Adobe Creative Suite. The current version is called InDesign CS—and is actually version 3.0 of the program.) Book layout with PageMaker, Quark XPress, and other professional page layout programs is similar in most respects, although there are some procedural differences in each program. When there is a significant difference, we'll mention the PageMaker approach. Quark XPress will not be specifically addressed. Quark XPress is heavily used by professional graphic artists, so it is assumed that a Quark user will be familiar with the program. See chapter 7 where we discuss the page layout programs and suggest those that are most appropriate for book design and layout.

File clean-up

Our first assumption is that the basic manuscript has been prepared in a word processing program. Due to the substantial market share of Microsoft Word, that is the most likely program used to write the manuscript. Computer commands in the following paragraphs are specific to Microsoft Word 98, a slightly older Macintosh version of the program. (I haven't felt the need to upgrade my copy of Word at this point—the commands in the most current version are the same or quite similar.)

The manuscript file should be carefully proofread, spell checked, printed out and proofread (again) from hard copy and otherwise cleaned up to ensure that your time with the page layout program is spent with design and typography rather than fixing errors that should have been caught at an earlier stage in the production process. We also make the assumption that the manuscript has been professionally edited. (If not, please go back and read chapter 2 to understand the importance of the editing step!)

There are a few things that cause problems and make the typography less attractive. First are extra spaces between sentences. (I know, you were probably taught, "period–space–space" when you learned to type.) There are also instances of other flaws.

Before you begin preparing your document as a book, you'll need to execute a series of find/change procedures to "clean" the file:

First, go to the menu: **Edit>Replace...** and enter **<space><space>** (hit the space bar twice) in the "**Find what?**" box. Click into the "**Replace with:**" box and hit the space bar once, i.e., search for two successive spaces and replace them with one space. When the first instance is found, select the "**replace all**" choice, unless you have specific places where multiple spaces are intended (if that's the case, you probably should have used tabs instead). Re-run the Replace procedure as many times as necessary until you get a "not found" response.

Next, select the contents of the "**Find what?**" box and enter **<space>^p** and then select the contents of the "**Replace with**" box and enter **^p**. This removes any single spaces that may remain before the (non-printing) paragraph ending character. While this situation is usually benign, there are cases where mysterious problems might be related to the presence of these extra spaces. Usually, one "**replace all**" is all that's necessary to remove these extra spaces if you've already done the first step and removed all instances of multiple spaces.

Repeat the find/change procedure, but this time, make the first entry **^p<space>** and replace with **^p**. This eliminates spaces at the beginning of paragraphs. These sometimes appear during the writing/editing phase of a project. They become very obvious when the book is printed, but are frequently missed when proofing on screen or even on printed proofs.

We will assume that the file only contains a "hard return" at the end of each paragraph. If not, you must go through and eliminate the extra returns. If you have two returns at the end of each paragraph, we can begin the search and replace procedure. If not, go through the document and manually insert two returns after each paragraph. Then make the following search and replace steps:

1. Search for **^p^p** and replace with ~ ~ (2 tilde characters or some other character not used anywhere in the manuscript).

2. Search for **^p** and replace with **<space>** (Then repeat the remove extra space procedure described earlier). The whole manuscript will now look like a single paragraph.

3. Search for ~ ~ (2 tildes) and replace with **^p**. This will restore the paragraphing throughout the manuscript.

Another issue is the removal of extra blank lines. Often writers prefer to put two returns at the end of each paragraph. Correct typographic practice is to either have a line (or half line) space between paragraphs or paragraphs may be indented. Never should there be both an indent and extra space between paragraphs. Assuming that you want to use indents, you can do a find/change by entering **^p^p** in the "**Find what?**" box and **^p** in the "**Replace with**" box. You may want to go through the document with the **change/find next** button rather than the **change all** choice if you have extra returns used to create the spacing on your chapter opening pages. (These larger spaces are better created by using the "space after" function in the style sheet.)

Excess use of tab characters may cause annoyance when you import your Word file into the InDesign document. Go through the Word file and note where you've used tab characters. If you have used a tab at the beginning of each paragraph, that's more properly handled by a first line indent in the style definition. So, do a search for **^p^t** (end paragraph followed by a tab character) and replace with **^p** (end paragraph) only. If you have other instances of tabs—such as in a simple table—you may wish to convert the text to a table—which imports into InDesign reasonably well. If you're using PageMaker, change all tables to plain text, as PageMaker doesn't "recognize" Word tables and will ignore them (failing to import the table material). The main point, with tabs, is to minimize their usage as they may cause confusion in sorting out the imported material.

Finally, if you have any extra indents, or other specialized formatting (that isn't obvious in the context of the text), then you should remove the formatting from the Word document—first "flag" the specially formatted material with a note or other reminder so that you can create a correct format for the element in the InDesign file. For example, when I have a long quote that is given an extra indent (left and right), I'll insert a tag, "<quote>" at the beginning of the passage and "</quote>" at the end of the passage. This follows a convention used in HTML, the mark-up language used to encode pages for display in a web site. Enclosing a "command" inside greater than/less than symbols and using the "slash" character (/) to indicate the conclusion of a particular element. All the page layout programs do have various tagging methodology that will interpret tags (in the format required by the specific program) and will encode the tagged text with the associated stylesheet definition. We are not attempting to do this here—this is simply a way to make specially formatted elements obvious as we go through the text in the page layout program paragraph by paragraph.

While we're discussing tags, if your document has external elements, such as photos or illustrations, it's a good idea to insert tags showing the location in the text where they should be placed. Again use the greater/less than symbols to set off the tag. Organize and number your photos, illustrations, or other items (including the file names) so that they can be matched to the tags. Then you can put a note like this: "<insert photo #23 at or after this position>." That reminds you (or tells the typesetter) where the photo may be positioned. It's also good to note any restrictions on placement that may apply to a particular element. Sometimes, a photo or other illustration merely needs to be near a point in the text. In other cases, you may rather have elements appear at or after the reference point, thereby allowing the element to be discussed in the text before the reader is distracted by the photo or illustration.

Having completed tagging any special elements, select all the text and open **Format>Paragraph...** from the menu bar. Once the dialog displays, using the **Indents and Spacing** tab, set **Alignment** to Left, **Indentions** (left and right) to **0** (zero) and **Special** to **(none)**. In the Spacing section, set **Before** and **After** to **0** (zero) and **line spacing** to "**single**." This removes all extra formatting from the Word document.

Next, go through the document and remove all forced page and section breaks. These will be invisible in the page layout program, but will probably cause incomprehensible mid-page breaks. They are almost inevitably confusing once the document is in the page layout program.

Removing specialized formatting is primarily to avoid having annoying coding in the file that may get imported into the page layout document. For example, if you've set a section of the letter sized manuscript to have a two inch indent on both left and right, you'll discover that the text will take up numerous pages with a half-inch line width when being set into a book-sized document with a 4½ inch line width.

The last step is to do a final spell check, then save the document as a Rich Text Format file. While most page layout programs can directly import a Microsoft Word document, saving as RTF reduces the amount of unintended coding that remains in the file that might cause unexpected problems in the page layout program file.

Preliminaries

Before we actually import the RTF document into the page layout program, we need to do some preliminary set up with InDesign. Start InDesign and select **Edit>New> Document...** from the menu. This will bring up the "New Document" dialog box. Starting at the top, Ignore the "Document Preset" drop down menu. Set the "Number of Pages" to **10**. Check the box "Facing Pages." If selected, unclick (or do not select) the "Master Page Frame" option. (Master page frames are available for compatibility with Quark XPress. They are actually superfluous to InDesign.) In the "Page Size" section, ignore the drop down menu and set "Width" to **6 in** and "Height" to **9 in**. Be sure to click the portrait orientation if it is not already selected. In the "Columns" section, set the number to **1**. (The Gutter is not relevant.) Finally, in the Margins section, set the "Top" to **.835 inch**, "Bottom" to **.985 inch**, "Inside" to **.75 inch** and, "Outside" to **.875 inch**. (See pages 35–36 for a discussion of margins.) You should have a resulting dialog box that looks like this:

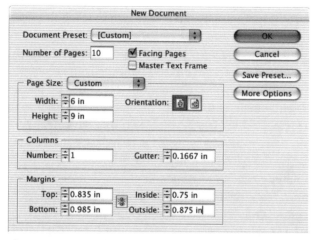

You can save this as a preset (Click "Save Preset..." and give it a name of "**6 x 9 book**") if you wish. We do not need to view or change any of the settings under "More Options," so you can click the **"OK"** button. [PageMaker's new document dialog is quite similar, but features exclusive to InDesign are absent.] A new document will appear with the first page (a right hand [recto] page) showing. This will become our Book Half Title page.

On the Book Half Title page, drag out a text frame to the size of the margins (click-drag the text tool from the top left to bottom right margins) and type the main title of your book. Position the title about ⅓ of the way down the page from the top margin. Set the title in your chosen typeface at a size that's reasonable and attractive considering the length of the title and size of the page. Break up the title over several lines, if appropriate. (See the Book Half Title page of this book as an example.) The title should be centered (left and right) within the frame created by the margins.

Move to the next spread (pair of pages). The back of the Book Half Title page is usually blank. On the right hand page, create a new text block. Type in the title (or copy it from the Book Half Title page) then type the sub-title (if any) below the main title. Set it somewhat smaller than the main title (breaking it over several lines, if appropriate). Space down several lines and type the author's name (no need for the word "by"). Just above the bottom page margin, type in the name of the publisher and (on the next line) the city (and state, if appropriate) where the publisher is located. (Again, see the main title page of this book as an example.) If the publishing company has a logo, it may be placed just above the publisher name.

Now, move to the page 4-5 spread. The left (verso) page is the copyright page. Here you will enter all the copyright information along with the name and address of your publishing entity. CIP data will also go on this page, if available. Otherwise, only the LCCN and ISBN would be shown. Finally, the statement of the country of origin should be put here. Usually that will be "Printed in the United States of America."

THE BASELINE GRID. Let's assume that we've decided to use Minion for our body typeface set at 11 points on 14 points of leading. We'll need a baseline grid spaced at 14 points to support this leading. Get out some scratch paper as we'll now need to do some calculating to figure the starting point for the baseline grid. (It needs to start 2 lines above the top of the text block margin.) We take the top margin (.835 inch) and subtract 2 lines (14 points x 2 or 28 points) from .835 inches (28 points equals .389 inches) to get a result of .446 as our starting point for the baseline grid. Go to **Preferences>Grids** to enter our information.

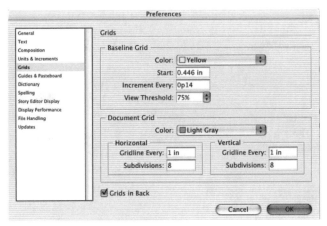

Your grid preference dialog should look something like the example at the bottom of the previous page. Note that I've entered the increment spacing in points **(0P14)** rather than as a decimal fraction of inches. That reduces any rounding errors. I have also set the grid lines to appear in yellow as I personally find that less distracting than the default light gray. Once you've set your grid values, click **OK**. (We can ignore the Document Grid settings as they aren't used in most book layouts, unless there are many photos to contend with.)

Below, we have a table showing some of the most common grid starting points at various leading measurements. The measurements are calculated first in points, then converted to the decimal inch equivalents to avoid compounding rounding errors. Fortunately, InDesign allows you to enter picas and points interchangeably with inches (and metric) measurements. When working with type, use points (and picas) for most calculations, then convert to the decimal inch equivalents, if necessary. Often it's better to simply enter the pica or point measurement directly in the dialog. (Pica/point measurements are expressed as NNpNN where the first number is the pica measure (if any) followed by a "p," which is followed by the point measurement. So, 12p27 would be 12 picas and 27 points. You do not have to convert points to picas or picas to points. In this case 27 points is actually 2 picas and 3 points (as there are 12 points to the pica).

Leading in Points	Equivalent decimal Inches	2 lines (decimal in.)	Starting point for grid with .835 inch top margin
12	.167	24 pt. or .333 in.	.835 - .333 = .502 in.
13	.181	26 pt. or .361 in.	.835 - .361 = .474 in.
14	.194	28 pt. or .389 in.	.835 - .389 = .446 in.
15	.208	30 pt. or .417 in.	.835 - .417 = .418 in.

After you've entered the starting point, go to the menu **View>Show Baseline Grid** to make the grid visible. Use the magnifying glass tool to zoom in to the top left margin so you can ensure that the baseline grid starts two lines above the top margin and the baseline exactly aligns with the top margin. If it does not, check your math and check your entries in the **Preferences>Grids...** dialog and make any adjustments necessary to get the grid to properly align with the top margin.

Now, we'll take a look at the bottom margin. With this example, we can see that the bottom margin is very slightly above the baseline for the thirty-seventh line on the page. We can leave this as it is and when we 'flow in' the main body of the book, every page will end on the thirty-sixth line. If we prefer to have a standard page with thirty-seven lines, we'll need to make a small adjustment to the bottom margin.

First, we'll measure the distance between the bottom margin and the baseline of the thirty-seventh line. The easiest way to do this is to select the rectangle drawing tool and simply draw a box from the thirty-seventh baseline to the bottom margin line. Then, with the box selected, look at the Transform palette and read the "**H:**" measurement as (in this case) **.0143** inches. Take the bottom margin of **.985** and subtract

.0143 to arrive at a new bottom margin of **.9707**. If the bottom margin and a baseline exactly coincide, InDesign's auto-flow feature will fail to utilize the last line before the margin. Therefore, let's round the **.9707** to **.965**. To implement this change, go to the Pages palette and select the master page spread entitled "A-Master" (the alternative is "none" which you can't select). Now go to the **Layout>Margins and Columns...** menu and enter .**965** in the bottom margin field. Zoom in and look at the bottom margin and ensure that it is slightly below the thirty-seventh baseline. Look at the Pages palette and make sure that all pages have an "A" in them. [PageMaker doesn't have a visible baseline grid. However, you should set "leading method" to "baseline" in the preferences. In the Paragraph dialog, do *not* select the "align next paragraph to grid" as I've found PageMaker to be somewhat inconsistent in the way it handles the baseline grid. It's better to carefully calculate position of non-paragraph elements in multiples of the primary leading measure.]

Paragraph styles

Now, return to the page 4-5 spread (double-click that spread in the Pages palette). Next, display the Paragraph Styles palette (if it's not already visible). Delete any default styles listed. (Select them and drag them to the trash container at the bottom of the palette.) Use the pop-out menu on the Paragraph Styles palette and select **New Paragraph Style....** The paragraph style options dialog will appear. This is a rather complex, multiple page dialog. In the first window, you are asked to give the paragraph style a name (let's call it "**BodyText 11/14**" for body text of eleven points on fourteen points), then there is a drop-down menu to set what paragraph style the new style is based on. In this case, we'll select **[No paragraph style]**. The next dropdown menu allows us to specify what the paragraph style of the next paragraph should be. For this style, we'll select **Same Style**. Now, click the **Basic Character Formats** page in the window on the left side of the dialog box. That will display the typeface choices. For this example, we'll select the Minion Pro font family, set the **Font Style** to **Regular**, **Size** to **11 pt** and **Leading** to **14 pt**. Set the **Kerning** to **Optical** (that usually gives better letterspacing than "metrics"), **Case** to **Normal**, **Tracking** should be **zero** (in most cases) and **Position: Normal**. Check the box for **Ligatures** and make sure the other check boxes are unchecked.

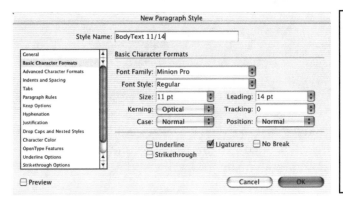

SHORTCUT: Set some text with the characteristics you prefer, select it, then go to the style menu and select **New Paragraph Style...** The dialog will reflect the selected text. Input a style name, check or adjust any of the optional settings, and click **OK**.

Click to the next page in the dialog, **Advanced Character Formats** and check that all values are set to the default values (**Horizontal Scale: 100%**, **Vertical Scale: 100%**, **Baseline Shift: 0 pt** [zero points], and **Skew: 0°** [zero degrees]). The dropdown menu **Language** should read **English:USA** (presuming that is the language you're using).

Now, click to the **Indents and Spacing** page. Set **Alignment** to **Left Justify**, set the **First Line Indent** to **.25 inch** (leave the other indents and space before/after set to zero). Finally, set **Align to Grid** to read **All Lines**.

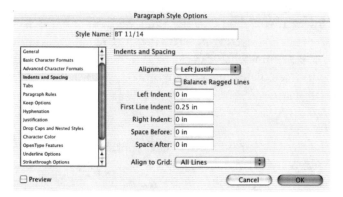

Now select the **Tabs** page of the Paragraph Style Options dialog. Note that the left margin shows a split with the lower triangle at the zero point on the ruler line while the upper triangle is at the ¼ inch mark. This was carried over from the Indents and Spacing page. Position the cursor at the .25 mark and click to insert a tab. Repeat on the .50 mark, then click the **Repeat** button to fill out tabs every quarter inch along the normal reading line. (We may never use these tabs, but they're helpful if special formatting is needed later on.) Leave the **Leader:** and **Align On:** fields blank.

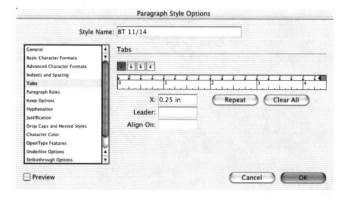

Now, click to the **Paragraph Rules** page of the dialog. Here we just want to be sure that the **Rule On** button is *not* selected. Note that the dropdown menu will show you either the **Rule Above** or the **Rule Below** attribute.

Next, click on the **Keep Options** page of the dialog. For now we can leave this page alone. We could set a parameter that we need to keep at least 2 lines together in any paragraph, however it's usually less confusing to handle this task manually—and changes you make to one page aren't as likely to ripple through the whole book. In the lower portion of the box, see the **Start Paragraph** parameter. It should read **Anywhere**. The other choices include: **In Next Column**, **In Next Frame**, **On Next Page**, **On Next Odd Page**, and **On Next Even Page**. This will be useful when we are setting up our new chapter opening styles.

Now, click to the **Hyphenation** page of the dialog. Click to select the "**Hyphenate**" check box (if it isn't already selected). Conform the settings to: **Words with at Least: 5 letters, After First 2 letters, Before Last 2 letters, Hyphen Limit: 3 hyphens**, and **Hyphenation Zone: 0.5 inch**. Some of the more fastidious publishers prefer to set the Hyphen Limit to 2 hyphens or may wish to set **Words with at Least: 6 letters, After First 3 letters, Before Last 3 letters**—make those settings if you wish. The continuum line from **Better Spacing** to **Fewer Hyphens** defaults to the center. I usually like to move it two or three notches toward **Better Spacing**. Finally, select the check box **Hyphenate Capitalized Words**.

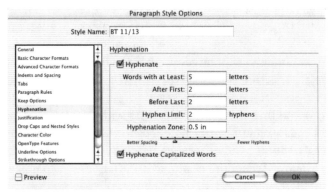

Next, click to the **Justification** page of the Paragraph Style Options dialog. The defaults here are relatively conservative and are generally acceptable as a starting point. The wordspacing parameter is, however, just a little too generous. For **Wordspacing**, set the minimum to **85%**; desired to **100%**; and maximum to **125%**. For **Letterspacing**, set the minimum to -5%; desired to 0%; and maximum to **15%**. (These values may be adjusted slightly for different typefaces.) For **Glyph Scaling**, set the minimum to **98%**; desired to **100%**; and maximum to **102%**. (Glyph scaling is generally best avoided, but these small values are barely perceptible and may give the software just enough flexibility to "save" a difficult paragraph. Set **Single Word Justification** to "**Full Justify**." (This refers to a single word on a line by itself. This may occur with a long word on a short line, perhaps where text is flowed around an image.) And, set the **Composer** to "**Adobe Paragraph Composer**."

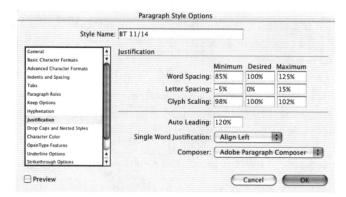

Next, look at the **Drop Caps and Nested Styles** page in the dialog. We don't need to make any changes to this page once we're sure that there's nothing set.

We can skip the **Character Color** page as we will be setting our text in the default, black, so now select the **OpenType Features** page in the dialog. The usefulness of this page depends on the features in your selected typeface. We're using Minion Pro, an OpenType face from Adobe. The "Pro" variants of Adobe OpenType have many useful and clever features built in. Using the "OpenType features" we can activate and control these automated functions. (They have no effect on regular PostScript or TrueType faces.) The OpenType file format uses "unicode" to allow more than 65,000 characters in each type file, although such a large number of characters is rarely used in Western/Roman typefaces. Before the OpenType format, to have the many typographic features in a "Pro" font required using several font files that were often called "expert sets," which would include separate font files with true (not computer generated) drawn small caps, oldstyle figures, alternate type forms, alternate figure styles, the less-common ligatures, fractions, and decorative symbols. These multiple font files made it difficult to utilize many of the more desirable variations and also make spell checking a document particularly complicated.

Here, we will select the "**Contextual Alternates**" OpenType feature and we'll set the **Figure Style** to "**Proportional Oldstyle.**"

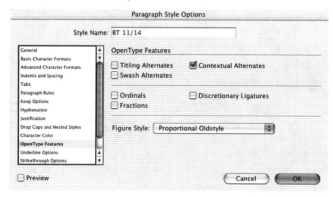

The next page in the dialog is **Underline Options**. As we've previously discussed, underlines are not normally part of typeset work. In the few instances where an "underline" is required, it's better to use a "rule" (see the Paragraph Rules page in the dialog). For general emphasis, italic or bold are normally used. Check to be sure that the underline options are not selected. (The checkbox should be unchecked.)

The next page is the **Strikethrough Options** page. This is an uncommon typographic feature. The only place where it's regularly used is in voter pamphlets where details of a proposed change to a law are detailed and the text to be removed is shown as strike-through text. Hopefully, you'll not be using this feature in your document. Again, check to be sure that the strikethrough options are not selected. (The checkbox should be unchecked.)

The final dialog page is **Bullets and Numbering**. This page is used to set up automatically bulleted or numbered lists. This is a new feature in InDesign CS that I haven't experimented with sufficiently. For now, I prefer to set up styles for these features manually, so we can also skip this page.

You can now click **OK** to accept our settings and close the dialog. Once you've done this, you've completed your first paragraph style. (Frankly, this discussion has taken far longer than the actual setting of the various options.)

Our second paragraph style. Fortunately, we can create a series of useful paragraph styles based on our first style. With the pointer tool (be sure no text or text blocks are selected) go to the **Paragraph Style** palette, highlight **BT 11/14** and select **duplicate style...** from the flyout menu. The **Paragraph Style Options** dialog will display. Our next style to create is for body text that does not have a first line indent. We'll call this "**BodyText 11/14 NI**" (for body text 11/14 no indent). Change the name in the style name window at the top of the dialog by selecting "**copy**" and typing "**NI**".

Now, select the **Indents and Spacing** page. In the **First Line Indent** field, highlight the **.25 in** and enter **0** (zero). You've completed all the changes necessary for this style. Click **OK** to accept the settings and close the dialog.

Our third paragraph style. Now, we'll want a paragraph style featuring a drop cap for the first paragraph of each chapter. We could use the BT 11/14 NI and manually insert the drop cap, but we might inadvertently set different parameters in different chapters. It's better to create a paragraph style to cover the need. Again, using the pointer tool, with no text or text blocks selected, highlight the **BT 11/14 NI** style on the **Paragraph Styles** palette and choose **Duplicate style...** from the flyout menu.

When the Paragraph Style Options dialog displays, change the **Style Name** to read "**BT 11/14 DC**" for body text 11/14 drop cap. Further down, set the **Next Style** option to read **BodyText 11/14**. The next style option only applies to text that you type into your InDesign document. It has no effect on imported (placed) text.

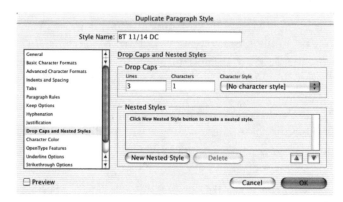

Now go to the **Drop Caps and Nested Styles** page. Enter **3** in the **Lines** field and **1** in the **Characters** field. This will give us a three-line drop cap consisting of the first character in the line. You may now click **OK** to accept the changes and close the dialog.

We can go on in a similar manner to create additional paragraph styles. I suggest you create a "page break" style: duplicate the BodyText 11/14 NI style, rename the style **Page Break**, then go to the **Keeps Options** page and set the **Start Paragraph:** to **On Next Page** or **On Next Odd Page** if you want your chapters to always start on a right-hand page.

You will also need paragraph styles for the Chapter Number (if used), Chapter Title, and for heads of the various levels used. We'll look at an alternative way to create these styles later.

> **DROP CAP AND QUOTE MARK**
> If you use drop caps or raised caps as a paragraph opening style and you encounter a chapter that starts with a quotation mark, what do you do? A super large quote mark looks rather strange. Making two letters "drop caps" doesn't help much.
>
> The practical solution is to simply eliminate the opening quote mark. Readers are rarely confused (or even notice). This practice is "approved" by the *Chicago Manual of Style* in item 11.41 in the 15th edition.

Character styles

One of the great features of InDesign (not available in PageMaker and some other page layout programs) is that it supports character styles. These are styles that apply to letters or words within a paragraph. Once set, these character styles override the basic paragraph style and will remain even if the paragraph style is changed. Now, let's create a few useful character styles.

Most documents will need to have italic and bold character styles. So, display the **Character Style** palette (if it is not already displayed) and, using the pointer tool with no text or text block selected, use the flyout menu to select **New Character Style...**. The Character Style Options dialog will display:

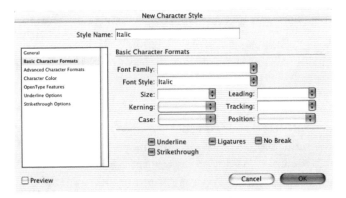

Here, we've input **Italic** for the **Style Name**, then moved to the **Basic Character Formats** page and selected **Font Style: Italic**. We've been careful to not make any other changes. Click **OK** to save the new character style definition.

You may double click the italic style in the **Character Styles** palette to re-open the **New Character Styles** dialog. Then, click on the various pages to explore the options available. You'll see how you can effect most aspects of the way type is displayed. This book uses many character styles to generate the type effects used. This time, click **Cancel** to leave the dialog without making any changes.

Now, use the flyout menu to create a new character style. This time, give it the name "semibold" and go to the **Basic Character Formats** page and set **Font Style:** to **Semibold**. [Minion Pro also has medium, and bold weights. Semibold is appropriate for use with the regular weight.]

If needed, you can also create character styles for small caps, captions, swash italic capitals, oldstyle figures or lining numbers, or other features that your font may offer.

Flowing text

We have finally reached the point where we're actually going to bring the text of the book into the InDesign document. Remember, all this time the document has been open to the pages 4–5 spread.

First select the pointer tool, ensure that no text blocks are selected, then go to the main menu bar and select **File>Place...**. The **Place** dialog will appear. Its appearance will depend on your computer operating system, but it will consist of a file navigation section and will have a couple of checkboxes labeled **Show Import Options** and **Replace Selected Item**. The checkboxes should normally be unchecked. Use the file navigation features to locate your book manuscript word processing file that you saved as RTF (rich text format) in an earlier step. Select that file, then click **OK**.

If the text suddenly appears in a box somewhere, you can actuate the **Edit>Undo** item. If all has gone well, you'll have a "loaded cursor." (The cursor will have changed appearance from an arrow to a symbol that looks like a corner of a page with lines of type). Hold down the **<shift key>** and the "loaded cursor" should change its appearance to look like a corner with a solid squiggle line. (See the drawings in the sidebar.)

The default loaded cursor will only flow from your insertion point to the bottom of the page. The squiggle-line loaded cursor will flow the complete file into the InDesign document—*creating as many new pages as necessary to contain the file*—which is exactly what we want to do. So, holding the shift key, position the loaded cursor within the text block area in the upper left corner and click the mouse. If everything is done right, a lot should soon start happening as the InDesign document is filled with the text from your RTF file. You may need to wait for a couple of minutes before everything settles down. Once this has occurred, save your file! *[We haven't mentioned it earlier, but you should be saving your file at least every half-hour or more frequently if you are making many changes. See the InDesign program documentation for details.]*

If the source file has page breaks or section breaks, the "flow" may stop when one is encountered. Look at the last page where text has flowed and look to see if there's a small red plus sign (+) in a little box near the lower right corner of the text block. If so, that's an indicator that there's more text to flow. You may need to use the pages palette flyout menu to add 3 or 4 pages and manually flow the text to get by the page (or section) break embedded in the input file. Once the page or section break is passed, you may start the autoflow again. If you have encountered one page or section break, it's usually best to check the last page yet again to ensure that all the text has flowed into the document.

InDesign is quite competent (few or no unexpected quitting incidents, program doesn't feel sluggish) in handling books of 300+ pages in a single file, if they are all text. Significantly longer books, or books with many photos (or other illustrations) are better broken up into smaller segments. (Use the "book" feature of InDesign to coordinate the segments.) With InDesign 2.0, I found that a book of 586 pages was best handled in two files; a book with hundreds of photos (the project was a "tour guide book") worked best when broken into 50 to 75 page segments. PageMaker similarly becomes less stable when documents are very long or have many images. Likewise, breaking the document into reasonably sized segments is recommended. (All the page layout programs claim that they can handle documents far longer than practical reality. The claimed document size limits are based on internal programming counters that can't be exceeded. In actual practice, the programs all become sluggish or unstable when more modest file sizes are reached.)

Paragraph styling (coding)

Once the manuscript is placed within the InDesign document, you then need to apply the style sheet to the paragraphs. I like to apply character styles first (so that I don't

---

Loaded Cursors

Above is a loaded cursor with a text file waiting to be placed. This indicates that it will fill the page from the point where it is clicked until it reaches the bottom margin.

This is a loaded cursor that indicates that it will flow the text continuously until the complete file is imported into your InDesign document. This setting will create all the pages necessary to contain all the text.

inadvertently obliterate any styled text). Starting at the beginning of the placed text, select each word or sentence (or any continuous group of words/sentences) that have special styles (such as bold or italic), then click the appropriate style in the Character Styles palette. Go through the whole book, or if you wish, process each chapter in turn.

Next, click an insertion point in the first text paragraph of the first chapter, then click the "**BodyText 11/14 DC**" style in the Paragraph Styles palette. The paragraph should immediately reflect the style that you earlier described. If a small plus sign (+) appears after the style name in the paragraph styles palette, this indicates that some of the paragraph does not reflect the defined style. If there is bold or italic type in the paragraph, you should ensure that the appropriate character styles have been applied. Once you are certain that everything has been "fixed" then (with your insertion point in the paragraph) again click the style name in the Paragraph Styles palette, this time while holding the option key (Alt key in Windows). This should clear the plus sign and change any previously unchanged text to match the specification in the style.

Proceed through the book applying the desired styles to each paragraph in a similar fashion. For longer blocks of text with the same style, you can make your insertion point in the first such paragraph, then move to the last paragraph and while holding the shift key, select all the text before applying the appropriate style from the Paragraph Styles palette. (In some books, I go through and apply character styles to the bold or italic material, then select all the text in the "story" (that's all the text imported with the "place" command), then apply the body text paragraph style. Then, starting from the beginning of the book, I apply page break, chapter number, chapter title, and drop cap paragraph styles as appropriate. This method ensures that there are no "stray" characters left over from the imported file that may cause problems later.

Take care, when selecting text, to select the "invisible" paragraph end character at the end of each paragraph. Sometimes these invisible "characters" are left unstyled or incorrectly styled and can cause improper leading or other odd behavior.

Other stray characters that remain "unstyled" can also cause some problems, especially if they are in a font not elsewhere used in the text and/or use a font not available on your computer. These unstyled/wrong font characters are very difficult to locate in PageMaker files, but InDesign has a "find font" feature (from the menu bar: **Type>Find Font…**) that simplifies correcting this annoyance.

OPTICAL MARGIN ALIGNMENT In general, more attractive typesetting results if the Optical Margin Alignment feature is selected. Go to the menu and select **Type>Story** to display the Optical Margin Alignment option. This feature applies to all text in the same thread (a "story"), so it only needs to be applied once to the main story and any other text blocks where justification is used. Simply make an insertion point in the target text block and click the check-box and set the point size to the primary size (body text) used in the story palette. (In our example case, that would be 11 points.) When selected for the first time, the whole document will be recalculated, so be patient, it may take anywhere from a few seconds to a few minutes depending on your computer and the size of your document to complete the calculations.

Master pages and running headers and folios

Books have running headers (sometimes footers) that feature the page numbers and (sometimes) information about where you are in the book, such as a chapter title. Fiction may only show the title of the book and the author's name. Nonfiction, especially a "how to" book, might show the book title on the left and the chapter title on the right. A book divided into sections may include the section title on the left with the chapter title on the right. These running headers (or footers) are created on the master page(s).

All the page layout programs allow you to have multiple master pages. InDesign allows master pages to be based on other master pages. For example, I'll create a master page with the left header (that is usually the same throughout the book), then create master pages for each chapter (with the chapter title on the right) based on the master page with the left header.

All the page layout programs also provide for automatic page numbering. InDesign further refines this feature by allowing multiple sections in a book. (This simplifies having front matter numbered with lowercase Roman numerals with the body of the book numbered with Arabic numerals. To accomplish this numbering method with PageMaker, the front matter must be in a separate file from the remaining pages. Use the "book" feature to link the two [or more] files.)

In general, I create master pages for those with a left running header, a drop folio (page number at the bottom of the page, usually used on chapter opening pages), and a master page for each chapter. Additional master pages for index pages or other pages where special formatting is needed may also be created. Once the first layout of the book is nearing completion, go through (starting at the beginning) and apply the master pages as necessary throughout the book.

Running heads should be positioned on the second line above the main text block—a single blank line should separate the running head from the main text block. Drop folios are usually positioned about ⅔ of the way toward the bottom of the page from the defined text block. (Folio position doesn't vary even if the page happens to run short or long).

Some designers place a line the width of the text block at the top or bottom of the page (or both). While this sometimes adds a note of completion to a book, I suggest avoiding the practice. Often, due to unbalanced lines or other typographical effects, such lines may cause the *optical appearance* that the book is cut incorrectly. In most cases, this is merely an optical illusion. But it can cause considerable consternation in the author/publisher when he/she first examines the books from the printer.

Finalizing the layout

Once all the text is styled and any photos or other graphic elements are placed into the document, then you need to go through and make adjustments to perfect the typography and layout. (I usually wait until after proofing the initial "first pass" layout, as proofing changes may affect page endings and other aspects of the book.)

Starting from the beginning of the book (always "work" the formatting issues from front to back, as changes on earlier pages may affect pages further along in the book), look for "bad breaks" and stacked words. Bad line breaks are where the hyphenation programming has failed, often due to words not being in the program dictionary. Look for overly loose or tight lines (where word/letterspacing appears to be outside the acceptable range. InDesign can show you those lines that are outside the parameters set in the spacing dialog. Go to **Preferences>Composition** and select **H&J Violations** and click **OK**. After a few moments, InDesign will highlight lines that fall outside the preference parameters with one of 3 shades of yellow. The darker the yellow, the more serious the violation.

Despite the program's assistance, a visual examination may also find instances that can be improved that are otherwise within the minimum parameters set in the preferences. Even if the program hasn't identified these situations, you should still fix overly tight or loose lines by manually inserting hyphens or rewording the text to make a better line break.

Another situation to watch for are "stacked" words. (The hyphen limit set in the style preferences should have avoided any stacked hyphens.) Stacked words (or hyphens) are cases where the same words are repeated at the beginning or end of two or more lines. This is usually very distracting to a reader, and may cause them to lose their place. If the wording can't be changed to eliminate the problem passage, it may be possible to insert forced breaks to split up the alignment of the stack.

Other more common situations are to eliminate "widows" and "orphans." While typographers have many definitions for these situations, basically you want a page to begin with a full line or, if that isn't possible, at least a line that extends at least ¾ of the line width. Although less serious, it's preferred that the first line of a paragraph should not be the last line on a page. In addition, the last line of a paragraph should be at least a single word of more than four characters. (A paragraph with a final short line consisting of two or three characters from a hyphenated word are best avoided. InDesign will usually allow you to select the "force justify" setting to eliminate such occurrences. Also, hyphenated words at the end of a right-hand (recto) page are not acceptable.

The primary method to fix "bad" page breaks is to run the page one line long or one line short. Where this gets tricky, is that both pages of a spread (facing pages) should end at the same point. To fix some bad break situations, it's often necessary to go back 2 or 3 pages and run one or more of those pages long or short to eventually fix the bad break. (This becomes an interesting puzzle at times.)

Another technique, especially helpful in InDesign is to use the "force justify" paragraph alignment setting on a paragraph that has only two or three words in its last line (these can be anywhere on the page). Often, if the paragraph is long enough, the words will "carry up" to the line above, and the paragraph will rejustify into an acceptable arrangement. [In PageMaker, use the "tracking" settings to readjust a paragraph to achieve a similar result. It usually takes several experiments, involving differing groups of lines and words, to find an acceptable arrangement.]

The force justify paragraph setting can be quite powerful with InDesign. The program attempts to have most paragraphs end in the middle half of the line, that is, more than ¼ and less than ¾ of the line length. Often, if a paragraph ends with less than

¼ of the last line, you can select the force justify setting and the paragraph will reformat, usually drawing up the last line so that it ends on the right margin. Care should be exercised, however, as InDesign will attempt to implement your selection with the force justify setting even though the word/letterspace is inappropriately tight. At some point (where InDesign can't squeeze any more letters into the existing area the force justify setting will cause the paragraph to drop down a line and excessively open up the letter- and wordspacing to bring the last line to the right margin. So, if you use the force justify setting, be sure to carefully inspect the paragraph to ensure that it still has reasonable word and letterspacing applied.

To fix a hyphen at the end of a recto page, check the next page, and based on the length of the word and the amount of the paragraph before and after the break, input a forced line break before or after the hyphenated word. InDesign usually rearranges the letter- and wordspacing to make an acceptable accommodation to the solution.

Once you have completed making all fixes, return to the beginning and go page by page through the book looking for any missed typographical problems, ensuring that the running headers are correct (from your master pages), and that completely blank pages do not have a running header or drop folio.

Summary

It's hard to over-check your document. It's easy to miss many of the small typographical elements that enhance the quality of the finished book. So, I always check these elements as I page through the (almost) finished book (always from front to back).

Check for bad page and line breaks. Are hyphenated words correctly broken? Are any lines of type set too tight or loose? (Often a problem where the software doesn't recognize a word that should be hyphenated—insert a soft hyphen to fix.) Are there stacked hyphens or stacked words? (Adjust word/letterspacing or re-write the passage to fix.)

Pages, especially right pages, should not end with a hyphenated word (use forced line breaks to fix). Pages should not start with a short line (unless it's a complete paragraph—as might occur with dialog). If possible, avoid having a page end with the first line of a paragraph that carries on to the next page. This is a less serious error than a short line at the beginning of a page.

Running headers should appear on all pages except for chapter opening pages or blank pages. Chapter openings should have a drop folio or other graphic treatment to lead the reader's eye to the beginning of the text. Blank pages should be completely blank.

If all chapters start on right pages, check to make sure that no chapters start on a left page.

Finally, run a spell check and (with InDesign) use the Find Font feature to insure that there are no unintended fonts used in the book. Print the book (with crop marks) and carefully proofread it. Use a professional proofreader or a very reliable friend/relative to proofread. (A typo can be very embarrassing!)

- 10 -

Selecting a Printer

THE NUMBER ONE MISTAKE that a new independent publisher can make is to go to the local "Mom & Pop Quickie Print" shop or to a large franchise printer to get price quotations on printing a book.

Printers are all taught: *Never* turn down a job (if at all possible). The printer will smile, tell you, "Sure, we do this kind of job all the time." And then take down all the specifications and information so they can find a printer who can really do the job. Of course, Mom & Pop will add a 30-50% markup to the price quoted by the actual printer. You will think that book printing is terribly expensive! Of course, you may want to establish a relationship with a local printer for flyers, mailers, and stationery. Your local printer can efficiently help you with many projects that you will need for your business—just be aware of their capabilities, staff skills, and equipment. You can also use this good relationship when you need some general printing advice or to check out ink colors on a PMS (Pantone Matching System) color chart.

The reality: Out of some 50,000 printers in the U.S., there are approximately 100 printers that either specialize in printing books or have book printing as a significant portion of their revenues. Since these printers are specifically equipped to print books (and usually, little else) they can print your book very efficiently and deliver them to you at a price far lower than anything that a local print shop can ever do—even with the transportation charges across the country.

How much? On any given day, almost any printer can beat the price of any other printer. Every print job is a custom production run. Printing prices vary by the nature of the project and the amount of work in the shop. If the shop is fully scheduled, the print price is likely to be somewhat higher than if their schedule is empty. Printers have high overhead and are also in danger of losing skilled workers if they don't keep them adequately busy (i.e., paid). Therefore, you should submit a Request for Quotation (RFQ) to a reasonable selection of printers specializing in books. Drop only those printers from your list that you know can't perform efficient or satisfactory work on your book.

(For example, don't submit RFQs for a book with a 10 x 14 trim size to printers who specify nothing over 9 x 12.)

Economics of printing

It's helpful if you understand the basic economics of printing. The two most common methods of book printing (digital and offset) have widely different economics. Digital printing is characterized by minimal expense to prepare the job for printing, but relatively high unit costs to run the job. Offset printing has rather high cost to make the job ready to print (outputting printing plates, mounting them on the press, printing "make ready" copies until the press is running correctly).

Below is a chart showing the relative unit costs of printing a project digitally and with offset. (Unit cost is on the vertical axis and the quantity printed is shown on the horizontal axis.) As you can see, there is some "economy of scale" with digital printing, but (for this project) it never quite reaches a $2.50 per copy unit cost no matter what the total number printed. However, with offset the unit cost becomes less and less as the total quantity increases. Our experience, with typical quantities printed by small publishers, is that the most dramatic savings in unit costs are achieved at about the 2000 quantity.

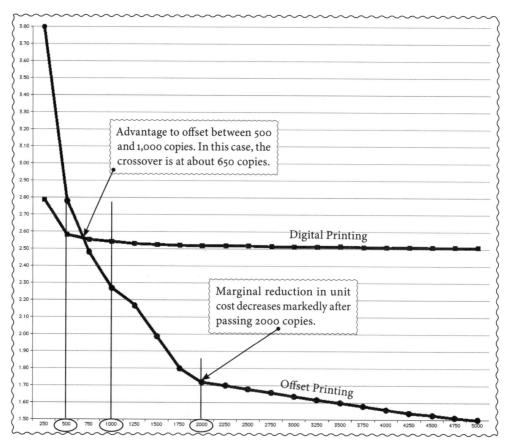

Advantage to offset between 500 and 1,000 copies. In this case, the crossover is at about 650 copies.

Digital Printing

Marginal reduction in unit cost decreases markedly after passing 2000 copies.

Offset Printing

Covers

While it's generally more convenient to print your cover with the same shop printing the interior pages of your book, there are some circumstances where you may wish to print your covers at a different shop. If your cover has special features, ask the book printer to bid on the cover as a separate item so you can compare the cost with non-book printers who may specialize in 4-color work. If you have an elaborate cover (perhaps with foil stamping, embossing, and/or other features), the book printer may not be able to efficiently produce the cover for you. Always schedule the covers to be completed before the book and allow a generous time for shipping, if necessary. The book printer (usually) will perform the bindery operations and will need to have the covers on hand before the books can be finished. Do not have the covers trimmed to size before bindery! Let the printer (or other sub-contractor) performing the bindery trim the covers to the appropriate size for the machinery being used. Print a generous extra supply of covers. Some covers will be lost during the bindery process. The leftovers can be used in your media kits and/or can be printed on the back by your local print shop (and trimmed) to be used as (oversized) post cards or hand outs for book shows, etc.

If your book is bound in hard cover, you will want to print about 20% extra dust jackets. Often returned books will look somewhat shopworn. A new dust jacket will often be all that's necessary to return a hard cover book to saleable condition.

What if there are problems?

No printer ever sets out to do a bad job. If problems do arise, first let your Customer Service Representative (CSR) or sales representative try to solve the problem for you. If your primary contact is unable to give you satisfaction, then be sure you talk to an owner or manager who has decision-making authority. Usually a negotiation will result in a satisfactory outcome. Before you complain, think about the situation and what it will take to fairly settle the issue. (Prepare yourself with facts and information. Avoid allowing too much emotion into the negotiation over an issue.) Printers are generally reluctant to give a refund, especially beyond the pro-rata unit cost of damaged/unacceptable books. Often a printer would rather re-run all or part of a job rather than give a refund—and this is usually the preferred outcome in any event.

Sometimes a publisher will think of ways to use damaged or substandard books. Keep in mind, there is a saying in the printing business: "if a job is good enough to keep; it's good enough to pay for." So, be prepared to return any and all unacceptable books. Only if the printer suggests that you might find a use for "bad" copies should you agree that is a possibility. Remember, you're trying to maximize the quantity of books of acceptable quality—you don't need to solve a problem for the printer. (Usually, unacceptable books will be recycled with other print shop waste paper. In some cases, unacceptable copies might be used in equipment setup or staff training before recycling.)

Danger signs

Be very careful when dealing with new CSRs or sales representatives (ask how long they've been with the printer)—they may make promises they simply can't keep.

(After the job is accepted, call and talk with the production manager to verify the schedule and ensure that everything is in order.) Be sure to get a written confirmation of the price and specifications for your job. If you change *any* specification, no matter how small, be sure to get a new written confirmation reflecting the change. Sometimes a seemingly small change may cause the price to change by a substantial amount—you may need to re-bid the project in these circumstances as other printers may be able to handle a book with that change more efficiently. Start by checking with the two or three "next best" choices from those who responded to your original RFQ.

Any printer that has recently changed owners or senior managers may have problems. (It takes time for an owner or manager to understand a particular operation and it takes time to build the staff teamwork relationship that results in good quality.) Get recommendations from other independent publishers; but be aware that you will get both good and bad references on any given printer. Also, be aware that dissatisfied customers tend to be more likely to respond to a request for comments placed through a neutral forum such as the online Pub-Forum or at a local publishers group meeting. (See the links page at the Æeonix web site for information about Pub-Forum.) No printer is perfect and even the best occasionally have a "job from hell" where everything that can go wrong, does.

Print brokers

The general rule is "don't work with a print broker." A good print broker may get you a better deal than you would get by yourself. A bad print broker may cause you to wish you had never lived. Tread with caution. Some print brokers, rather than checking with their "usual" printers, take in the job, promise you a good price, then try to place the job with a printer who will do the job at the lowest price. Brokers may purchase odd lots of bargain paper that may result in a poor print job. A broker may emphasize the wrong factors—for example, the broker may demand the printer turn the job around quickly, when you'd rather they take a little more time to ensure top quality. Brokers may place your job with a distant printer then find that the cost of transportation is (much) higher than they assumed or you may be billed for transportation without warning. Brokers will blame all problems on the printer, then disappear when the printer blames the problem on the broker—you may end up dissatisfied and you may have an expensive problem with little chance of getting satisfaction. Generally, you should avoid brokers, except, as discussed below, with overseas printing orders (usually 4-color jobs). If you shop carefully for a printer and develop good relationships with your printer(s), you can usually do much better on your own than by using a broker.

The exception to the "don't work with a print broker rule" is with overseas printing. Often, overseas printers work with selected brokers to represent their interests in the U.S. market. These brokers may work with several printers or may provide exclusive representation for a single printer. When this relationship exists, the brokers are more like "staff" sales representatives and generally handle all the details with you before the job is submitted to the plant. Other overseas printers may have an exclusive arrangement with a "manufacturer's representative"—an independent businessperson who provides services for a particular printer. And, lastly, a few of the larger overseas printers have U.S. based sales office(s) where the staff are all direct employees of the printer. In actual practice, I have seen little difference in the customer service provided by overseas printers with any of these variations in local representation. Partly, this is due to the complexity involved with a full color project (those most likely to be printed overseas), and to the nature of an out-of-country entity doing business in the U.S. market.

A word about "Printing On Demand" (POD)

The often confusing term "POD" has been applied to several printing/publishing processes. First there are publishing firms (usually) with a strong Internet presence. These companies may offer full service POD and eBook publishing services. Often they will offer to publish a book for you and will provide production services in addition to printing. These *POD Publishers* frequently offer some sort of (minimal) marketing and distribution services (most often through a web site operated by the publisher and through Amazon.com). While this has a certain appeal, and may well be appropriate for some projects (perhaps fiction or poetry), you need to exercise caution to ensure that you are paying for and receiving the services you really need. Read all agreements carefully as they sometimes have nasty aspects that may cause needless headaches later on—such as a claim on future royalties if a book is later sold to a large, traditional publisher. (Check the National Writer's Union web site *(www.nwu.org)* for commentary about publisher contracts. Use the link "Contract Advice" to learn more about publishing contracts.)

The typical online POD agreement will have up-front charges of $300 or more (sometimes way more) and will offer a variety of services, including page layout and cover design. I will note that the book interior design and typography is often of mediocre quality. Covers are often a formula design (you get to choose "style 'A-6'"). If you follow this path, be sure that you can continue to use a cover design (if it's effective) if you decide to graduate from an online POD publisher. (See also Ivan Hoffman's excellent web site *[www.ivanhoffman.com]* that discusses these and many other legal issues for small publishers.)

These notes aren't intended to be a complete critique of the online POD publishers. The intent is to suggest that you carefully check publishing agreements, do your homework, and check any agreement with the National Writers Union or the intellectual property lawyer of your choice to ensure that you aren't signing up for something that's unreasonable or doesn't actually allow you to accomplish your publishing goal.

The second meaning of "POD" is a true *print-a-book-for-each-order-received* paradigm. The largest purveyor of such service is Lightning Source, a sister company to Ingram Book Wholesalers. With Lightning Source (and other similar services) you are the publisher, while Lightning Source is a printer with an associated distribution function. Books prepared by Lightning Source will be automatically listed in the Ingram book distribution database, so any retail bookseller can easily find and order your book (which will be printed to fill the order since no books are warehoused except as an electronic file). The downside of this is that "POD" books produced by Lightning Source are (usually) not returnable (by the booksellers) and therefore are only ordered when the bookstore has a confirmed order from a customer. This does nothing to get your book on display in bookstores. You can possibly use Lightning Source as a part of your marketing plan, if for no other purpose than to simply be assured that your book is in the Ingram database (but there may be better alternatives). Lightning Source charges various set-up fees based on the preparation work required to get your book into the system. Typically, these charges are around $200 to $300.

The third meaning of "POD" is as a description of a digital printing service (usually using a toner-based duplication process) in contrast to a printer using a traditional offset press. Digital printing has minimal set-up cost as compared to offset printing, however the unit cost of production does not decrease significantly after the first copy is printed. With offset printing, much of the cost is in the set-up and plate making *prepress* work. Once the press is running, the incremental cost of each additional copy is actually quite low. While a particular job might cost $3.50 per copy from a digital printer, it might cost only $0.85 per copy on a press. But the digital printer may only have $50 (or so) in set up costs to absorb in the job while the traditional printer is likely to have over $1,000 in set-up costs on the same job. These economic differences generally suggest that it is more cost effective to print 500 or fewer copies with digital (POD) printing and that 1,000 or more copies should be produced with offset printing equipment. Between 500 and 1,000 copies, you'll need to do some comparison shopping as factors such as book length, trim size, and binding, may encourage a choice for either method of production.

Dan Poynter, author of *The Self Publishing Manual,* has coined the term "PQN" (Print Quantity Needed) as an alternate acronym to apply to this third meaning of POD. PQN follows the model of printing only a very small number of books, as needed, to fill immediate demand rather than printing a "full" run of books in the traditional manner. Depending on the aspects of the particular project, this may prove more cost effective when you calculate all the costs of the investment in producing, shipping, and holding inventory. At the very least, a "PQN" run at an early stage will allow you to establish what, if any, market interest there might be for your book without committing to the large investment that printing several thousand copies would involve. Most POD (or PQN) printers either do not charge a set-up fee or have such fees of less than $100. If you encounter a POD/PQN printer with a set-up charge, be sure to compare the total cost of the order to a printer without such a charge. Also, determine if the charge is made for *exact reprints* of a book, since it is quite likely that you'll make frequent reprints if you're using the POD/PQN method.

List of book printers

To see a good selection of printers who specialize in books, please visit the Æonix Publishing Group web site at *www.aeonix.com* and select the link: "list of book printers."

The list consists of printers specializing in books that I've identified. Some of them may offer digital printing services (black ink) in addition to offset printing services. Those printers that specifically offer digital printing have been flagged with "Digital" printed in red after the company name; however, always be sure to inquire, as other printers may also offer digital printing, but available information did not specifically mention it. Generally, however, printers handling offset printing of books do not have *inhouse* capability for digital printing. The prepress workflow of digital and traditional printing are sufficiently different that it becomes difficult to efficiently offer both production options at a single printer. Many offset printers use third party digital printers for their digital production, although some of the larger printers may have sufficient digital printing volume to justify establishing a separate department (and workflow) to handle the work. (In general, you'll find that the prices quoted will reflect the extra cost associated when the digital production is outsourced.)

No recommendation is intended by the presence or absence of any particular printer from my list; nor is any recommendation meant by any comment associated with any particular printer. Some of the listed printers are located in Canada, others are located in Asia. Be aware that there may be currency, transportation, and customs issues involved in using a printer outside the United States—be sure to explore and understand all such issues before you place an order.

Four color projects may be printed in Spain and Italy as reasonable alternatives for an East Coast publisher (due, in part, to transportation costs or transit times from Asia). The general quality of European printers is excellent. Relative changes in the value of Euros to Dollars may have a significant impact on the economic practicality.

You may get suggestions to try a printer in countries outside the United States, Canada, China, Hong Kong, Singapore, Taiwan, or Korea (who have established reputations in working with U.S. market requirements). Printers in Third-World/lesser developed countries (who don't regularly prepare work for the U.S. market) may present special problems. Many of these printers may not be sufficiently aware of our market expectations to produce acceptable product for the U.S. There may be serious language translation problems. I've heard enough sad stories that I suggest that printers outside the established areas should be avoided by the beginner. (The success stories generally involved people who were very close to the printers. For example, a publisher of a cookbook featuring recipes from Central America was born in Nicaragua and normally spent several months of each year in the city where her book was printed. Another publisher, whose book was printed in Puerto Rico, was at the print shop daily to supervise the production. (The shop was also a "quick print" operation that would not normally be able to be competitive in book production. The shop owner was a family friend which may have also impacted the price and quality.)

There are lesser developed areas that are trying to break into printing for the U.S. market (in particular, India, Mexico, and Thailand). In time, printers in these locations may become highly competitive. My suggestion is that only the most savvy publishers should deal with printers outside the more established markets. Aside from a house or a new car, your book may represent the largest single investment you may make in your lifetime. It just makes good sense to not take excessive risks to gamble on saving, at most, a few hundred dollars.

Preparing a Request For Quotation (RFQ)

Below is a discussion of a typical Request For Quotation with common specifications and options for common trade books. Some projects may require more elaborate specifications. Full color books (mostly children's picture books) aren't fully addressed in this discussion.

Start out with the formalities of who you are:

[Company Name]

[Address]

[Voice and Fax numbers]

[Email address where responses may be sent]

[Date of RFQ]

Request For Quotation

Please quote your best price and delivery time for printing and binding the following book:

Title: [title of book]

Author: [author(s) name(s)]

Publisher: [name of publisher]

Specifications

QUANTITY. Please quote: 1,000; 2,000; and 3,000 (or insert the quantities you wish). Generally print estimating software can handle 3 quantities with one set of entries. Most beginning self-publishers should print no more than three thousand books. (You can always go back to print more, if they are eventually needed.) Indeed, 1,000 copies is most likely reasonable for most titles. Due to the economics of printing, the 2,000 quantity is often quite attractive as that quantity generally spreads the set-up costs over a large enough quantity that the unit cost is quite reasonable. (The unit cost will drop most dramatically between 1,000 and 2,000 copies. The unit cost reduction will not be so dramatic with the increase in quantity from 2,000 to 3,000 copies.) The alternate choice is digital printing. Digital is generally only cost effective with very short runs; perhaps from 100 to 500 copies. Set up costs are quite low with digital printing. However, the unit cost does not decrease significantly as volume increases. For any given project, the crossover favoring one printing method over another is between 500 and 1,000 copies. The chart (page 128) shows typical charges for an offset and a digital print job. You can easily see that the digital price doesn't vary much over the total volume, while the offset job drops rapidly as quantity rises, then levels out after passing 2,000 copies. The best economic quantity for digital printing is between 200 and 500 copies. Unless you're using digital printing only for a specific need (such as bound galleys) where the quantity is easily estimated, it is most reasonable to request a quotation for 200, 350 and 500 copies. I note that some digital printers price on a gradually declining basis, while others are constant until passing a threshold, such as 300 copies. In the latter case, 300+ copies may have a significantly better unit cost than quantities below 300 copies. When discussing a project with a digital printer, be sure to discuss their pricing policy so you can take advantage of such threshold pricing.

NUMBER OF PAGES. State the number of pages, including all blank pages, but do not count end papers if hard bound. Generally, the page count should be evenly divisible by 8 (although 16 or 32 are better) as most printers print books in "signatures" of 32 pages. Do not try to count the sheets or prefigure anything to "help" the printer. Simply give the total number of pages in the book. Be sure to include the page count of front matter if it was numbered with Roman numerals separately from the body of the text.

If your page count doesn't divide evenly by 16 or 32 you should ask the printer if a page count that is evenly divisible by 16 or 32 might prove more economical. For example, your book has 244 pages—this does not divide evenly by 8. You would need to add 4 pages, making a total of 248 pages, to divide evenly by 8. (248/8=31) However, 248 does not divide evenly by either 16 or 32. To divide evenly by 16, you would need another 8 pages, making a total of 256 (256/16=16). This allows the printer to print 8 signatures of 32 pages. (256/32=8) If you find you can reduce your 244 page book to 240 pages (perhaps by running a few pages one line longer to eliminate mostly blank pages at the ends of chapters), then the printer can print your book as 7½ signatures of 32 pages. (The half signature is usually prepared as two sets of 16 pages to make up a full plate, so the press only prints half the number of copies as for the full signatures.)

The added pages to fill out a signature are usually placed at the back of the book. These can often be utilized to promote other works or other services offered by the publisher. If there are more than four or five added pages, it might be better to add two or four pages (one or two leaves) to the front of the book to make room for a display of review quotes ("blurbs") in a subsequent printing. By making this allowance in the first printing, it eliminates dealing with the issue of either sacrificing the half-title for the blurbs or paying for the printer to re-paginate the signatures. (Page numbering should ignore the extra blank pages and start numbering with the half-title as Roman "i" or Arabic "1.")

Digital printers usually can print any page count, so long as the page count is an even number, although there may be a few situations where the page count must divide evenly by 4.

TRIM SIZE. Trade books are most commonly 8½ x 5½ or 6 x 9 inches. (The choice of trim size is driven by the market—look at books in the same genre and select a similar size for your book. See chapter 3 for a full discussion of common trim sizes.) The trim size may have significant impact on the printing cost. Some presses, particularly digital presses used for POD/PQN printing, are more efficient with 8⅜ x 5⅜ inch trim. (If you leave sufficient space in the margins, you can adjust an 8½ x 5½ book to this slightly smaller size, if necessary.) Indicate which edge, long or short, is to be bound—most books are bound on the long edge. A book with "landscape" orientation would be bound on the short edge. Your entry here might read, "6 x 9, bind on long edge."

INTERIOR COPY. This might read: "Provided as electronic files in InDesign CS (ver. 3.0) and Adobe Acrobat PDF. Files will be on Macintosh (or Windows) CD-ROM."

You need to tell the printer what software was used to produce the book and what media will be used to send it. Most printers also have "file submission" sheets to fill out with detailed information once you've made your printer selection. The use of Acrobat PDF files has become almost universal—and has some cost-saving advantages. I always provide the underlying program files "just in case" there's a problem; however I haven't encountered any in the past few years.

ILLUSTRATIONS. This might read, "contains 12 photographs, included as part of the electronic file." Describe any photographs or illustrations and how they will be provided to the printer. Generally, it is best to include all images as part of the electronic file. If you wish to have the printer prepare photos or illustrations, there will be additional costs. (Note: always "link" images to your layout program file, do not "embed" the images in the file. It is preferred to "embed" all text files, however.) See chapter 13 for a discussion about scanning images.

BLEEDS. Bleeds, where photos, illustrations, or other artwork run off the edge of the page, can add significantly to the cost of printing. Often, a larger sheet size is required for the press, depending on the trim size of the book. Normally, the entry here is "none." If you do have a bleed, the art should extend ⅛ inch beyond the edge of the page. Important elements should be no closer than ¹⁄₁₆ inch from the edge of the sheet. (Cutting devices generally have an accuracy of about ¹⁄₁₆ inch.) A smaller trim size may offset added cost for use of bleeds (e.g. use 8⅜ by 5⅜ or 8¼ by 5¼ inches instead of 8½ by 5½ inches). Many printers specify a safety zone (usually ¼ inch from the edge) where no type should intrude.

PAPER. This might read, "your house stock in natural, 55 or 60 lb. book, vellum finish. Please provide spine measurement with selected paper stock. Paper shall be acid-free and neutral pH if available." Paper is approximately 30 to 40 percent of the cost of manufacturing a book. *House stock* is the normal paper that the printer uses. For small publishers, printing in typical quantities, it is cost prohibitive to select a specific paper other than one normally stocked by the printer.

Common choices of weight are 50, 55, 60, and 70 lb. weights. 50 lb. paper is the same weight as 20 lb. bond commonly used as copy paper. It is usually too thin for highest quality two-sided printing, as it often allows type from the back side to show through resulting in a lower quality, and thinner book. 55 lb. paper is a special grade made especially for books and is usually only available in natural, off white, or ivory colors (not bright white). Most 55 lb. papers are quite thick for their weight and make a good quality book with a nice heft. 60 lb. paper (equivalent to 24 lb. bond) is the workhorse of printing. Many 60 lb. papers are physically thinner than the special 55 lb. book paper. 70 lb. (and heavier) papers might be used where extra bulk is required or for other special purposes. The 55 lb. high-bulk paper is generally not available to printers on the West Coast.

Printing papers are offered as "regular" or "opaque." Opaque grades are more expensive than regular, but offer greater resistance to show-through from printing on the reverse side of the sheet. Unless you've used a particularly heavy, dark type, opaque grade is unnecessary in the 55 or 60 lb. weight. Photos that may be adversely impacted by type on the reverse might reproduce better with an opaque grade. If in doubt, ask your printer for guidance. If you use a 50 lb. paper, you might be inclined to use the opaque grade. However, for most books it's probably more desirable to have a thicker book by using the 55 lb. high bulk paper or a 60 lb. regular paper. If you have a particularly thick book, where extra bulk is undesirable, then a 50 lb. opaque may be exactly what you need. Otherwise, 60 lb. regular paper costs about the same as 50 lb. opaque.

Color choices are either an off-white, cream, ivory, or natural (unbleached) shade or "bright" white. Most books are printed with one of the off-white colors as they have less glare and are easier on the eyes. If many photographs or other graphics are used, a bright white might be preferred. How-to and reference-type non-fiction are often printed on bright white, while novels, memoirs, and other works without diagrams or photos are frequently produced on one of the off-white shades. White papers are graded with brightness numbers. Typical book printer stock is usually graded in the eighties.* Especially bright white paper might have a rating over ninety. (Papers with the lower brightness rating usually look gray or slightly yellow when compared with a higher rated paper.) Some specialty papers (for laser printers or photo quality ink jet), are rated above one hundred. Since the rating is based on the amount of light reflected from a standard source, a rating above one hundred seems impossible, but some papers add a fluorescent additive, so they actually "reflect" more light than received from the source. It is unlikely that papers with an extra-high brightness rating are either economic or desirable for printing books.

Uncoated book printing papers come in vellum (slightly rough) or smooth finishes. Vellum is usually thicker and the slight roughness further reduces glare. Again, if photographs or other illustrations are used, it may be better to use a smooth finished paper. Coated papers are usually reserved for books with many photographs or color printing. The basic choice is between gloss coated and matte coated finishes. Gloss coated paper will give the maximum quality to photos—at the expense of potential glare when reading the text. An art book that has little text is an example of the kind of book that could take good advantage of gloss coated stock. Matte coated paper is usually a better choice if there is significant text. The primary advantage of coated paper is that it reduces dot gain—the ink doesn't soak in and spread through the paper—giving a more accurate reproduction of the image. (This is less of a problem with digital printing, as the toner doesn't soak in to the paper—but the methods of applying toner are less accurate than offset printing. Digital printing also suffers from significantly lower overall resolution.)

Acid free and neutral pH is specified so that the book will last for a significant time. Most papers used for book printing are now made using an acid free process. Wood fiber based papers with high acid content tend to self-destruct after 30–50 years. If you wish, you may specify "recycled" paper, however recycled stock usually costs more than regular paper. Regular papers already contain a significant portion of wood fiber recovered from other forestry production by-products (sawdust and wood chips). Adding post consumer recycled material adds cost due to the expense of collecting and processing the material. Recycled papers may be weaker than corresponding regular products.

---

*Recently, the standard white printing papers were upgraded to a brightness of 92. One manufacturer began producing only this grade and all other paper manufactures matched it to remain competitive. Brightness ratings are not used with colored (natural, off-white, ivory) papers.

There are, however, some book titles competing in markets where recycled papers are required by consumer expectation. Most printers have a reasonable house selection of suitable recycled paper stocks.

INK. "Black throughout" is one example of an ink specification. You may specify "soy-based ink, if available." Due to environmental regulations, printers are using products with much lower impact on air quality than those used a few years ago. There is no particular benefit to the environment for soy-based ink vs. other inks now generally used. Indeed, petroleum-based ink was never a significant contributor of volatile organic chemicals (VOCs) generated in printing plants. The primary source of VOCs in printing came from alcohol, used in the fountain solution, and from solvents used to clean unwanted ink from the press. Printers have (generally) migrated to other low VOC products for the purposes formerly filled by the alcohol and organic solvents. (On a personal note, I was shocked to discover (during a trip to Peru) that some of the Amazon rain forest is being destroyed to make fields for growing soy bean crops. To me, this is hardly a positive environmental trade-off. See the January 2007 issue of *National Geographic* for an article on the topic.)

If your work includes full color material, the specification would be for "process color" or if you have spot color elements, then you would specify "black plus PMS XXX" (where the "XXX" would be the specific color number from the Pantone Matching System color chart). A color project with two spot colors might read (as an example), "black plus PMS 187 and PMS 485."

PROOFS. This might read, "complete bluelines or electronic alternative." Electronic proofs are the standard as printers have mostly moved to all electronic work flows. The main frustration with electronic proofs is that you can't determine the actual quality of half-tone photos as the output is printed on an ink jet (cover) or laser printer (interior). It is also pointless to look for broken type or other "hard" flaws. However, the proofs should be checked for page make up errors (Are all pages present? Are pages in the correct order?) or are there errors with the typeface(s) used.

These filmless production techniques, where the electronic file is output directly to printing plates, save the expense of generating film negatives and eliminate the environmental concerns of the film processing chemicals. However, the cost of an "exact reprint" is no longer as significantly lower in cost when compared to the original when negatives are used. Some of the cost savings of the direct to plate workflow are offset by more expensive materials used in that process. At this time the majority of book printers are using a direct to plate workflow—and the remaining printers are adopting the new workflow as they upgrade equipment during their normal course of business. (Printing plates are rarely stored as they have a limited lifespan and are quite fragile (especially to damage from scratches) and awkward to manage. They are usually made from recyclable aluminum.)

COVER. Typically, this might read, "print one side only. (Front, spine, and back of book.)" There are only rare cases where you might want to print on the inside of the cover. If printing is done on the inside, be sure to keep all ink at least ⅛ inch away from the spine area where glue is applied. The binding glue will not stick to the ink, weakening the binding.

If you're printing a hard cover book, you might specify foil stamping. Usually the spine is stamped with the title or the title may be on the front and the spine. In some cases, the author name is also stamped, but it's usually not cost effective for modest-length press runs. (The larger the stamping die, the more expensive it is.) The foil stamping serves the buyer if the dust jacket is lost. It is a convenience and isn't required at all.

You may have seen hard cover books where the cover is printed with the design. This process is usually called "lithowrap." This approach is usually limited to text books or children's books. With text books, there may not be a dust jacket at all. Dust jackets on children's picture books tend to quickly become damaged, so it is important to maintain the appearance of the overall presentation with the more durable lithowrap cover.

If a dust jacket is used, then you add an entry to describe it. The entry may read, "Print one side only, 4 color process, full bleed."

EXTRA COVERS. This might read, "please specify cost per hundred of extra covers at each quantity." It is often helpful to have extra covers for your press kit and for other promotional purposes. It's considerably cheaper to print a couple hundred extras with the main production run rather than running promotional copies at another time.

With a hard cover book, you will refer to "dust jackets" rather than the cover. It's usually prudent to print ten to twenty percent more dust jackets so returned books can be refreshed with an unused dust jacket and sold as new.

Once you've selected a printer, check to see what the cost of running bookmarks or business card sized cover images might cost. Often such artwork can be "piggybacked" on the cover run for a nominal charge, *if there is room on the press sheet.* (A particularly efficient printer may use a press sheet that minimizes waste and therefore may not have any excess space for such extra artwork.)

COVER COPY. See the comments after "interior copy" above. The cover designer should use the spine measurement provided by the selected printer to adjust the spine width on the artwork before it is submitted to the printer. This is a common adjustment and is usually done at no additional cost. For initial preparation of cover art, the spine width can be estimated at 400 pages per inch as a compromise between common paper thicknesses. The 400 ppi measure is reasonable, but 55 and 60 lb. papers are often thicker (with some 55 lb. stocks at 360 pages per inch), while 50 lb. papers might be somewhat thinner.

Remember that a hard cover book is ¼ inch taller and wider than the nominal trim size of the book. Be sure that the dust jacket artwork includes the extra size. Likewise, the spine width is significantly larger than it is for a soft cover book. For a 'working estimate' add about ¼ inch to the spine measurement as estimated for a soft-cover book.

BLEEDS. This might read, "all 4 sides; ⅛ inch allowance made in artwork. Artwork is not trapped. You should make necessary adjustment or use 'in RIP' trapping." It is best to leave trapping (see glossary for explanation) to the printer, as trapping values vary by press used. ⅛ inch bleed is a common allowance for most printers. In some circumstances, the bleed allowance should be more generous. Be sure to ask the selected printer for their specifications. It is generally up to the printer to adjust for trapping or to provide trapping information to your designer.

COVER PAPER. The most common specification is, "10 pt C1S stock. Laminate with a lay flat gloss plastic lamination." Choice of cover weight is usually 10 or 12 point Coated One Side (C1S) paper. The majority of perfect bound book covers are printed with the 10 point paper. You might choose 12 point paper if your book is particularly large (over 6 x 9 inches), will be heavily used (e.g. a reference book), or is particularly thick (spine more than 1½ inches). I have encountered one printer who only offers 12 point paper. 10 point paper is .010 inch thick and 12 point is .012 inch thick. The cost difference between 10 point and 12 point paper is fairly small. If you use plastic film lamination, discussed below, it also adds thickness and strength to the cover, reducing the need for using the thicker, stronger 12 point cover stock. Digital printers will offer an equivalent, but probably thinner, cover stock that will run through their equipment.

Choices of finish include: no finish, UV or aqueous coat, or film lamination. It is wise to specify a finish as unprotected ink can easily be damaged, possibly making books unsalable. Plastic film lamination is slightly more expensive than the liquid UV or aqueous coatings. The advantages of film lamination outweigh the additional cost. "Spot" varnish, embossing, foil stamping, or other special treatments can be used with laminated covers in most cases. Other times plastic lamination may interfere with desired special effects. (Of course, such special features will add to the cost of the printing.)

Plastic film laminate is available with matte (dull) or gloss (shiny) finish. Lay flat laminate should always be specified if film lamination is used. The least expensive plastic film, polyester, tends to exaggerate the tendency of covers to curl with changes in humidity. Though more expensive, polypropylene and nylon films are more flexible than polyester and tend to allow book covers to adjust to humidity changes with significantly less of the curling associated with the cheaper polyester film. Most book printers only use lay flat film, but specifying the lay flat property is a wise precaution in case a printer makes an error. Note: If you live in Florida (or another very humid environment), some cover curling is unavoidable. Film laminate is essential for digitally printed covers, as the toner tends to crack on the spine corners without laminate holding it in place.

Cover curling can also be caused by the basic manufacture of the book. Paper has a grain (due to the fibers becoming aligned while being drawn into the paper making machine). The grain should be parallel to the spine in a properly made book. Some combinations of page size and press configuration force the book to be printed with the grain perpendicular to the spine of the book. Books printed in this manner will tend to have greater problems with cover curl and may have a weaker spine. It's certainly reasonable to discuss this issue with the potential printer and to select one who will properly manufacture your book, even if it's somewhat more expensive. (The heartbreak of having a garage full of curled books is unimaginable!)

If the book is to have a hard cover, you would specify the weight and finish of the dust jacket stock. You might also specify a foil stamp on the cover and/or spine—or a "lithowrap" cover. (Lithowrap is a printed cover, usually the same as the dust jacket or instead of a dust jacket. If you wish to do a lithowrap, get instructions from the printer to create the art as the design must (usually) extend an additional ¾ inch in all dimensions to fully wrap around the cover boards. Hard covers are usually ¼ inch taller and wider than the nominal trim size of the book block.

Also, if the book is hard cover, you should specify the thickness of the "binder's boards" (the heavy cardboard stock used for the cover) and the fabric used to cover the book. Typically, 80 or 88 point binder's board is used. Depending on who does the binding (the printer or a subcontractor) and the size of the printer, you may have more choices. Discuss the binder's boards and the fabric choices and colors with the printer. Fabric is often described as "equivalent to Arrestox B grade" in your desired color. (Aresstox is a particular brand of book cover fabric.)

COVER INK. A typical specification is, "standard 4-color process." One- or two-color printing is an alternative, but usually doesn't result in a significant savings at typical run lengths. Printers will charge extra for the time it takes to clean the press for the specified inks, reducing the impact of any savings. The marketing benefits of full color normally far outweigh any modest savings by using fewer ink colors.

When preparing artwork, be sure that artwork for the 4-color process is converted to or designed in the CMYK color space. (Computers often default to the RGB color space.) If the printer receives artwork with RGB colors, the translation to CMYK may give very unexpected results (see chapter 13). If possible, print a trial separation on a PostScript printer before submitting the job to the printer to ensure that the colors will separate as expected.

If printing a cover with spot color(s), it is essential to print trial separations before the file reaches the printer. Most printers prefer that spot colors be "defined" as the specific color used. However, due to technical differences between various programs you may have unreliable results with specific spot color definitions. In these circumstances, the "spot" color can be defined to be the same as one of the non-black CMYK colors. This eliminates outputting "bad" separations where the software creates multiple plates for a single spot color. If you (or your designer) must do this, be sure to make it very clear to the printer that the "yellow" plate (for example) is to actually print with "PMS 359."

PROOFS. This might read, "color match print and complete bluelines or electronic alternative from same RIP used for plates." (See the discussion with the interior proofs, above.) Digital proofs are commonly being used for interior and cover proofs, however color match prints tend to more accurately show what the cover looks like. Some printers are using calibrated proofing printers with various degrees of success. If color is critical, then ask for a color match print as the electronic alternatives are less likely to be as accurate.

If color is especially critical, you may request a press proof. This is actually printed on the press. This requires complete setup of the job, and is likely to be quite expensive.

BINDING. Usually, this will read, "soft cover, perfect bind."

Other binding choices:

RepKover or Otabind for a "lay flat" binding suitable for work books or cook books where the user may wish to have the book remain open at a particular page without breaking the spine. (Do not confuse this "lay flat" binding method with the "lay flat" plastic film lamination.)

Various metal or plastic spirals, Wire-O, and plastic comb bindings (often referred to as mechanical bindings) are also available. These alternatives have their advantages for certain books, but these bindings usually eliminate being able to print the title on the spine. A book without a printed spine is at a significant disadvantage when sold in book stores where the spines are all that's displayed on the shelves. Plastic combs might be screen printed with spine information, but that usually isn't practical in ultra-short run (PQN) quantities. There are techniques with the Wire-O binding where a cover paper can wrap around the metal coil to create a cover with a printed spine. Nevertheless, this binding method is somewhat awkward and potentially expensive.

Another option is saddle stitching. This method places a staple in the spine of the book and is usually limited to a maximum of about 100 pages. Saddle stitching does not normally leave a spine for printing, although there is a square back saddle stitch method I've seen advertised (by an equipment vendor). I haven't seen the square back saddle stitch offered by book printers, yet.

Common hard cover bindings are adhesive bind, notch bind (similar to adhesive bind) and smyth sewn. The smyth sewn binding is the traditional hard cover binding method and results in the strongest and most durable book that will open and lay flat. It is more labor intensive and is more expensive than the adhesive-based binding methods. Adhesive notch binding may require extra margin at the inside of the pages depending on trim size of the book. The notch binding cuts triangular grooves across the spine of the book, increasing the contact area for the glue. The resulting bands of glue also act as a flexible hinge. Notch binding is more durable than a straight adhesive bind, but less durable than a smyth sewn binding. Be sure to ask the printer about binding allowances if you choose a notch bound book cover. Adhesive bindings (regular or notch) tend not to open or stay open as easily as a smyth-sewn binding, so consider the likely usage of the book when choosing a binding.

Hard cover books may also be made with a "library" binding. This is a modified smyth sewn binding where an additional fabric tape is used to strengthen the connection between the spine and the covers. It adds to the expense, but may be worthwhile if a book is expected to have significant sales to libraries (particularly for children's picture books). It is generally only worthwhile to do as a combination job with regular hard covers. It's not suggested if the book is otherwise only to be produced in soft cover. If books are available to library customers with a library binding, you can generally charge a little more for the book (I've seen a $1.00 up-charge for library binding vs. regular smyth sewn.) Mixed runs of hard and soft cover books inevitably have a surplus of one binding and a shortage of the other. Given a choice, most consumers will opt for the lower cost of soft covers. Libraries often choose hard covers. It's nearly impossible to accurately estimate the split to the market. Unless you have a good estimate, you might be better off simply running all copies as hard cover or soft.

When selecting an appropriate binding for your book, you need to look at similar books already in the market. If most (or all) similar books are hard cover, then your book should be hard cover. Likewise, if most (or all) similar books are soft cover, you should select an appropriate soft cover binding technique. The various mechanical bindings and the Rep Kover/Otabind are selected when it's necessary for the book to remain open when placed on a table or counter. As I'm sure you've experienced, the normal soft cover book snaps shut if not held open, unless you crease and break the spine. These "lay flat" bindings are most often used with cookbooks or students' workbooks.

Understand, too, that traditional large publishers first release a book in hard cover only. That forces buyers (who want the book) to purchase at a price with the highest margins for the publisher. Later, if the book was successful as a hard cover, it is then released as a "trade paperback." This is the high quality soft-cover book that is the primary focus of *this* book. Once sufficient trade paperback sales have been achieved, then the book is often released as a "mass market" paperback (the rights are usually sold to a specialist publisher for this step). While there is some cost savings shared between the hard cover and trade paper back (the same typesetting is used for both), each version of the book is printed separately and has the usual start up costs. The marketing and trade practices of mass market books (those are the ones commonly sold in drug stores and supermarkets) are inadvisable for small publishers to emulate due to the economics involved.

When a small publisher produces a book in hard cover, they are likely to print the trade paperback version at the same time to get the benefit in production cost of a longer press run. But, the small publisher has both soft and hard cover books on hand and for sale at the same time. Then, the two versions compete for buyers, who most often opt for the lower cost book.

The glue used in the binding can also strongly impact its strength and the durability of the book. However, smaller publishers usually are only able to accept what's offered. Bindings using "PUR" adhesive are significantly stronger, however the PUR adhesive requires specialized application equipment that has not found its way into most shops. (PUR cures by contact with air. Therefore, the adhesive must be applied with a pressurized, closed system that does not allow air contact until it is applied to the book block. Other adhesives are either hot-melt or cold applied. Hot-melt glues are most common and are similar to the hot-melt glue sticks sold in hardware stores. The cold adhesives are most closely related to Elmer's white glue.

PACKAGING. "Shrink-wrapped in groups and packed in a 275# burst test carton. Cartons shall be tightly packed and sealed and shall weigh no more than 35 lbs. each." The 275 lb. burst test cartons are somewhat heavier than those commonly used by many printers, however they protect the contents better. Expect to pay a small premium. The 35 pound carton weight is to facilitate handling and re-shipping of books to wholesalers. Heavier cartons, besides being difficult to lift, are subject to more shipping damage when shipped individually. Baker & Taylor, a large wholesaler, requires that cartons weigh less than 40 lbs. Optionally, you can specify shrink-wrapping in convenient groups, usually 5 or 6 books. Consider how the books might be marketed and sold—you may want books shrink wrapped as "singles." The negative of shrink wrapping your book as a single is that many booksellers don't unwrap the book, discouraging browsers from handling and buying the book. Shrink wrapping is intended to reduce scuffing damage in transit. If the book covers have plastic film lamination (as recommended) and the cartons are tightly packed, potential scuffing damage should be well controlled. Shrink wrapping can add about 20 cents per bundle, so is not an insignificant expense, especially if you shrink wrap as singles. It's possible to have most of an order shrinkwrapped in groups and have a specified number as singles. If you're working with a trusted printer, who you know will properly pack your books, you can probably not bother with shrink wrapping.

SHIPPING. Books are normally shipped on wooden (or plastic) pallets and secured with "stretch wrap" to hold the shipment together. You will need to be prepared to receive a palletized delivery. If you don't have a loading dock (or have arranged to use a fulfillment service with one), you may wish to ask the carrier to use a "bobtail with lift gate" for delivery. There may be either a slight delay and/or there may be a modest extra charge for this service. Otherwise, the truck driver is likely to "flip" the pallet of books off the back of the trailer and drop them about 4 feet to the ground. It's neither good for the books nor for your nerves. Often, a driver will agree to hand unload the cartons, in which case, it's helpful to have a group of friends and neighbors handy to help out. Although it will generate a slightly more expensive shipping charge, you can specify "inside, residence delivery" in your shipping instructions. Most printers will not quote this in an initial response to an RFQ, so it might be better to bring up delivery options when you have narrowed the field to a couple of printers.

"Please advise cost of shipping to your city, state, and ZIP Code." Or, if you are going to be using a fulfillment or other warehousing service, the city, state and ZIP Code of the delivery location. Be sure to mention any delivery issues, such as "residential delivery, no loading dock" or "second floor or inside delivery," if applicable. In bidding the project, the shipping cost can make a higher priced printer more attractive once the shipping is taken into account. While many printers pass on the large discounts available from the truck lines, remarkably, others do not. Also, you may check for discounts available to you through trucking company agreements with publisher organizations like Publishers Marketing Association and SPAN. (See the Resources section for information.)

Books should be stored to allow air circulation around the cartons. Remove the stretch wrap (if any). Leave (or place) the cartons on their wooden pallets. **Do not** allow the books or cartons to be in *direct contact with concrete*. (Concrete is not water proof and moisture will wick through the concrete into the books causing severe damage in a relatively short time.) Also, be sure that the cartons aren't tight against a wall as this prevents free air circulation and may allow condensation and mildew to form. The proverbial "cool, dry, place" is best for storing books. Try to avoid a garage or shed that gets excessively hot. Excess heat can cause or increase cover curl and can dry out the glue in the spine causing books to "crack" when handled—followed by large sections of the book falling out of the cover. Extreme changes in humidity should also be avoided as much as possible.

Terms. "Please specify your terms." Most printers will ask for ⅓ to ½ of the printing cost up front with the balance due when the books are ready to ship. If you've done business with the printer before, you can apply for a "30-day" account. If you are granted credit, then you would have thirty days from the time of invoice to pay. Printers are usually quite conservative in granting credit. (Think about it. If you buy a car and don't pay, it can be repossessed. But if you don't pay the printer, what can they repossess? Books are a custom product and have no value to the printer.)

You should always carefully review a sample book page by page before making the final payment. When the books are received, you (or your agent) should carefully check several books randomly from different cartons in the shipment. If books with unacceptable printing are found, then a greater number of books should be checked. In "worst" cases, all books in a shipment may need to be inspected with unacceptable books segregated from those that are usable. *Do not wait to inspect books* and if a significant number of unacceptable books are found, *immediately report the problem in writing* (a letter, *not* an email) to the printer. The printing contract has specific (and short) deadlines and requires *written* notice of unacceptable printing. Of course, you should also call your Customer Service Representative and/or salesperson to discuss the problem and begin negotiations to resolve the issue. But *a telephone call does not protect your rights* under the printing contract. When you call, it's appropriate to mention that you'll be sending a letter to "confirm" the report of the problem.

Additional "Boiler Plate". "Please give a detailed quotation, including cost of overruns, reprints, and delivery charges. Please provide your best estimated production schedule. Please provide details of any other miscellaneous charges. Any item in this RFQ takes precedence over any industry trade customs and conventions. The CD-ROM and electronic files remain the property of the customer and are to be returned on completion of the job. Any fonts provided with the electronic files will be promptly removed from your system(s) after the job is successfully ripped and plates made in keeping with the terms of software licenses covering their use.

"Please return quotation to [insert your name, address, and e-mail address.]

"Note: If you are emailing your response, please be sure your company name appears in the first line of your message. Thank you."

These last three paragraphs of boiler plate take care of a few details to avoid any misunderstanding with any conflicting provisions of the standard printing trade terms and conditions. I found, to my surprise, that some printers would send an e-mail copy of a quotation letter (to be printed on letterhead) without specifically indicating their identity (assuming that the company name would be in the letterhead).

Underruns and Overruns. This is the surprise element of most offset printing jobs. The standard printing contract allows the printer to deliver *and charge for* up to 10% over your requested quantity. The reasoning is that there are unknown potential processing losses during bindery, so the printer needs to print extras to ensure that there will be enough to meet the order quantity. This reasoning isn't all that convincing, as the number of set-up copies can easily be estimated and serious bindery machine errors that destroy a significant number of copies is extremely rare. However, while a few printers deliver exact quantities, most will simply deliver the extras and add them to your bill (it is a "sale" without any sales effort). Either you should plan for the extra quantity and related charges or you should discuss this with the printer to work out some accommodation.

Sample RFQ

To assist you in setting up your RFQ, following is a sample RFQ without the discussion. You may download this sample RFQ from the Aeonix web site. [Items in square brackets are reminders and should be deleted from the final submission.]:

[Enter your company name, address, phone, fax, and e-mail address.]

[Date of RFQ]

Request For Quotation

Please quote your best price and delivery time for printing and binding the following book:

Title: [enter title of book]

Author: [enter name(s) of author(s)]

Publisher: [enter publisher's name]

Specifications

Quantity: Please quote: [enter up to 3 quantities; 1,000; 2,000 and 3,000 suggested.]

Number of pages: [enter number of pages]

Trim Size: [enter height x width; cover type; binding type; bind on (long or short) edge.]

Interior Copy: [enter program(s) and media used]

Illustrations: [describe photographs, illustrations, or screening used in the book, if any. Otherwise enter "none."]

Bleeds: [Indicate bleeds in main body of book, if any. Otherwise enter "none."]

Paper: Your house stock in [enter weight, color, finish, and recycled content, if desired]. Please provide spine measurement with selected paper stock. Paper shall be acid-free and neutral pH if available.

Ink: Black throughout. [Mention "soy based" if you wish.]

Proofs: Complete bluelines or electronic alternative.

Cover: Print one side only. (Front, spine, and back of book.)

Extra Covers: Please specify cost per hundred of extra covers at each quantity.

Cover Copy: [enter program(s) and media used]

Bleeds: [adjust terms as necessary] all 4 sides; ⅛ inch allowance made in artwork. Artwork is not trapped. You should make necessary adjustment or use "in RIP" trapping.

Cover Paper: [10 pt or 12 pt] C1S stock. [enter finish, UV or aqueous coating, or plastic (matte or gloss) film lamination.]

Cover Ink: Standard 4-color process

Proofs: Color match print and complete bluelines or electronic alternative.

Binding: [enter cover and binding type]

Packaging: [adjust entry as needed] Shrink-wrapped in groups and packed in a 275# burst test carton. Cartons shall be tightly packed and sealed and shall weigh no more than 35 lbs. each.

Shipping: Please advise cost of shipping to [city, state, and ZIP Code of destination where books are to be shipped. Note, too, if residential delivery.]

Terms: Please specify your terms.

Please give a detailed quotation, including cost of overruns, reprints, and delivery charges. Please provide your best estimated production schedule. Please provide details of any other miscellaneous charges. Any item in this RFQ takes precedence over any industry trade customs and conventions. The CD-ROM and electronic files remain the property of the customer and are to be returned on completion of the job. Any fonts provided with the electronic files will be promptly removed from your system(s) after job is successfully ripped and plates made in keeping with the terms of software licenses covering their use.

Please return quotation to [enter where and by what means (letter, fax, email) you want the quotation to be sent.]

Note: If you are emailing your response, please be sure your company name appears in the first line of your message. Thank you. [Strangely, I've received a number of quotes where it was not obvious who was sending it.]

Printer's response

A sample quotation, as received, is on the next page. This quotation, from Central Plains Book Manufacturing is for a project I handled while finalizing this book. The name of the project and other personal data has been blurred. This quotation is reproduced as a sample only for educational purposes. While I have printed many projects over the years with this printer, this is not a recommendation to use (or not use) them. The specifications for your job are likely to be different and they may result in costs and prices quite different than the quote shown. For a list of printers who specialize in books, please visit the Æonix Publishing Group web site at *www.aeonix.com* and follow the link to "list of book printers."

Note that this quotation refers to terms and conditions "on the back" of the form. They are not reproduced here as we have a discussion of printing terms and conditions in the next section of this chapter.

When you receive a quotation, you should immediately check to see that it reflects the RFQ you submitted. While infrequent, errors are made in preparing a quote. It might prove annoying (or worse, you may not have the budget) if the specifications don't match. If you do see an error, don't assume that you can take advantage of a 'better price'— what will happen is that the printer will compare the job as submitted against the quote, and if the specifications don't match, then the printer will revise the quotation. (The terms and conditions as explained in the following discussion allow this.)

Date: 04/06/2005

Pete Masterson
Aeonix Publishing Group
~~address line~~
~~address line~~
~~address line~~
~~address line~~

Central Plains
BOOK MANUFACTURING

RE: ~~Birth Mix Failure~~ (CPBM #122636)

Dear Pete,

Thank you for the opportunity to serve your book manufacturing needs. Central Plains Book Manufacturing is pleased to present the following qua
components and pricing to Aeonix Publishing Group.

DESCRIPTION	CUSTOMER SELECTION	COMMENTS
Quote Number	122636	New
Trim Size	6 x 9	
Page Count	160 Pages.	
Quantity	1,000; 1,500; 2,000	
Overruns/Underruns	Up to 10%	
Binding Type	Perfect Bound	
Printing Type	Offset	
Turn Time	see comment	15 in house working days after final approval of both proofs
Text Copy	Pdf files	Based on file(s) that conform to our digital file guidelines. Requires a hard copy of the te & a completed disk to film submission sheet to start a job.
Text Stock	60# New Life Opaque	
Text Ink	Black/Black	No Bleeds
Text Proofs	Digital proof	
Text Halftones	None	If scanning or stripping in of halftones is required, charge is @ $11.00 each.
Cover Copy	Electronic File(s)	Based on file(s) that conform to our digital file guidelines.
Cover Stock	10 pt C1S	
Cover Ink	4 color process	
Cover Finish	Gloss Lamination	
Inside Cover	Blank	
Cover Proofs	Laser	

PRICING OPTIONS

(PRICES AND DELIVERY ARE SUBJECT TO PRINTERS ABILITY TO OBTAIN PAPER AT CURRENT COSTS)

QTY	PRICE	60# OFFSET	EST FREIGHT
1,000	**$2,406.0**	$2,253.00	$134.00
1,500	**$2,741.0**	$2,541.00	$171.00
2,000	**$3,076.0**	$2,829.00	$228.00

SHIPPING

(UNLESS OTHERWISE SPECIFIED CHARGES FOR RECONSIGNMENT, DROP SHIPMENT, RESIDENTIAL OR INSIDE DELIVERY ARE NOT FREIGHT COSTS.)

Ship to zip code - ~~94595~~
Freight Paid By Customer.

PAYMENT TERMS AND CONDITIONS

50% plus signed bid to start job; 50% with return of proofs; any additional charges, including freight, due prior to shipping. Price proposal is good fot
days. PRICES & DELIVERY ARE SUBJECT TO PRINTERS ABILITY TO OBTAIN PAPER AT CURRENT COSTS. All job materials must be received
at CPBM within 60 days after receipt of signed proposal. BY SIGNING THIS QUOTE YOU HAVE AGREED TO PAYMENT TERMS & CONDITIONS
AND TRADE CUSTOMS NOTED ON REVERSE SIDE OR ATTACHED PDF FILE.

By: _~~signature~~_____   _____

~~Name, title~~   Signature/Title/Company Name/Date

Vice President, Sales - Central Plains Book Mfg   Are you sales tax exempt? ☐ Yes ☐ No

Printing Trade Customs

What are *Trade Customs?* As the name suggests, Printing Trade Customs reflect the common business practices of the printing industry. However, Trade Customs are neither recommended nor required practices. Some printers may elect to follow them; others may choose not to. The use of any Trade Customs is an independent, individual business decision. Each printing company drafts its own contractual provisions, and will consider its business interests, relationships with customers, and other competitive issues.

Trade customs, as they apply to the printing industry are no longer binding to users of printing services. An earlier set of trade customs was legally binding to both the printer and the customer (in the absence of other specific contractual provisions) with the assumption that the customer (as a commercial client) was aware of the customs. Federal Trade Commission rules require that 51% of an industry group must ratify a set of Trade Customs before they become the *assumed contract provisions* in the absence of a more specific contract. At this time, there is no single or even group of printing trade organizations that encompasses the necessary 51% of printers to ratify the proposed trade customs, so there are presently no official printing trade customs accepted by the Federal Trade Commission. This is actually good news, in that there are no trade customs that might impact your printing contract *without your realization.* (Oddly, there may be printers who don't realize that there are no currently applicable trade customs—and they continue to submit quotations without reference to any underlying contract.) However, with some minor variations, the trade customs are likely to be incorporated into your printing contract and you should read and understand the contract before you sign it or make any payment. Often the printing contract is printed on the back of the quotation form. [When I owned a print shop, these terms and conditions were printed in light gray ink in small type on the back of a quotation form. To say the least, they were difficult to read. It was surprising that in the several years I operated the print shop, nobody ever actually read the contract nor complained about the difficulty in reading it.]

The trade customs as set forth below are those that were proposed to replace earlier trade customs that became out of date. While not having the full impact of officially recognized Trade Customs, most printing contracts cover each of the detailed elements and many will incorporate the proposed trade custom items verbatim.

DISCLAIMER. The following is a general discussion intended to help you to be aware of the elements of a printing contract. This book does not endeavor to offer any advice with respect to any specific agreement you may negotiate with a printer and does not offer any legal advice. **If you need legal advice you should seek the advice of a qualified attorney.**

The darker sans-serif type is the Printing Trade Custom and the regular text (serif type) are the author's comments.

Printing Trade Customs (as proposed)

1. QUOTATION. A quotation not accepted within 30 days may be changed.

Prices for materials may change and other circumstances may impact the ability of a printer to produce a project, so it's necessary to limit the time period a quotation is valid. Publishers should wait until a book is almost ready to print before seeking quotations. If a printing price is needed for planning purposes (perhaps at the outset of the project), let the printer know that is the case. Then, allow the printer(s) to re-quote the project when it is ready to print. Keep in mind, if the final specifications are different (even in small ways) from those submitted for planning purposes, that the final quotation may be substantially different from the original quotation. Submit preliminary specifications to several printers to get an idea of the range of likely quotations. Printers have different equipment, workflows, and staff skills, so there can be a 200% difference (or more) between the high and low quotation on any particular job. When the final specifications are ready, submit them to the "best" three to five printers from the preliminary quotations. If the specifications are significantly different (e.g. different trim size or hard cover instead of soft cover), then submit a new Request for Quotation to an expanded list of printers to ensure that the project and printer are well matched. If the printer has previously quoted on the project, note at the beginning of the RFQ "Changed Specifications" or some other indication to explain why the "same" RFQ is being resubmitted.

2. ORDERS. Acceptance of orders is subject to credit approval and contingencies such as fire, water, strikes, theft, vandalism, acts of God, and other causes beyond the provider's control. Cancelled orders require compensation for incurred costs and related obligations.

A contract isn't made between the publisher and printer until the printer accepts your order against their quotation. There may be circumstances that occur between the time a quotation is issued and when the order is submitted that may make it impossible for a printer to accept the order. If you cancel the order after the printer has started work, you'll have to pay for any work performed by the printer up to the point of cancellation. You should be aware that offset printing generates significant costs before the job gets to the press. A good policy is to never submit a project until you are certain that it is viable and that the necessary funding is available to complete the project.

3. EXPERIMENTAL WORK. Experimental or preliminary work performed at customer's request will be charged to the customer at the provider's current rates. This work can not be used without the provider's written consent.

See the comment with item 4, below.

4. CREATIVE WORK. Sketches, copy, dummies, and all other creative work developed or furnished by the provider are the provider's exclusive property. The provider must give written approval for all use of this work and for any derivation of ideas from it.

The printer has invested time and materials in the project and is entitled to reimbursement for his work. The printer may have copyright to creative aspects of parts of such work. If you wish to use or re-use experimental or creative work produced by your printer, be sure to make appropriate arrangements before proceeding.

5. ACCURACY OF SPECIFICATIONS. Quotations are based on the accuracy of the specifications provided. The provider can re-quote a job at time of submission if copy, film, tapes, disks, or other input materials don't conform to the information on which the original quotation was based.

If the order as submitted doesn't match the specifications in the RFQ originally submitted, then the pricing needs to be re-estimated. Even minor changes to the specifications may cause significant changes in cost of materials or time involved. Something as simple as a bleed, not in the initial specifications, might force a project onto a larger sheet and possibly a different press with highly significant changes to the cost of production. (Yet, the same change, in another printer's shop, may result in no change in costs.) Before proceeding, be sure to review the final specifications against those in the quotation. If there are differences, have the printer (or the printer and several of the closest competitors) re-bid the project before moving ahead.

When you receive a printer's response to your RFQ, always carefully check the specifications as quoted against the RFQ. There have been frequent cases where the printer has mixed up portions of the quotation. Common areas of confusion are split runs of hard and soft cover books—it's not infrequent for the printer to always assume that the smaller quantity will be printed in hard cover (but that's not always the case). Another possible confusion might occur when unusual trim sizes are involved. One RFQ I made was for a book with a trim of 5⅜ x 5⅜ inches (square). Several printers provided quotations for a trim size of 8⅜ x 5⅜ inches. This occurred on enough quotations that it had to be re-submitted—with a specific notation following the trim size that "this book is square." While several printers were then able to correct their quotations, a couple of others had to withdraw their bid as they were unable to manufacture a book with those specifications due to limitations in their (binding) equipment or work procedures.

6. PREPARATORY MATERIALS. Artwork, type, plates, negatives, positives, tapes, disks, and all other items supplied by the provider remain the provider's exclusive property.

Printers generally prefer to maintain ownership of all the intermediate materials used. However, items specifically billed as a separate line item on an invoice generally become the client's property. Preparatory and intermediate materials invoiced as part of another item generally remain the property of the printer. If you have any need for preparatory and intermediate materials created during the printing process, be sure to discuss this with the printer and make the appropriate arrangements *before* proceeding with the project.

Disputes under this section usually involved negatives required for making printing plates. It's becoming a non-issue as printers move to electronic workflows that result in direct-to-plate methods of prepress preparation. While negatives might have some value if you use another printer, printing plates are highly unlikely to be desirable: they are difficult to store without damage and most plates can only be preserved in a usable state for a relatively short time. In addition, plates are specific for particular brands and sizes of press. There's no guarantee that you can find another printer who uses the exact same size or brand of plates. (Brand is important as inks and other press chemicals are generally optimized for particular physical characteristics of the plate.)

7. ELECTRONIC MANUSCRIPT OR IMAGE. It is the customer's responsibility to maintain a copy of the original file. The provider is not responsible for accidental damage to media supplied by the customer or for the accuracy of furnished input or final output. Until digital input can be evaluated by the provider, no claims or promises are made about the provider's ability to work with jobs submitted in digital format, and no liability is assumed for problems that may arise. Any additional translating, editing, or programming needed to utilize customer-supplied files will be charged at prevailing rates.

Make backup copies of your work, be sure that a virus does not lurk among the files being submitted, put no extraneous items on disks submitted, and test any disk or CD for readability before sending it out. If the printer's prepress staff must do extra work to prepare your files for printing, you can expect that they will charge for that service. Cost is usually higher than if the problems had been corrected before the project was submitted to the printer. Once a job is in the hands of the printer's prepress staff, the press scheduling may force the publisher to pay for expensive work rather than take it back for correction by the typesetter/designer or inhouse staff. The caution here is to be sure that a job is correct and meets the printer's needs before it is submitted for production. (Most printers will provide a checklist or other information about how to prepare files for them. While nearly all printers needs are quite similar, the work processes of each may cause variations. For example, one printer requires that Acrobat PDF files be produced from PostScript files generated using a specific "PPD" file that supports their particular high resolution imagesetter.)

8. ALTERATIONS/CORRECTIONS. Customer alterations include all work performed in addition to the original specifications. All such work will be charged at the provider's current rates.

Put requests for alterations or corrections in writing. Often, the printer will provide instructions for documenting alterations or corrections on proofs or accompanying forms. Any other changes to the quantity or other specifications (paper, bindings, etc.) should be submitted in writing with a request that the printer provide written estimates of the cost. You don't want any nasty surprises when the bill is presented. In my experience, printer staff may not be very forthcoming with reminders that an "innocent" change will result in significant extra expense.

9. PREPRESS PROOFS. The provider will submit prepress proofs along with original copy for the customer's review and approval. Corrections will be returned to the provider on a "master set" marked "O.K.," "O.K. with corrections," or "Revised proof required" and signed by the customer. Until the master set is received, no additional work will be performed. The provider will not be responsible for undetected production errors if:

- **proofs are not required by the customer**
- **the work is printed per the customer's O.K.**
- **requests for changes are communicated orally.**

All I can say is *check all proofs very carefully.* Once you give your "O.K.," you are committing yourself to be financially liable for any work printed, even if it has an embarrassing error right smack in the middle of the cover (it's happened)—as long as the printed piece matches the "approved proof." Also, most printers will not accept an approval by phone. They will require that any and all proofs be signed and in their hands before a job will move forward. Some printers may allow a fax with a signature for an approval of a minor change.

10. PRESS PROOFS. Press proofs will not be furnished unless they have been required in writing in the provider's quotation. A press sheet can be submitted for the customer's approval as long as the customer is present at the press during makeready. Any press time lost or alterations/corrections made because of the customer's delay or change of mind will be charged at the provider's current rates.

Press proofs are usually provided during a session called a "press check." You have to be present at the printer's facility to perform this function. It's rarely done for book interiors printed with black ink, but may be worthwhile for an important work printed in color. (There are limited color adjustments possible once a project is on press.) If you decide to do a press check, bring a book to read and expect a long day usually spent in a dull waiting area. Be careful what you ask for, as you could generate significant extra costs if the press check generates serious changes. It is fair to expect that the press proofs will be reasonably close to the prepress "match print" or "calibrated print" as provided. The press proof is not the time to begin requesting significant changes! (Note: the "charged at provider's current rates" can involve some *very substantial costs*. Larger presses may have an operating cost of $300 per hour or more.)

With color printing done overseas, the proofs submitted for approval are often true press proofs. This allows the best evaluation of the color that will be produced. It is also an acknowledgment of the considerable cost of "match prints" as it's less expensive for the overseas printer to actually print a few copies of the project rather than prepare substitute analog or even digital proofs on larger projects. (It's also a reflection of the lower labor cost associated with overseas printing.) Recently, overseas printers have begun to provide ink jet proofs as they convert to electronic prepress technology, but you can usually obtain true press proofs for a moderate extra charge. If the color work is very important, then it might be money well spent.

11. COLOR PROOFING. Because of differences in equipment, paper, inks, and other conditions between color proofing and production pressroom operations, a reasonable variation in color between color proofs and the completed job is to be expected. When variations of this kind occur, it will be considered acceptable performance.

While *extreme* variations are unlikely, there are always *some* variations between "calibrated prints" from a proofing device, "match print" proofs, or even true "press proofs" and the printed output. It is important to ensure that you are satisfied with the prepress proofs provided. If the prepress proof isn't acceptable, the on-press output is likely to be unacceptable as well. There are many factors that can impact the appearance of printing, including the weather, humidity changes, lighting—you should try to view the proofs and final output in open shade with natural daylight or, at least, with lighting that simulates natural light. (In my studio, I have installed fluorescent lights with special "full spectrum" 5000ᴋ color temperature tubes. These lighting specifications are the standard used by printers and designers when evaluating color.) Even printing the same artwork on different days may cause small color shifts. This last factor impacts overseas color printing projects, so if you're having a color project printed overseas, be aware that there can be variations from the press proofs that are simply beyond control of the printer. However, such variations *should be* relatively minor.

12. OVER-RUNS OR UNDER-RUNS. Over-runs or under-runs will not exceed 10 percent of the quantity ordered. The provider will bill for actual quantity delivered within this tolerance. If the customer requires a guaranteed quantity, the percentage of tolerance must be stated at the time of quotation.

Although unusual, there are several steps in book manufacturing that may destroy book interiors once they have been printed but before the book is finished. **Most printers will automatically print 10% "overs" to avoid coming up short,** so expect to be billed accordingly. The "over/under" price is generally stated as a unit (per book) price on the quotation. Always clarify this point with your printer before accepting a quotation. Some printers will agree to "exact count," although most will charge more for that service. Some printers simply use this Trade Custom as a means to "upgrade" the order by 10%. Keep in mind, any of the Trade Customs are negotiable and the total quantity is, as well. While 10% over may not be unreasonable on an order of 1,000 books, it surely is unreasonable for an order of 10,000 books.

13. CUSTOMER'S PROPERTY. The provider will only maintain fire and extended coverage on property belonging to the customer while the property is in the provider's possession. The provider's liability for this property will not exceed the amount recoverable from the insurance. Additional insurance coverage may be obtained if it is requested in writing, and if the premium is paid to the provider.

While printers normally carry business insurance comparable to other businesses that may hold customer property, the printer is not accepting any liability beyond the value of the materials. As intellectual property often far exceeds the value of the paper or other media, the only real "insurance" is to maintain multiple backup copies in safe places. The printer also does not take on any liability for *lost profits* if, for example, a shipment of printed product should be lost before an important event or holiday. Consider, for example, that you have a little gift book about Christmas trees and the shipment is damaged beyond recovery in the printer's warehouse before the carrier picks it up. Time and circumstance prevent replacement books to be ready for "this" season. The printer is only responsible for the cost of materials and production, not for any profits you may not realize due to receipt of the books after the holiday in January. If you have some critical situation where extra insurance is appropriate, you may wish to discuss this with your insurance agent and/or to make arrangements with the printer for coverage.

14. DELIVERY. Unless otherwise specified, the price quoted is for a single shipment, without storage, F.O.B. provider's platform. Proposals are based on continuous and uninterrupted delivery of the complete order. If the specifications state otherwise, the provider will charge accordingly at current rates. Charges for delivery of materials and supplies from the customer to the provider, or from the customer's supplier to the provider, are not included in quotations unless specified. Title for finished work passes to the customer upon delivery to the carrier at shipping point; or upon mailing of invoices for the finished work or its segments, whichever occurs first.

Freight charges are paid by the publisher either directly to a carrier or as added to the printer's invoice. Some printers, if they handle the shipping arrangements, do not pass through carrier discounts. You might get a better transportation price with discount arrangements through a publisher's trade organization such as the Publishers Marketing Association (PMA) or the Small Publishers Association of North America (SPAN). (See the Resources section for details about these organizations.) If you plan to receive the books at your home or a business office without truck unloading facilities (a "dock"), you should specify "residential delivery" and/or "second floor delivery." Some locations will require "inside delivery." Discuss your situation with a carrier representative and/or the printing salesperson and make the appropriate notations in your Request for Quotation sent to the printer. The passing of title makes you responsible for filing a claim for freight loss or damage regardless of who pays the transportation charge, although most printers, if they are paying the carrier, will assist you with this process, should it be necessary. *Transportation moves under its own contract with its own set of rules, so react quickly to any problem to avoid loss of your rights.* Contact the carrier for copies of transportation agreements so you can review them. If appropriate to your business needs, secure additional insurance for business losses in excess of actual costs of the physical product.

15. PRODUCTION SCHEDULES. Production schedules will be established and followed by both the customer and the provider. In the event that production schedules are not adhered to by the customer, delivery dates will be subject to renegotiation. There will be no liability or penalty for delays due to state of war, riot, civil disorder, fire, strikes, accidents, action of government or civil authority, acts of God, or other causes beyond the control of the provider. In such cases, schedules will be extended by an amount of time equal to delay incurred.

Most book printers schedule their presses weeks in advance, so delays in returning proofs or other requested items can cause production of your book to be delayed as "your time" is reallocated. So, be prepared to review proofs promptly and to provide any other requested items as quickly as possible. A printer's reputation will let you know the likelihood of delays occurring from situations on their end. Of course, every printer may have unexpected difficulties or equipment failures, but such things are usually infrequent and the "better" printers can usually respond to such difficulties in a more timely and organized manner.

16. CUSTOMER-FURNISHED MATERIALS. Materials furnished by customers or their suppliers are verified by delivery tickets. The provider bears no responsibility for discrepancies between delivery tickets and actual counts. Customer-supplied paper must be delivered according to specifications furnished by the provider. These specifications will include correct weight, thickness, pick resistance, and other technical requirements. Artwork, film color separations, special dies, tapes, disks, or other materials furnished by the customer must be usable by the provider without alteration or repair. Items not meeting this requirement will be repaired by the customer, or by the provider at the provider's current rates.

For the small publisher, it's usually better to let the printer provide all the materials used for production of your book. While larger publishers may be able to save money by negotiating directly with paper manufacturers, keep in mind that a minimum order from a mill is a truckload or rail carload quantity (thirty thousand to one-hundred thousand pounds). That's a *lot* of paper. And, the paper must meet the technical requirements for the specific printer. It is best to simply specify "your house stock" of whatever color and general type of paper you wish to use. Most printers have a reasonable selection of bright white and off-white/ivory colored stocks as well as a selection of coated papers for projects requiring them. There are likely to be choices of virgin and recycled papers. Most printers will, upon request, provide you with a sample kit of house stock printing papers and cover materials.

17. OUTSIDE PURCHASES. Unless otherwise agreed in writing, all outside purchases as requested or authorized by the customer, are chargeable.

This is somewhat obvious. If you ask the printer for an outside service or material, you can expect to pay for that service. Be aware that the printer is entitled to charge a mark up to cover his handling/administrative costs and for profit. Often, when requesting quotations from competing printers for a service that only one can provide with their own equipment, the one who depends on an outside provider will have a much higher price on that aspect of the service. A good example is the cost of case binding books into hard covers. A printer with inhouse capability will generally offer the service at a significantly lower cost than a printer that may need to package and ship the book blocks to a bindery that may not even be in the same state.

18. TERMS/CLAIMS/LIENS. Payment is net cash 30 calendar days from date of invoice. Claims for defects, damages or shortages must be made by the customer in writing no later than 10 calendar days after delivery. If no such claim is made, the provider and the customer will understand that the job has been accepted. By accepting the job, the customer acknowledges that the provider's performance has fully satisfied all terms, conditions, and specifications.

The provider's liability will be limited to the quoted selling price of defective goods, without additional liability for special or consequential damages (i.e. loss of profit or loss of opportunity). As security for payment of any sum due under the terms of an agreement, the provider has the right to hold and place a lien on all customer property in the provider's possession. This right applies even if credit has been extended, notes have been accepted, trade acceptances have been made, or payment has been guaranteed. If payment is not made, the customer is liable for all collection costs incurred.

Make sure that you immediately inspect a reasonable number of the books received. Select several cartons and examine sample copies from each carton. Don't take samples from the top of the carton. (Printer staff has a human tendency to put the "best" on top, this is not an effort to deceive, but more a natural desire for the customer to see the very best work first.) If "bad" books are found, enlarge the sample you review. In any print order, it's possible that you may find a copy or two that fail to meet specifications. If there is a significant problem, you may need to inspect all copies and segregate the bad from those that are acceptable.

Once you have determined the extent of the problem, you should then contact the Customer Service Representative and explain the problem. Be prepared to move up the chain of supervision until you eventually discuss your problem with a manager or owner who has authority to adjust your problem. Always document your complaint *in writing* within the time limit (usually 10 calendar days) in your printing contract.

What often sets *good* printers apart from those who aren't as good is the way that they respond to a problem. *Good* printers will usually take an active interest in reaching full customer satisfaction.

Since printing is a *process,* there are errors that are transient. A smear on the cover, or a blotch of ink on a page may only appear on a relatively few copies. These are inherent to the process, and unless the problem persists on a large number of copies, are not considered a material failure in the project. If you're really annoyed, set such copies aside. You may need them to note editorial changes for a future edition or you may prefer to sell them at a discount as hurt books through your web site or other means.

When you find a problem be sure you have adequately surveyed the shipment so that you can fully describe the issue and the number of copies affected. Printers will usually try to determine if the problem affects the salability of the product. Minor flaws, while annoying, don't often have an adverse impact on sales. (For example, one publisher was quite upset that the back cover text was not properly centered. A test subject didn't even notice it—and once the problem was pointed out, the subject responded, "Oh yeah, it happens…") This example is also illustrative, in that the problem was not the printer's fault—it existed in the artwork and should have been caught, at the latest, when the printer provided proofs. (Of course, this type of problem should have been caught before the project was even submitted to the printer.)

Decide what it would take to solve the problem (in your mind) before you even discuss it with the printer. Be aware that printers generally prefer to reprint rather than reduce the price or refund money. Printers also have a saying, "if it's good enough to keep, then it's good enough to pay for." So, if you think it's "good enough to keep," then think carefully about what you need to be made whole for an error.

19. LIABILITY.

1. Disclaimer of Express Warranties: Provider warrants that the work is as described in the purchase order. The customer understands that all sketches, copy, dummies, and preparatory work shown to the customer are intended only to illustrate the general type and quality of the work. They are not intended to represent the actual work performed.

2. Disclaimer of Implied Warranties: The provider warrants only that the work will conform to the description contained in the purchase order. The provider's maximum liability, whether by negligence, contract, or otherwise, will not exceed the return of the amount invoiced for the work in dispute. Under no circumstances will the provider be liable for specific, individual, or consequential damages.

Again, the printer is notifying you that they will only be responsible for producing the project to the specifications provided in the contract. And, if there are claims, the printer is only responsible to a maximum of the actual cost of the project and *is not responsible* for any lost revenues or other damages beyond the cost of the printing.

This contract language is found in many purchase contracts to deflect the more stringent common law warranties that would otherwise apply. (If you have questions about this topic, refer to a business law text or consult with an attorney.)

If a publisher has business needs that require additional protection, then the publisher should obtain insurance to cover those business situations, paying a premium appropriate to the situation. Some of the liability issues may be covered under a *printer's errors and omissions* insurance policy. However, such policies seem to be quite narrowly drawn and don't cover the most common situations. So, if you need extra insurance, don't depend on coverage that printer may or may not have. Contact your business insurance agent and arrange the coverage you may need.

20. INDEMNIFICATION. The customer agrees to protect the provider from economic loss and any other harmful consequences that could arise in connection with the work. This means that the customer will hold the provider harmless and save, indemnify, and otherwise defend him/her against claims, demands, actions, and proceedings on any and all grounds. This will apply regardless of responsibility for negligence.

1. Copyrights. The customer also warrants that the subject matter to be printed is not copyrighted by a third party. The customer also recognizes that because subject matter does not have to bear a copyright notice in order to be protected by copyright law, absence of such notice does not necessarily assure a right to reproduce. The customer further warrants that no copyright notice has been removed from any material used in preparing the subject matter for reproduction.

To support these warranties, the customer agrees to indemnify and hold the provider harmless for all liability, damages, and attorney fees that may be incurred in any legal action connected with copyright infringement involving the work produced or provided.

2. Personal or economic rights. The customer also warrants that the work does not contain anything that is libelous or scandalous, or anything that threatens anyone's right to privacy or other personal or economic rights. The customer will, at the customer's sole expense, promptly and thoroughly defend the provider in all legal actions on these grounds as long as the provider:
- promptly notifies the customer of the legal action
- gives the customer reasonable time to undertake and conduct a defense.

The provider reserves the right to use his or her sole discretion in refusing to print anything he or she deems illegal, libelous, scandalous, improper or infringing upon copyright law.

With this provision, the printer, quite reasonably, requires the publisher to defend the printer and accept all responsibility for copyright infringement, libel, invasion of privacy, misuse of celebrities' image or right of publicity, or any other of the many liabilities associated with copyrights and trademarks. A careful publisher will thoroughly fact check projects and, if appropriate, have a suitably qualified intellectual property lawyer vet the project before it is submitted for printing. The printer retains the right to reject the project if the printer is uncomfortable with it.

If you are working with a potentially controversial project, it's best to be up front and indicate the situation on your Request for Quotation. For example, when I produced *Confessions of a Dope Dealer,* by Sheldon Norberg, I realized that some printers might feel uncomfortable with a book that had so many references to drugs, sex, and rock & roll. I noted the controversial nature of the content directly on the Request for Quotation and indicated that any printer who did not wish to bid on the project should simply let me know and that they would be considered for future projects without penalty.

Printers rarely read the content of the books they print. However, when certain problems occur, they may need to read some of the material to determine where corrections might need to be made. During one such situation, another publisher discovered that the printer (or the printer's staff) was unwilling to work with "that kind of material." The job was pulled off the presses in the middle of the run, and the publisher was told that the printer was rejecting the project due to the "nature of the material." While there was no charge made for any of the work done by the printer, the book was put well behind schedule and the publisher was not able to have the book in the time previously promised to various buyers. (Another printer was found who did not object to the material.) The point is that it is surely better to advise the printer to be aware of the potential for controversial material before the project is submitted to avoid any unpleasant circumstances. Also, publishers should understand that many of the printers servicing independent publishers are located in smaller towns in the Midwest, where "community values" may be rather different from those of a publisher located in a major metropolitan area.

21. STORAGE. The provider will retain intermediate materials until the related end product has been accepted by the customer. If requested by the customer, intermediate materials will be stored for an additional period at additional charge. The provider is not liable for any loss or damage to stored material beyond what is recoverable by the provider's fire and extended insurance coverage.

This term indicates that the printer will not store negatives or plates beyond the time that the project is accepted by the publisher (without payment of an extra fee). Printers realize that a large number of projects never have a second printing and it would otherwise quickly become a serious storage problem if the printer were to attempt to keep intermediate materials for every project. In some cases, upon request, a printer will ship negatives to the publisher. However, the evolution in production methods has resulted in many printers going "direct to plate" or even "direct to press" eliminating the preparation (and cost) of producing negatives. These changes in production methods has eliminated the need to store intermediate materials as they no longer exist. (The initial printing is more efficient (and less expensive) but the downside is that discount for an "exact reprint" is no longer as large as was once the normal case.)

22. TAXES. All amounts due for taxes and assessments will be added to the customer's invoice and are the responsibility of the customer. No tax exemption will be granted unless the customer's "Exemption Certificate" (or other official proof of exemption) accompanies the purchase order. If, after the customer has paid the invoice, it is determined that more tax is due, then the customer must promptly remit the required taxes to the

taxing authority, or immediately reimburse the provider for any additional taxes paid.

This is a common disclaimer in many wholesale contracts. Sales taxes are collected by sellers of tangible products. The rules vary from place to place, but generally sales to out of state customers are exempt from sales taxes. A publisher should obtain a "sales tax permit" from their local taxing authority and prepare suitable "certificates" (often a postcard form) as required by the printer to document that no sales taxes need to be collected by the printer on the manufactured books. Of course, the publisher will need to obtain similar "exemption certificates" from wholesalers and retailers and should collect sales taxes from retail purchasers as required by the rules in the local area. (In most cases, the publisher will need to pay sales (use) taxes on copies used for promotional or office purposes.)

23. TELECOMMUNICATIONS. Unless otherwise agreed, the customer will pay for all transmission charges. The provider is not responsible for any errors, omissions, or extra costs resulting from faults in the transmission.

This reflects the new data transmission possibilities via the Internet and other private telecommunications services. It is not usually an issue in book production.

Well, what does it all mean?

It is very important that you read the terms and conditions that apply to your print project—if you only get a memo form with a basic price quote without the full terms of the printing contract, be sure to ask for the complete contract so that you can read it. The salesperson can say what they will do, but if it isn't written in the contract, the promises of the salesperson are worthless.

In most cases, a job will be printed to your specification as set forth in the printing bid. Be sure to review all bids to ensure that they actually reflect what you want and expect from the project. (It might seem odd, but I have encountered many printer quotations that misstated the dimensions of a book, the page count, or the kind and quality of the paper to be used. These quotations were non-responsive and I either set them aside or had to request the printer re-bid the project.)

If a job goes bad, then the quality of the service and responsiveness of a printer becomes a major factor. Keep in mind, despite the best intentions, even the very "best" printer can have a bad day and spoil a job. Hopefully, the printer's quality control system will find any problems and the job will be reprinted or otherwise remanufactured before you ever see any errors. However, printing is a *process*. As such, quality can vary through the press run or other transient errors (such as ink spotting, or over/under inking) can occur resulting in less desirable copies. Some errors can come and go before the press operator is even aware that a portion of the job is of poor quality. A "good" printer will make some accommodation (reprint, credit bad copies, offer a general price reduction) to reach a reasonable level of satisfaction. A "less than good" printer may ultimately be quite hostile to making an effort to satisfy a customer complaint. It is a wise precaution to research the general level of satisfaction of customers of a printer.

- 11 -

Book Covers

MOM USED TO SAY: "DON'T JUDGE A BOOK BY ITS COVER." While that may be good advice when it comes to avoiding mistakes caused by false assumptions or accepting stereotypes about people or things, it isn't particularly reliable advice when it comes to books! Before your book even gets in front of a potential reader, it will first be judged by reviewers, distributors, wholesalers, librarians, and bookstore buyers—with their opinion strongly influenced by their reaction to the cover. If your book gets past all the gatekeepers, then your potential reader will form an impression of your book while looking at the cover—before they even decide look inside.

So, what can you do to make sure that your book is the one that gets a fair viewing?

We start with the obvious: Select a title and subtitle that get to the essence of the book as briefly and succinctly as possible. Keep in mind who you're trying to reach and what words they will understand. If you are in a field that has its own well known special jargon, it's okay to use it. If you are trying to reach folks who do not know the "in" language of a field, avoid jargon.

The example (on the right) of *Making Rain* refers to the jargon on the legal and accounting professions to refer to those who secure new business as "rain makers." The book uses an adventure novel to illustrate various ethical considerations in "making rain" for lawyers.

Select an image, photo, or other design element for the front cover that further communicates understanding about the content—there is more truth about "a picture is worth a thousand words" than not. But, exercise care that the picture or other image is quickly understandable, particularly if the title requires a frame of reference to make sense. (A short, succinct title may fail if it is misinterpreted; for example, there are reports that the novel, *Fear of Flying,* was placed with books on aviation in some bookstores (before it became a best seller).

Consider, too, if your book would be best served by an all-text cover. Some books can best communicate all they need with the use of text. If your book fits this category, then it may be counterproductive to try to include elaborate artwork. The all text example, *Viatical & Life Settlements: An Investor's Guide,* is a discussion of various investment advantages and pitfalls associated with buying life insurance policies from elderly or ill people at a discount (the policyholder gets an early payment, the investor may earn a greater than average return). Frankly, there's no photo that can explain that idea.

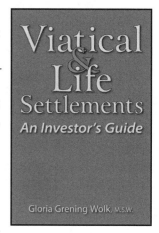

The spine

Nearly all books are displayed in bookstores with only the spine exposed (spine out), although a few titles may be displayed with the front cover out. This is usually a choice made by the store staff (with priority given to strong selling titles), but may also be influenced by extra payments made by the large publishers (a tactic that's usually unaffordable to smaller publishers). It's important that the title on the spine is clear and readable. It's usually a good idea that the spine reflect the color and style of the front cover, since your promotional material will likely feature the front cover. It might cause confusion if, for example, the front of the book is mostly colored blue, then the spine is some other color. If the design on the front/spine is inappropriate as a continuation to the back cover, then make a style change at the crease where the back cover joins the spine.

Dan Poynter suggests 'stacking' the letters of your title vertically, rather than turning them to one side. This works with some book titles—but often doesn't work all that well. If you do need to rotate the title, remember that the tops of the letters should be closest to the front cover. Your book is likely to be completely ignored if its title is the opposite of all the rest. (Think how you read book titles in a store (or library)—you stand back a bit and turn your head to make the reading easier. If one is upside down or facing the opposite direction, you're likely to skip on by rather than move your head around to read it. (I understand that this alignment of spine text is the opposite of the usual practice in non-English speaking countries—go figure.)

With a very thick book, you might be able to run the title left to right on the spine. Notice how dictionaries and the *Chicago Manual of Style* manage to display their titles in this manner. However, even if this is a possibility for your book, you need to balance the advantage of larger type against the easier reading of the title.

A well designed, attractive cover will not guarantee success. But a poorly designed, unattractive cover is likely to cause your book to never be selected and picked up. Your book is much less likely to be considered by reviewers, librarians, distributors, bookstore buyers or potential readers with an unattractive, ineffective cover. Finally, it's best if the cover achieves its marketing objectives. If it's a successful work of art, so much the better.

Back cover

The back cover text is the most important writing you'll ever do in connection with your book. It must answer the question: Why should I buy this book? (Right now!)

You need an arresting headline addressed to potential buyers, so they can relate to the book and find themselves in it. Don't just repeat the title (it's on the front and spine, after all).

For example: The back cover copy of the book, *The Best Medicine* reads: "The consumer guide that protects you against cut-rate medicine and guarantees you the best health care available." Notice that this headline tells you several things about the book right off. It also makes a promise to you (the reader).

Selling message. Concisely state what the book is about. Explore how the reader will benefit from reading the book. Dan Poynter's, *The Self-Publishing Manual* states "all authors should read his book even if they do not plan to self-publish" (the inference is that they will benefit by learning more about the publishing process).

Make your readers a promise: Promise health, wealth, entertainment or a better life. Promise to make readers better at what they do.

Focus on benefits. Remember, a fact is not a benefit. A fact is a statement with no obvious connection to the potential reader's needs. You have probably stated a fact (without a benefit), if you are left with the question, "So?"

"This dictionary has 165,000 entries." That's a fact—it's obvious that you can just ask "So?" Looking at Merriam-Webster's new 11th edition back cover it gives both facts *and* benefits: "The words you need today—anytime, anywhere." (Followed by an explanation about how print, CD-ROM, and Internet formats are supported.) "The most up-to-date dictionary available." (Sub-points include several facts about the number of new listings, etc.) "The features dictionary buyers ask for." (Followed by facts supporting this benefit); and so on...

People often don't immediately make the connection between a fact and a benefit, so you must anticipate the connection for them. "This book weighs 10 pounds, so it will work as a great doorstop, even in a windy location." Fact followed by benefit.

> ### Facts and Benefits
> When I owned a print shop, I discovered that people were very unlikely to immediately make the connection between the activity of a printer and the printed products they wanted. Over the years, our most effective ads included lists of the products we produced: newsletters, labels, flyers, post cards, brochures, business cards, etc.
>
> When making sales calls, I often had people ask, "Do you print X or Y or Z?" [Yes, of course we do!]

The back cover should include testimonials and endorsements (commonly called blurbs). As you prepare your mock-up cover, write two or three different endorsements from people you would like to quote. If "'this book changed my life.' —Grey Davis, (former) Governor of California," would look good, try it. Use names or titles recognizable in your field—sources that might impress potential buyers. Remember, this is just a draft. Go ahead, dress it up.

The whole point is to make the author look like the ultimate authority on the subject. Just two or three sentences will do.

Finally, call for action. The last lesson of salesmanship is "ask for the order—and shut up!" Close your sales message with a call for action: that is, to "take this book home with you now." This is the hardest point to accomplish—and is usually the most frequently skipped. But Dan Poynter, on the back cover of his *The Self Publishing Manual,* closes quite well with: "This best-selling manual on self-publishing has shown thousands of people the faster, surer way to break into print. What are you waiting for?" Now, *that* is a call for action!

Required Elements

BISAC SUBJECT HEADING. A category designation, usually placed on the top left or near the price. The official source for these categories is the Book Industry Study Group's "BISAC Subject Heading Package." They'll gladly sell you a copy for about $25 from their web site at *www.bisg.org.* You can also obtain a discounted copy if you are a member of the Publishers Marketing Association. Another approach is to use your favorite search engine on the Internet and look for "BISAC Subject Heading." I ran across a site with the whole listing online. Another way is to go to a large bookstore and to look at the shelves where your book might be displayed. Look at the categories on similar books and look at the category designations of the shelves. You want to list a category on the back cover of your book so it will help the minimum wage bookstore clerks to put your book where your potential readers will have a chance of finding it. It *is* best to use the official list ... and take note of the BISAC category number, as your distributor will ask for that. (Some booksellers and libraries want to be notified of new books in certain categories.) [BISAC stands for Book Industry Standards and Communications.]

PRICE. Bookstores like a price on the back cover. Remember, the check stands in larger stores are often staffed by students or other part time employees who aren't necessarily interested in books. Make it easy for them to not make mistakes.

Don't get creative with the price. End it with 95, 98, or 99 cents. Market research has shown that people simply add the dollar whether it's there or not. An even-dollar amount won't gain you anything, because the buyer either isn't aware or won't pay any attention. An intermediate price (say $19.50) simply gives up 45 cents that the buyer has already "assumed" they are paying.

BAR CODE with International Standard Book Number and price extension. The bar code on a book identifies the ISBN which, in turn, identifies the publisher, title, author and edition (soft cover, hard cover, etc.) The price extension identifies the currency (the number 5 means U.S. dollars) and a four digit number for the amount including cents.

Recently, a new 13 digit series of ISBNs started to be released. If you have ISBNs from the old 10 digit series, by 2006 you will need to convert them to the 13 digit format. It's really quite simple. Look at the bar code and read the numbers below

Sample of ISBN and bar code

the bars starting with "978." This is the 13 digit ISBN. The change is that 978 is added in front of the old 10 digit number (becoming, in the example 978-0-9474331-0-3). The last digit is the "check digit." It is derived by a mathematical calculation. The result of the calculation may change the check digit, in this case from "x" to "3." (A check digit of "X" actually stands for ten, but only a single character is used in the ISBN.) I then typeset the 13 digit ISBN and placed it above the standard barcode in the example. You may start seeing books with this arrangement in the near future*.

Cover layout

When setting up the cover in your page layout program, remember that the printer wants a single page with all the elements positioned in the proper arrangement. As you look at a cover wrapped around a book, keep in mind that it's printed flat, then folded around the book. Here is a sample of the rough layout. Note that the front cover is on the right. (For a layout of a dust jacket, see chapter 11, Children's Books).

BISAC Category Headline Description/Sales statement Facts and Benefits Testamonials, "blurbs" Author qualifications Call to action (close the sale) Price and ISBN bar code	S P I N E T I T L E Author Name Publisher Name or Logo	Title SubTitle Author Name Foreword by (name)

In this typical layout, we've shown the spine title set in a vertical format. Most titles should be rotated ninety degrees, with the top of the type toward the front cover. Often, only the author's last name is shown on the spine. Sometimes the author name can be rotated to the same orientation as a rotated title, especially with longer names. If the publisher has a logo, it would be shown on the spine (as well as on the main title page).

If you look at books in the stores, you'll see many with the author's name at the top—and often larger than the title. This is appropriate when the author's name is likely to be more of a "draw" than the actual title of the book. If your name is Tom Clancy, Steven King, or Tom Wolfe then it pays to have your name very large and above the title. For lesser authors (that's nearly everybody else), a more modest display of your name is

*As we go to press for our second printing, the 13-digit ISBN is now the standard. It's no longer necessary to show the 10 digit ISBN on your book.

appropriate. That also applies to photos of the author on the cover, front or back. Where there may be marketing reasons for an author's photo to appear on the cover, usually it's best to leave such photos for an "about the author" page in the back of the book or, possibly, on a dust jacket flap.

SPINE WIDTH. The printer will advise the width of the spine once you've selected the paper. Since paper thickness varies significantly, I usually start with a 400 pages per inch estimate for 60 lb. paper and 500 pages per inch for 50 lb. paper. The cover art will need to be adjusted for the actual spine width once it is confirmed by the printer. The formula to calculate a spine width for a paperback book is quite simple: number of pages divided by pages per inch (for the paper being used) plus twice the thickness of the cover stock to be used. For example, a 224 page book using high bulk 55 lb. book paper with 10 pt. C1S cover would be calculated thusly: $(224/360)+(2 \times .010) = .6422$ inches. Note that pages per inch includes both the front and back of the sheet. Paper is usually specified by paper merchants by its "caliper" (the same as the device used to measure the thickness). So to recalculate the caliper of a paper into "pages per inch" (PPI), divide one (inch) by the caliper to arrive at the number of sheets per inch, then multiply that by two to get the number of pages per inch. For example, a paper with a caliper of .0042 will have about 476 PPI; calculated thusly: $1/.0042 = 238.095$; $2 \times 238 = 476$.

Some thoughts on cover design

Some folks may have the idea that they can design their own cover. Be realistic. Can you create and render a design that looks completely professional? If not, let an experienced designer do the work. The cover is simply too important for halfway measures. Either ensure that it is done right or you should completely rethink your project.

Now that I've said it, let's consider what a cover design might cost. A competent cover designer might charge between $600 and $3,000 for a completed cover design. Most experienced designers charge between $1,200 to $1,800 to develop a cover. If you have a *well developed* concept, an experienced designer may be able to charge a little less. (There can also be additional costs for photos or special illustrations on top of the designer's charges.) The lower price range may come from a designer who is not particularly experienced with cover designs. An inexperienced designer may be a reasonable choice, but beware of designers who charge significantly less than $600. (I've heard quotes of as little as $300—but these usually end up as incomplete designs and you'll end up having to hire another designer to finish or fix the design and make a complete, ready for the printer, file. At the other end (above $2,000 or so) you're probably getting into a "name" designer situation. Most small publishers don't need to pay such fees and there's little advantage to be gained by using a "big name" designer for a cover on a book that may not sell more than a couple thousand copies.

When planning your cover, you have to realize that there is a very big difference between conceptualizing and designing. (Perhaps this is a sore point for me.) Often a non-designer will provide a designer with a concept. The designer will then render the concept into a design—using the training, experience, and skills that separate a designer from ordinary folks. The publisher may have acted as "art director" and may have conceived the basic idea, but (please) understand that the designer actually *designed* the piece.

Choosing a designer

Chapter 14 has a thorough discussion of this topic. Here are a few things to consider when evaluating a cover designer.

When seeking a designer, try to get referrals from other small publishers (if possible). You may be able to find some if a credit is printed in a book you like (usually on the copyright page), although books from larger publishers may have been done by in-house staff. Seek out those who have a portfolio of work that includes many (or is exclusively) book covers. Take note of the designer's style. A designer who does highly realistic images may not be the one you want to execute a surrealist design. If you see that all the covers use a similar palette of colors (that you don't like), you may have difficulties getting the designer to try other color combinations. Frankly, most of the "designer problems" I've heard about are actually very predictable failures of communication between the author-publisher and the designer.

The design process usually works by a thorough interview between the publisher and the designer. You should be prepared to explain any ideas you have and show samples of covers (on other books) that you like. (I've often explained that I can design "anything"—but it's nice if it's something that the client likes.)

If you don't like the direction in which the design is going, speak up sooner rather than later. Most designers will prepare two or three concepts as part of the quoted price—then they will begin to render a final design. Additional concepts will cost extra. You need to communicate your impressions earlier rather than later. It's fair enough to show concepts to friends, librarians, or book store staff (manager or book buyer), but keep focused on what you want and don't turn the project into a "design by committee" as that usually generates very unsatisfactory results.

Be sure that the cover designer understands that the marketing considerations come before the artistic merits of the cover. I have heard too many complaints from publishers who had a designer refuse (for example) to make a title larger or some other adjustment because, according to the designer, "it will hurt my design."

Copyright

Be sure to talk with your chosen designer about copyright issues *before* you do any substantive work on the project. Beyond discussing generalities about the design, be sure you understand the terms and conditions of the design agreement and understand exactly what reproduction rights you are buying. The typical design contract will give you the right to reproduce the cover on the book but you also need the right to reproduce the cover in advertisements, posters, catalogs, on the Internet, and possibly on other items such as T-shirts, coffee mugs or other promotional items.

Some designers do not license these other uses as a matter of course. You need to discuss your advertising and promotion plans and be clear with the artist on what you need to do. If the designer is not cooperative (or at least reasonable) on these issues, go somewhere else. (This caution also applies to commissioned illustrations or photographs. In one case, a client had a poster-size version of the cover made to put by a display of the books being sold at a concert. The photographer, whose work was prominently featured on the cover, was quite upset, "I didn't know you were going to do *that* with my photo." Eventually, the photographer was sufficiently calmed and agreed to the "unexpected" use of the photo.) Most designers will agree to these promotional needs beforehand; but a lack of prior understanding can lead to bad feelings and even lawsuits.

When selecting a designer, be sure to look at the style of work they have done in the past and discuss the look you are seeking. You may waste a lot of time trying to get a look that's completely different from what a designer normally creates. Designers (should be) happy to show you their portfolio and explain in what style they prefer to work.

Some final points to consider

Who is your market? Is a cover going to a very "macho" group or is it intended particularly for women—the design should reflect the intended market. The *Making Rain* book was intended for an audience of lawyers, as someone associated with the publisher said, "a very testosterone-soaked bunch—even the women!"

It shouldn't look like a magazine cover—it's a book cover! I've seen this error on books produced by small magazine publishers. A magazine has to "shout" from the rack in a store. A book has a less frantic background among other books.

Avoid designs that look like "variation 'C'" from a high volume publisher (such as one of the online, so-called "POD" publishers). Spend some time looking at the author package offerings at the online publishers. Then you'll know what to avoid.

If catalog sales are likely to be important—the cover should be readable in a very small size (about 1 inch tall) and in shades of gray. At the very least, the design should hold up for a low resolution image on Amazon.com. (Amazon wants an image of 648 pixels for the longest dimension. A typical 6″ x 9″ book gives a 432 x 648 image. Other size books should be scaled to the 648 pixel size—a 5½″ x 8½″ book results in a 419 x 648 pixel image.)

The title on a book cover should be readable from about ten feet away. Most books will have a less than ideal display situation (in book stores) so it has to be able to get the reader's attention. The browsing potential buyer should not have to "work" to read the title and get some feel for the book from the cover image.

- 12 -

Children's Books

NEARLY EVERY PARENT HAS THOUGHT ABOUT IT—writing a picture book for children. However, the reality is somewhat more difficult than you might perceive by looking at books in your local bookstore or at a library. Here, we'll try to help you understand some of the basics. And, if you wish to persist in this endeavor, we will direct you to some additional resources.

First, I have to caution that publishing children's picture books is not for the faint of heart. It's expensive and very risky. Children don't buy books—parents, grandparents, librarians, and teachers do. So, not only does the book need to appeal to children, it must also appeal to the adults in the child's life.

In self-publishing, it's best if a book can be marketed to a defined (and relatively small) group that can be reached through highly directed marketing and publicity efforts. Parents and children are a diverse and large group, making the marketing task just that much harder. (This marketing factor is shared with nearly all works of fiction.)

Another bit of bad news is that the large publishers have nearly insurmountable advantages due to the cost structure of printing and distribution. Due to the cost structure, it is highly unlikely that a small publisher can compete with a paperback book in this segment. Therefore, be prepared for the extra expense of publishing your book in hard cover.

Children's picture books also have very strict format requirements. They are always twenty-four or thirty-two pages in length. This is a matter of the product being designed to facilitate cost constraints of printing and the typical press set-up rather than any outside marketing issues. Unfortunately, it seems like the most common error by a beginning children's picture book author/illustrator is to create a book that has too few or too many pages.

The thirty-two page format means that the book must have room for all the required elements of a book: half-title (optional), title page (often a double page spread) and a "copyright page" where the publisher information, copyright notice, and cataloging-in-publication data will appear. It is possible, if extra space is needed, to use the back side of the end papers, gaining two pages, for the publisher information and/or for a final illustration. Keep in mind that production needs often dictate that the endpapers must be uncoated stock and they will absorb the ink differently, causing an unintended color shift.

Another common error is to have a too-long story. Children's picture books are rarely over a thousand words. If your manuscript exceeds that length, it's time to start cutting. (One hint is to avoid lengthy scene setting or exposition. Simply start the story where the action begins. Use the interactions of the characters [and the illustrations] to set the scene and give a sense of location to the story.)

Another useful idea is to make up a "dummy" of your book. Take a print of your manuscript and cut it into blocks and paste it into a thirty-two page blank book. This will give you a chance to play with it to see how it fits into the format and to find the best breaking points for pages or for insertion of illustrations.

Finding a publisher

Many author/illustrators initially seek a publisher for their work. Here, the most common error is that an author goes to the trouble and expense of obtaining illustrations before approaching the publishers. Rule number one: Publishers almost always prefer to provide their own illustrations for a picture book rather than using those of the author.

If you're an author, you may provide simple illustrations (if you have the requisite skill) to simply show a possibility with your story. Indeed, some stories need illustrations to enable the reader to understand what the book is about. (For example, if you've written a book on the growth and development of a frog's egg through tadpole to adult frog, there's no way to adequately describe the process without illustrations.)However, don't provide complete, fully rendered illustrations. Just provide the story along with simple illustration suggestions as you see fit. Let the publishers make their own choices for how to produce the illustrations.

If you are an illustrator, prepare a portfolio and offer it to publishers so that you may get a commission to prepare illustrations for a book already in the publisher's plans. The talents and skills that make a good illustrator are often not the same as the talents and skills that make a good writer. Obviously, particularly unusual and talented people (Dr. Seuss comes to mind) may write material and provide illustrations as a package. Even so, it's usually best to offer the words and illustrations as separate elements that the publisher may choose to combine.

Publish it yourself

Once you've collected a sufficient number of rejection slips from the larger publishers, you may decide to proceed with your project yourself. Now, you will need to ensure that you have sufficient funds to proceed with the project. You will have expenses for editing, typesetting, illustration, prepress color preparation, printing, and shipping.

A children's picture book needs to go through all the same steps of book production as does any professional-quality book. Even though the language of a children's picture book may seem simple, it requires careful editing with a view to the specific age group of the children who are expected to be reading the book. Editorial errors are particularly dangerous, as having a few "over age" words may result in the book being given bad reviews and negative recommendations by teacher and librarian groups.

Understanding the format

As we said earlier, children's picture books are either thirty-two or twenty-four pages. To make this understandable, let's see how the book is assembled:

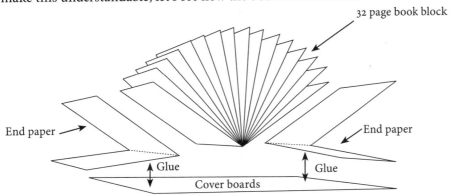

This above illustration shows a thirty-two page book block with end papers glued to the boards that make up the cover. The glue on the end papers extends partly around the edge to stick to the book block near the spine. This is the most common case bound method of binding a hard cover book.

The thirty-two page book block is printed with sixteen pages on each side of a single sheet (or side of a web). This same technique can be used for a twenty four page book, if the printer can efficiently print the pages twelve per side. Otherwise, a twenty-four page book is assembled thusly:

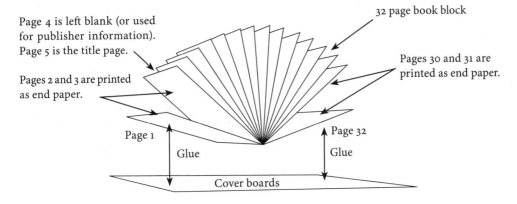

In this case, no separate end paper is used. The glue is applied to pages 1 and 32 of the thirty-two page book block. The end paper design is printed on pages 2–3 and 30–31. The first page of the book is printed on the fifth page of the book block. If you then count the available pages, you'll find that there are 26 available pages for the illustrations and text (running from page 4 through page 29 of the book block). As you can see, there are opportunities, with this method, to create a book of 29 pages if you start with the title page on the third page of the book block—but that isn't recommended. It's better to keep the book to twenty-four or twenty-five pages and retain the expected end paper look of the first and last double page spreads.

As for durability, end papers are usually printed on heavier uncoated paper for better glue strength. The book block is usually printed on a coated (matte or gloss finish) for better color reproduction of the illustrations. When the second method is used to obtain a shorter (twenty-four page) book, the glue is applied directly on the coated paper. However due to the large coverage area (as compared to the standard method with end papers), there is sufficient glue strength to hold the book together. The book block itself is (should be) smyth sewn for maximum strength.

In an alternate method, called a "library binding," an additional two-inch-wide strip of fabric is sewn to the spine of the book block during the smyth sewing step. The fabric is then glued to the binder's board used in the cover. Library binding is only done with the traditional end-paper construction.

Preparing the book

In most respects, preparing a children's picture book is much the same as for any other book; however, various accommodations are made for the illustrations and for the audience.

TYPESETTING. Typesetting in a children's picture book should follow the suggestions in Chapter 4 and 5. In addition, care in selecting a typeface is important. Frequently, a traditional typeface with a large x-height is preferred, although for artistic purposes, a book with very few words might be set in something more decorative. Type is set larger than would be acceptable in an adult publication—14, 16 or even 18 points with two to four points of leading are not uncommon. (Older pre-school children may begin to read for themselves, so generous size and plenty of leading to assist with eye-tracking will help the young reader.) While traditionally-based fonts are preferred, selecting from those that are more "friendly" is appropriate. I suggest looking at New Century Schoolbook, Stone Informal, and Bembo. For a more casual look, Comic Sans or one of the similar handwritten-looking typefaces may be appropriate.

COVER. Due to the low prices ($6–$9 or so) charged by the large publishers for soft cover children's picture books, you'll probably want to publish a hard cover book. The large publishers can print ten or twenty thousand copies of a soft cover title while a small publisher usually can only afford to print three to five thousand copies. This difference makes it economically impossible to compete with soft cover books. Hard cover children's picture books are more often priced between $12 to $16, leaving some room for a possible profit.

When setting up the cover art, you will most likely want to print a dust jacket and to have a "lithowrap" (printed) cover. The basic artwork is the same for both. However, the lithowrap for the hard cover will have much wider margins for wrapping around the binder's board. Remember, too, that hard cover books are about ¼ inch larger than the nominal trim size. So, an 8″ x 10″ trim size will have a cover of 8¼″ x 10¼″. The dust jacket and lithowrap will need to be designed with these dimensions in mind. (Always get exact dimensions from your printer, once you have made your selection.) A typical dust jacket will have a layout as follows:

Flap	wraparound allowance	Back	Spine	Front	wraparound allowance	Flap

The flaps are usually about 3½ inches, the double line between the flaps and the front/back of the book represents the wraparound allowance (usually about ⅜ inch), and the front and back are the size of the covers. The spine is usually about ⅜ inch, but this should be confirmed with the printer and will vary slightly depending on the paper used to print the book and the thickness of the binder's boards used. A bleed allowance of an additional ⅛ inch should be added around the outside dimensions of the dust jacket for trimming.

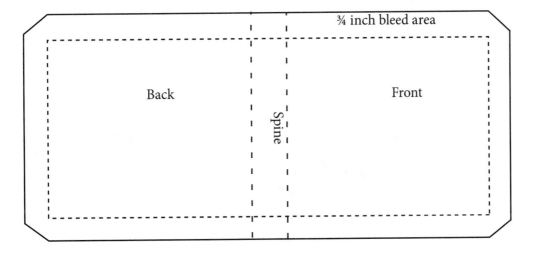

Plan for a lithowrap cover. The lithowrap uses the same art as the dust jacket, but is adjusted to allow for the large bleed. Corners are clipped to reduce bulk where the paper is folded around the corners of the binder's board. The end papers cover the edges of the wrap sheet, leaving about ⅛ inch revealed on the inside.

The lithowrap is the same dimension as the front/spine/back (without the flaps and flap wrap allowance) but with a ¾ inch *wrap allowance* added around *all four edges*. I usually work up the dust jacket, then duplicate the file and change the paper size to reflect the larger lithowrap, then adjust the artwork to account for the wrap area.

COLOR PRINTING. This is one of the most difficult aspects of a children's picture book. Since the issues of color printing also apply to almost all book covers or dust jackets and to adult books printed in full color, we discuss this topic in the next chapter.

- 13 -

Color Printing

This is one of the most difficult aspects of preparing artwork for printing. If an illustrator has created art on the computer—and when you scan hard-copy transparencies, photographs, or other artwork—the scans will be in the RGB (Red-Green-Blue) color space. Likewise, photographs taken with a digital camera are also in the RGB color space. (Don't run away—I'll make this clear in a moment!). The RGB color space means that the colors are represented by values of red, green, and blue that are mixed to generate one of the millions of colors that a computer can display. (The human eye is able to discern more colors than can be displayed on a computer or that can be printed on a piece of paper. However, human vision is a combination of the physical sensor of the eye plus the processing of the brain. We have an amazing capacity to interpret colors and even visualize colors that aren't present.) The RGB system of creating colors is quite similar to how our eyes physically react to colors. One caution—as we age, our perception of the intensity of colors tends to fade. As a result, older individuals may perceive rather saturated, brilliant colors as being rather less intense than they actually are. (That also accounts for a tendency of older women to over-apply facial make up.) If you are in late middle-age or older, try to obtain opinions of younger people with good color vision when making final decisions on colors.

Printing depends on light reflecting off a sheet of paper with inks used to create colors. Printing uses the CYMK color space (Cyan [a blue-green], Yellow, Magenta, and blacK). These four inks are used to approximate the colors of nature and are called the 4-color process method of printing color—sometimes simply called "process color."

One might hope that the color spaces would correspond exactly. If they did, life would be so very much easier for designers who work with colors every day! Unfortunately, each color space—and each device that displays/prints color—has a range of colors that they are capable of displaying/printing. This range is called the color gamut (or gamut for short).

In the diagram below, the largest circle corresponds to the visible spectrum. The human eye has receptors sensitive to red, green, and blue light. All visible colors are combinations of these three primary colors. As you can see, the RGB color gamut (what a computer monitor can display) is considerably smaller than what the human eye can perceive. A monitor uses electrons to stimulate red, green, and blue phosphors. Due to the physics involved, the CMYK color gamut used in printing is even smaller yet, somewhat different—and notably lacking—in the greens and blues (particularly purples) when compared with the RGB gamut. CMYK printing uses the subtractive primaries Cyan, Magenta, and Yellow (plus blacK for shades) to display the available range of colors.

What this means, is when you open a file in Adobe Photoshop (or similar, professional image editing program) and convert the colors from RGB to CMYK you'll often see a color shift—the on-screen image will appear darker, duller, and less vibrant. What happened is that the software has made relative adjustments to all the colors in the image with the "best intentions" of keeping each and every color separate from the others. If you're inexperienced in making these color-space conversions, you may end up quite frustrated as you manipulate various brightness, contrast, saturation, and hue settings to recover the pleasing colors you may have liked in the original image. To avoid this situation, it is best if you let your illustrator know that you will only accept a CMYK version of the illustration files (if drawn directly on the computer). If scanning original drawings/art works, you should fully calibrate the scanner/computer before working with the illustrations. Ultimately, you may wish to have the color conversions done by an experienced graphic artist or by the printer's prepress staff.

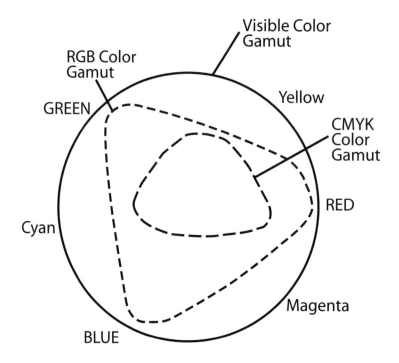

- 14 -

Scanning

While many books are all text or text and tables, some books can benefit from inclusion of photos and other illustrative artwork. In this chapter I will address the steps to prepare photos and line art for reproduction inside your book. However, this is not intended to explain every nuance, nor is it intended to give you an instant education in all the subtleties of Photoshop or how to prepare color photos. Those topics are beyond the scope of this book. See the resource section for additional references for books on scanning and preparing artwork for publication.

Equipment

There's a saying from the early days of computers, "Garbage in, garbage out." That truism also applies to the quality of photo reproduction. First you need a good photo. Next, you need to have a scanner capable of acceptable quality. Then you need to process the scan appropriately to maximize the quality of the reproduction.

We'll assume that you have a reasonably good photographic image—at least one that is in focus and close to being properly exposed. It doesn't matter if it's color or black and white.

But, what of the scanner? Over the years, the cost of "decent' scanners has come down by an amazing amount. These days, scanners fall into three broad categories. High-end professional models—prices range up (way up) from about $1,000. At the lower end of this range are scanners suitable for professional level work in moderate to high volume. Those at the top of the range (upwards of $30,000) can handle very large original and very high volumes of work with exacting quality. The middle of the range, roughly priced between $250 and $1,000 are the solid quality scanners that are used by professionals for less critical work and by more sophisticated consumers. The bottom tier are scanners that cost less than $200 and even as little as $50. These are the mass market, consumer models, not really suitable for professional level work, but certainly adequate to scan snapshots of the kids for their grandparents. Frequently these low-end scanners are packaged with a computer and a printer at a special price—and the package may prove to be an excellent value for the home computerist (if all the pieces work together).

So, what makes quality in a scanner? First, it isn't the claimed maximum optical scan resolution. Nearly all scanners on the market claim a resolution of 1600 x 3200 or more. While a scanner that can sample an original at such a high rate is "better" than one that can only resolve 600 x 600, the sampling rate isn't a key factor in quality. The real quality comes from the optical system that gets the image from the original to the CCD sensor and the smoothness of movement by the mechanical devices that move the optics inside the scanner. These are the factors that set the high-end and mid-level scanners apart from the cheap scanners at the low-end.

Both the high-end and mid-level scanners have excellent optics. A low-end scanner may have cheaply manufactured lenses that may not have the best alignment. So when you scan, a less-than-optimal image is transferred into that 3200 x 1600 sensor.

What about those non-square scan resolutions? Another marketing hype. In reality, the driver software has to convert the scan into square pixels before you can view or edit your image on the computer. So, if you set a 3200 x 1600 scanner at anything above 1600 x 1600, you're getting data that's been partially estimated by the driver software. Same goes for those 9600 x 9600 (and higher) "interpolated" resolutions that scanner marketers love to throw out. In reality, "interpolated" simply means "guessed."

So, what scanners should you consider? There are many priced from $200 up. Most will do just fine. Personally, over the years, I've had good service from several Epson scanners—currently I have the Epson Expression 1680. However, there are several Epson units priced much nearer to $200 to consider—I'd certainly recommend one of them. Also, some Canon models have been given good reviews. There are always new models being introduced to the market—search the Internet for reviews (computer magazines often do a "scanner roundup" every year or two) and see what models are given good ratings.

Software

This one's easy. The program that sets the standard for photo editing and manipulation is Adobe Photoshop. It has the power and features necessary to do practically anything to a photo or scanned image. Most consumer-level programs may do the job for you as well. For example, Photoshop Elements (that comes with many scanners) is modestly priced at $99 (list price) and has most of the editing features of Photoshop CS, the "pro" version of the program. For professional designers, Elements is crippled by not being able to work in the CMYK color space—a requirement if you're preparing color artwork for printing. (This means you'll need to find a copy of Photoshop or a similar high-end program for the book cover image.) However, I believe that Elements will allow you to convert your photos to grayscale, which is required for printing inside a book. Other consumer-level programs will also work for processing photos, as long as you can convert your image to grayscale. Since nearly all the programs have similar adjustment features, the following steps (made with Photoshop) will give you the basics on adjusting your image for printing.

Types of images

Images fall into one of two main categories (and a subcategory): continuous tone (often called contone) and line art (sometimes called bitmap). A continuous tone image is most often a fancy name for a photograph. Line art images are those that only show black or white, such as a pen and ink drawing or a simple line-drawing. So long as the color values are either zero percent or one hundred percent, then it is a line drawing (so far as printing is concerned). You have to be careful, however, as pencil drawings often reproduce better as grayscale—pencil marks rarely reach 100% black and mostly fall into in-between values. Indeed, a pencil drawing with shading may have a wide variety of tonal values so that treating it more like a photograph actually begins to make good sense. The remaining subcategory is previously printed images. If you look closely (using a magnifying glass) you can see if the image is made up of solid black lines (then it's line art) or if it's made up of small dots. If you see dots, then you're looking at a previously screened original—technically, this is considered line art, but the reality of working with previously scanned art is usually not so cut and dried.

LINE ART. When scanning a line art image, you want to maximize the contrast. Set your scanner to an appropriate dots per inch setting, normally to result in a scan of 1000 to 1200 dpi at the size of reproduction (we'll talk about this more later). Select the bit map or line art mode, then make the preview scan. Looking at the preview, you should adjust the scanner brightness and contrast so that the image looks as good as possible. One hint is to try to position the image in the scanner as straight as possible as a slight angle can cause annoying variations in line thickness. If the image is small, but difficult to align, then try setting it at about a 45° angle for the scan (you can straighten it in the image editing program). Set at such an angle, you're more likely to have less problems with fringing along the edges of straight lines, if most of them are vertical or horizontal.

Once your scan is made, open the resulting image file in your image editor (often the scanner driver is designed to work directly with an image editing program). Then inspect the image carefully (zoom in) and delete any unwanted spots. Sometimes, you may need to carefully draw over some of the lines if they are too thin or became broken. Mostly, it's a matter of ensuring that there are no unwanted splotches and that the remaining lines look right. Usually line art, if it is a good original, will scan quite nicely. When finished editing, save the image as a TIFF. Then import the image into your page layout application and position it as you wish.

CONTINUOUS TONE. Photographs always need adjustment to ensure that the tonal values are reproduced satisfactorily. If you simply take a photo, scan it, and place it in your page layout document, it will probably look dark and muddy—and you'll be quite disappointed with the outcome. Here, we'll take you through a few steps to try to make the photos look better.

Set up your scanner for 300 dpi (for the image at its final size) and 24 bit color. (I prefer to change the colorspace to grayscale in Photoshop, but you may prefer to set the scanner to scan in grayscale). Then perform the preview scan. Using the scanner controls, try to get an even range of tones, including the nuances of detail in shadow areas without blowing out all detail in the brightest areas. The scanner driver may have a mode that gives you access to many adjustment features—or it may not, depending on your scanner—so do your best. Once you're satisfied with the scanner adjustments, make your scan and open it in your image editing program.

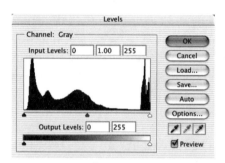

Here's a photo of Thomas Jefferson's home at Monticello as it was initially scanned. The image is less than exciting due to back lighting leaving the façade in deep shadow.

The levels dialog in Photoshop for this image is shown to its left. The histogram (graph) shows the frequency distribution of pixels over the range of tonal values from 0 to 255. Note that the frequency is biased heavily toward the darker shades with spikes in the light tones (the sky and roof structure).

This may not be the case for every photo, but our first step is to use Photoshop's shadows/highlights feature to open up the shadows. We might have been able to make this adjustment with the scanner driver if we had realized the problem before making the scan (or, if noticed at the time of the photo, opening the lens aperture by a stop or two).

Here's the histogram showing the new tonal distribution. Now it's more "normal" but the photograph still does not print as well as it could because we have not compensated for dot gain.

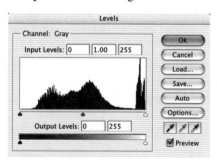

Ah, what is dot gain? When printing on an offset press, ink is placed on the paper under considerable pressure that "squishes" the ink onto and into the paper. With un-coated paper, the ink is also absorbed into the fibers and seeps into un-inked areas. This seeping isn't particularly noticeable with type because all the type has about the same

amount of spread. Photos, however, are another matter. As the ink spreads, it makes the halftone dots larger than they ought to be. This tends to block up shadows and shift the mid-tones. The usual result is a somewhat dark, muddy image. You'll need to get spe-cific dot gain data from your printer. Dot gain amounts vary with the press and paper used, so the required ad-justments will vary with your choice of printer and paper.

With this third image, we've ac-counted for dot gain. Using a "clouds" filter, we also enhanced the sky to make a more interesting background. In addition, we sharpened the photo very moderately. If I've done everything correctly, then this should look the best the photo possibly can under the circumstances.

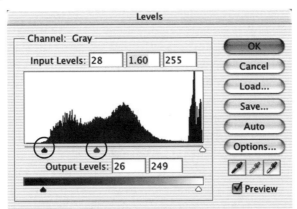

Looking at the levels dialog, here's what we've done. First we move the left slider under the histogram to the beginning of the data. This maximizes the tonal range of the photo. Next, to compensate for the mid-tone dot gain, we move the middle slider to the left—with the preview box selected, you'll see the photo appear to become lighter overall, but the shift is primarily in the mid tones. Note the center box at the top has changed to 1.60 from the normal 1.00 as shown in the earlier snapshots of this dialog. You should see that 50% gray in the photo has moved to about 40% (move your cursor over the photo and look at the "info" box).

Our printer has stated that the maximum desirable dot size should be 90% and that the minimum highlight dot should be 2%. Remember, the dialog shows the tonal range on a scale from 0 to 255, so the 10% lighter shadow dot is about 26 on the scale. The highlight dot at 2% is a reduction of about 6 from the maximum 255, or 249. As you can see, in the output area, we've entered 26 and 249 (or you can move the sliders accordingly).

Adjustments with curves

These same dot gain adjustments can be made with the curves dialog. After making all adjustments in levels to make the photo properly balanced with a full tonal range, we open the curves dialog and move 3 points on the graph. The graph starts as a straight,

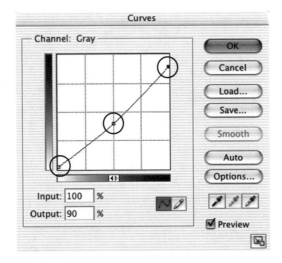

diagonal line. At the middle, we select the center spot and pull it down from 50% to 40% or so, depending on instructions from your printer. Next we move the left (white) end straight up by 2% which displays in the output box. Finally, we move the right (black) end straight down by 10% as shown in the output box (above).

Both the levels and curves methods generate the same result. However, you should first balance the tonal range of the image with the levels dialog before using curves to adjust the output. The tonal range of the image should be spread across the full range of available tones so that good separation of each tone is possible. The degree of this adjustment is, however, a matter of judgment and varies with each photo, so I can't give you any specific guidelines.

Digital cameras

The adjustment process is no different with a digital camera. The image is captured with a sensor in a camera rather than a sensor in a desktop scanning device. The main difference is that your camera photos are usually in color, so you'll need to change them to grayscale as a first step.

The grayscale conversion can be quite simple. Open the image in Photoshop (or a similar image editing program) and select **image>mode>grayscale.** Usually, this will give a satisfactory result (perhaps with some brightness/contrast adjustments). Other times you might get a better result with **image>adjustments>desaturate** followed by the mode change to grayscale. If the direct conversion doesn't provide a satisfactory result, then try the desaturate route. (First, using "undo" to revert to the color version, if necessary.) If the conversion with these methods isn't particularly good, you can spend some time with the levels or curves dialogs to try to improve the image.

In another technique, convert the image mode to Lab Color, delete the "a" and "b" channels, then convert to grayscale. In Lab Color, the "L" channel carries all the luminosity data in the photo. It's different from the simple grayscale conversion, so sometimes it may give a better outcome. (This assumes you're using Photoshop or another advanced image editing program. Lab color may not be available in the consumer-oriented products.)

With the last technique, we introduced channels. It's also possible to look at the R, G, and B channels individually. Sometimes selecting one of them (and deleting the other two) gives better results than the other conversion techniques. To avoid confusion since Photoshop changes the channel names when you delete one, I like to make a copy of the channel that gives the best grayscale result, then delete all the other channels. (Making a copy of the channel will give it a name of "copy of green channel"—or red or blue as the case may be.)

If you have originals that are dominated by a particular color, you may find that the individual channels that carry that particular color may have more definition of the elements than you can get with the straight conversion (that actually does a balancing between the values of all the color channels). For example, I was working with a photo of a room that was heavily saturated in red. The walls were red and the lights in the room were red. (This was an artistic display.) A conversion through the mode menu resulted in an image that was quite dark with little contrast. Using the red channel, we were able to get very good definition of the furniture and fixtures in the room. With more typical photos, the green channel often gives excellent results. Basically, don't be afraid to experiment a bit in finding the best grayscale conversion approach for each photo. Don't worry, 90% of the photos you're likely to work with will convert just fine with the normal approach—but the other 10% can benefit from some experimentation.

Sizing artwork

Earlier, I indicated that photographic artwork should be saved at 300 dpi *at the size of reproduction* and line art at 1,000 to 1,200 dpi *at the size of reproduction*. So, where did we get these dpi settings? What does this size of reproduction mean?

First, line art is sized to reproduce with 1,000 or 1,200 dpi because the acuity of human vision, though particularly good at seeing edges, can't resolve below $1/1000$ inch. Also, many imagesetters operate at 1,200 dpi, so matching the scan to the output resolution is a safe way to obtain good results. Indeed, whenever line art is involved, it's a reasonable practice to scan at the same resolution of the highest resolution output device likely to be used. Remember, file size isn't a problem because line art is (almost always) saved in bitmap mode, which generates much smaller files than grayscale mode.

The 300 dpi for contone art comes from a rule of thumb that the dots per inch scan rate of a photo should be about twice the lines per inch of the screen used to print. Most books printed on uncoated paper on an offset press use screens of 133 lpi; on coated paper 150 lpi is often used. So twice 133 is 266, rounded up to the next common scanner setting, usually is 300 dpi. The "extra" simply gives us some leeway in sizing the photo, especially if we need to tweak the size upward a little. Of course, it also covers use of common line screens up to 150 lpi or more.

The two-times-the-line-screen rule of thumb also has some leeway built in. In reality, there is usually sufficient data present in a typical photo with a factor of dpi to lpi of 1.5 to 1.7. That means we could potentially get away with a scan as low as about 200 dpi and still have an acceptable photo. However, (this is the zinger), not all photos will reproduce cleanly at such low resolutions. Therefore, to reduce the risk of a bad outcome, we simply use 300 dpi for contone artwork—that gives us leeway for moderate re-sizing and for screens with a higher than expected lines per inch value. For example, the selected printer may have upgraded to waterless printing, which often allows much higher line screen values. Or, we may choose to print on coated paper (with a 150 lpi screen) and then will need the extra resolution to ensure an even tonal range.

The photos of Monticello on pages 180–181 are 2.16 x 2.88 inches. This image originally measured approximately 5¾ x 7⅔ inches at 300 dpi. If I had simply placed the image into my page layout document and dragged the corner in to re-size it, the image would have ended up at 800 dpi—at the size of reproduction. Instead, I used Photoshop to re-size and reduce the resolution to reach the size I wanted to print at. While I could have just let the software take care of it, I would have needed more space on my computer, more space on the disc used to send the document to the printer, and the job would take longer to output on the imagesetter as it was processed. This last element could have a significant financial impact, as most printers will charge extra if "excess time" is taken while imaging a job.

As a practical matter, the document will be output as an Acrobat PDF before it is sent to the printer. Acrobat will "downsample" the images according to the job options you select—and that will most likely save you from excess time charges. However it's always better to make resolution changes on your own so you can knowingly make the compromises and choices necessary for optimum results, rather than taking on faith that the software will do the job you will like when the printing is done.

If you know in advance the size you want the image to be (a ruler can help here), you can adjust the scan resolution accordingly. Going back to the original of Monticello at 5¾ x 7⅔ inches, we could scan it at about 115 dpi to get to 2.16 x 2.88 inches at about 300 dpi. Taking the short edge measure and dividing the original 5.75 inches by the reproduction of 2.16 inches results in a 2.66 reduction factor. I chose the short side (the height) as that was the critical measure I used to size the image for the page. (That is, the width was not an issue with this photo.) Then I divided the desired 300 dpi by 2.66 to arrive at a scan rate (rounded) of about 115 dpi. (Now, my scanner would not allow me to select 115 dpi—so I would select the next highest scan setting, 150 dpi.)

With this example, we actually had an excess of data and had to discard well more than half of what was available to get down to the reproduction size. What about going the other way? In this situation, we need to be careful that we can scan the original at a high enough resolution so that the image will reproduce properly. Take, for example, a 35 mm slide (called a transparency by printers and graphic artists). Slides (35 mm) are one by one and a half inches. To reproduce a slide at 2.16 inches in height gives a factor of 2.16 (2.16 divided by 1). So, when we're enlarging the original, we multiply the scan rate by the factor; thus 300 x 2.16 = 648. My scanner allows either 600 dpi or 720 dpi. Since a little more resolution is better than too little, we'll choose 720 dpi. This also gives us some leeway to moderately crop the edges of the image to get rid of the rounded corners and bits of dust, debris, and fibers that usually appear at the edge of a slide.

Graphic arts supply stores sell "reduction wheels"—a circular slide rule that simplifies making enlargement/reduction calculations. You note the size of the original and the size of the area for the image. Then you find the measurement on the outer ring scale of the wheel corresponding to the original and then move the inner ring scale so that the copy size measurement aligns. Next you read the enlargement/reduction factor from a window. Remember to multiply the factor for an enlargement and to divide by the factor for a reduction.

Personally, I haven't used a reduction wheel in some time. I find it easier to use the box drawing tool in my page layout program to find the size of the reproduction area. Then (after cropping any extraneous edges) I open the "image size" dialog in Photoshop. After deselecting the "resample" checkbox (leave the "constrain proportions" box selected), I enter the desired size measurement into the appropriate document size field. (In a moment, the other fields will update with the corresponding information.) I can then determine if I have selected the critical dimension. (Either the height or width will be critical, depending on the layout and space available for the image. The second dimension will usually have some "play" allowing you to compensate if the image does not directly enlarge or reduce to the size wanted.) If the calculation looks satisfactory, you can select the resample checkbox and enter 300 in the dpi field. Click **OK** and the image will be modified as you need. (Hint: work from copies of the original—at least until you're comfortable with this technique.) Remember, enlarging a smaller image that is at a low resolution will not be improved by just changing the number in the dpi field. Photoshop (and other image editing programs) can do a good job, but when the data isn't present to support a higher resolution, then the outcome will be poor—and probably unacceptable.

Moiré patterns

When scanning previously printed material, for example a photo that was printed in a magazine or newspaper, interference patterns appear between the alignment of the dots in the screened original and the raster of sampling points made by the scanner. This interference pattern is called a moiré pattern. These can be almost imperceptible or they can be extremely distracting.

Many scanners (or their drivers) have a "remove moiré" or "descreening" feature that may work well or not at all. Usually it's worth trying the feature as a first attempt to rid a scan of the pattern. Setting the artwork at a slight angle (to be straightened in Photoshop) sometimes calms a moiré—and sometimes makes it much worse. (If you're getting the idea that these patterns are unpredictable and difficult to control, you're right!)

Another technique is to scan the image at 1,000 or 1,200 dpi in an attempt to "see" the individual dots—treating the image more as line art rather than a photograph. Again, this works some of the time, but not all the time. In part, it depends on the quality and screen value of the original and it depends on the ability of the scanner to resolve the individual dots. This is where low-end scanners demonstrate why they are so inexpensive. Printers often own "dot-to-dot" scanners, specifically designed to prepare artwork from prescreened originals as line art. Naturally, such scanners are very expensive.

Sometimes taking the high-resolution scan (above 1,000 dpi), then downsampling the image to a much lower resolution (300 dpi) will cause enough averaging of the data to tame the pattern. Once the image is scanned, careful use of the blur filter and possibly the dust and scratches filter may also bring a moiré pattern under control at the expense of making the photo rather blurry. Reduction of the reproduction size may bring the dots so close together that they will be treated as continuous tone data rather than as dots.

The last resort is to photograph the image with a regular camera. Sometimes this will soften the dots in the original sufficiently that the moiré patter will be tamed. Then, again, it may not work either.

Ultimately, the elimination of a moiré pattern may simply require considerable time and much patience working in Photoshop to selectively blur, average, and smear the dots together to eliminate their presence. Usually, before undertaking these extreme measures, you may want to consider the value of the image to the document. It may be that the presence of the image is simply not justified by the work involved to make it usable.

- 15 -

Hiring a Designer

I F YOU'VE MANAGED TO GET THIS FAR IN THE BOOK, you've either decided that you can do the design and layout yourself—or not. In either case, you have (I hope) learned enough about the process that you can converse with potential designers, printers, and other publishing professionals with confidence. Here we will go over the factors you should consider when you are seeking a designer to assist you with the cover design or the design, layout and typesetting of the interior.

First, just to be clear, there's a huge difference between having a concept or idea for a design and actually generating artwork that makes that design a reality. Even if you are heavily involved in decision-making with a design, you are truly dependent on the skills of the designer to take your ideas, comments, and directions to then turn it into a final, professionally produced piece of artwork suitable for reproduction. What I'm trying to say is, don't "claim" to be the designer if your contribution was the idea—even a fairly well developed idea. (Some years ago, I overheard a client say to one of my employees that "it was sure a nice cover that (the client) designed"—knowing full well that the designer spent many hours selecting appropriate photos, drawing connecting elements, and fine tuning the rendering of the design. Of course, the client had told us his ideas, and even provided some simple sketches of what he had in mind—but it was the talent, skill, and artistic vision of the designer that made the cover a reality. (I guess you might say that I was irked by the remark.) If you must, claim credit for the concept or say that you "art directed" the project—but give the designer due credit.

Cover design

The key skill in cover design is marketing sense with a good artistic vision. The designer may or may not have skills in photography or illustration—that's not much of an issue as illustrations or photography can be commissioned from an illustrator or photographer much better qualified than any cover designer. Indeed, most artwork appearing on a book cover is from the wide variety of clip art or stock photography that is available.

Clip art is pre-drawn illustration, often sold in collections of many similar styled drawings. Stock photography consists of photographs, taken by professionals, that may have been produced "on spec" (speculation) or as part of a personal or professional project that are, for any variety of reasons, available for sale.

Stock photography and clip art are available in a wide variety of offerings. At the high-cost end are "rights managed" items. Primarily photographs, these are fairly expensive, will be offered with the need to pay an initial fee and (probably) royalties or some limit of the quantity of copies reproduced without further payment. An example is a photo of a Rembrandt painting I used on a cover of the Farmer Street Press version of *What's Wrong With Dorfman?* The painting, entitled "The Anatomy Lesson of Dr. Tulp," was held in a museum in The Netherlands. Only photos offered through a stock photo agency were available for reproduction. Since the museum strictly controlled access to the paintings, only authorized photographers and photos were available. (That is, if a photo was used, it undoubtedly came from a copyrighted source that ultimately was licensed by the museum.) After negotiation, we paid $400 for non-exclusive rights to use the cover of the book up to a cumulative press run of 4,000. While the image is readily available on the Internet, the reproduction rights (even if you could get an image of sufficient resolution) are uncertain at best.

More commonly used by book designers are collections and individual photos of "royalty free" photography and art. These "better" ("best" being whatever you're actully willing to use) collections might charge $400 for a CD with 50 or 100 photos (usually scanned with top quality equipment by skilled operators) to more modestly priced "super collections" of hundreds or thousands of photos in multiple disc (CDs or DVDs) collections. These less expensive collections are often quite satisfactory. Although there is a risk of choosing a photo that may be used (or overused) in a variety of locations, it's rarely a problem as cover photos are most likely to be less "mainstream" than photos commonly used in advertisements or brochures. Indeed, a photo found in a collection of "fashion" photos became the basis of the cover of Diane Klein's, *In The Name of Help.* The photo (right) features a head shot of a young woman with frost in her hair, eyebrows, eyelashes, and on her face, as if frozen. It struck me as an ideal photo to visualize the psychological isolation which was an essential aspect of the main character in the  book. This photo was in a modestly priced stock art collection purchased as a "close out" from a dealer of graphic arts software and related materials.

An important aspect is that the photo collection offered rather broad reproduction rights, allowing the photos to be used for commercial purposes or in a product made for sale. Be sure to check the specific license terms of clip art or stock photography used in your project. Some contracts allow items to be used, for example, in company newsletters, but not in items for sale (such as a book). The other common restriction is that such "royalty free" artwork may not be "the major component of value" in an item for resale. So, a book cover may be okay (even though people do buy books based on their cover) as, presumably, the major component of value in a book is the content. However, a T-shirt wouldn't pass the test—T-shirts with photos or other artwork are sold primarily for the presence of the image, as T-shirts are readily available at lower cost with no artwork.

Interior Design

The key skill of a book-interior designer is the art and craft of typography. In many respects, this skill is very different in character than the skills of a graphic artist. Indeed, book design is usually an elective in typical graphic arts courses of instruction—or, worse, it's relegated to a 'unit' of a larger course. Typography is taught in graphic art/design schools, but often from the advertising/artistic viewpoint.

The main thrust of many graphic design schools is to produce "agency" designers. Admittedly, the large advertising agencies offer the greatest potential for high earnings. (Overall, graphic artist median salaries are about $30,000 annually, but top designers at a major agency might earn in excess of $100,000.) As a result, most graphic design education is focused on developing those seeking the highest paying positions.

So, what does this mean for someone who wants a book designed? Fundamentally, it means that 90% of the designers don't have much knowledge or any experience in designing books.

Rule number one: look for a designer with a talent for typography and who has experience designing and typesetting books. Your brother-in-law's nephew who just graduated from a graphic design course (even from an excellent school) may not be a good choice. Pay the kid to design some flyers or posters to promote the book after you've worked with an experienced professional.

Rule number two: see rule number one. All the sad stories I've heard usually boil down to a basic failure to find a designer with relevant experience doing similar projects. Book interiors are large, time-consuming projects. It takes a well organized designer to efficiently typeset the project and keep track of the requirements and details that make a successful book. In my experience, I've found that there is a relatively small contingent of designers who focus almost exclusively on books and related projects. Look for one of them. (Forget about your sister-in-law's daughter who just graduated from design school. No matter how little they charge, it will be no bargain.)

Interviewing the Designers

As you work on your project, you will probably come up with ideas on how you'd like it presented. Try to find examples. Make copies of pages you like—see if the typeface is listed, if you find one that attracts your eye. Gather photos and other concept elements. The designer will probably want to create something you like. (There are always designers like the one who worked [briefly] for me. One day she said that she thought she might take an "assertiveness training" class so that she could "get the clients to like her designs better." I suggested that she might be better off taking a "listening" class, so she could better appreciate what the clients were telling her rather than trying to force her designs on them.) As you discuss your project with the designer, take note of how interested they are in your ideas about the project. As they describe any concepts they may have, evaluate them with respect to your vision.

Keep in mind that designers (and illustrators, and photographers) have a style (or a range of styles) that they're comfortable working with. Do not seek something from the designer that is significantly outside the style they've established. Look carefully at their portfolio (whether in person or on their web site) and consider how the style of their work will conform with your vision of your project.

Questions to Ask

QUESTION: *Have you produced books in the past? Not brochures, not catalogs, not posters, not newsletters, but* **books?** *How many?* You don't want a beginning designer. It may seem fair to ask, how does a designer become experienced if nobody hires the inexperienced? First, a designer may work under the supervision of a more experienced designer or work as an intern with a book designer during their formal training. Second, your book project may be the third most expensive single investment you make in your life. (A home and an automobile are probably first and second.) So, considering the magnitude of your investment, don't you deserve the best? (Would you go to a surgeon for a heart bypass if you knew it was to be the first time they did the operation?)

QUESTION: *May I see two or three samples (for interior design) of the actual books or a reasonable selection of pages from them?* To be considered any further, they must give you a positive response. While I no longer send out sample books (they *never* get returned even when I've provided shipping labels and postage), I will arrange Acrobat PDF files that potential clients can download to see a representative sample of pages from books I've done. (Unless, of course, we can meet in person.)

In looking for a cover designer, you should view many samples, either at their web site or of images (or prints) they send you. I would have reservations about selecting a designer who could only show one or two samples—unless they're absolutely stunning! Again, the style of the cover and interior should be in a style compatible with your vision.

QUESTION: *What software do you use?* The correct answers are Adobe InDesign, Quark XPress, and (half-correct) Adobe PageMaker. Most book printers can work with the native file formats in these programs—a plus, although use of Acrobat PDF files in electronic workflows have reduced the necessity. There are a few other programs capable of professional-level work (such as FrameMaker and TEX), but the first two are used by more than 95% of book designers. (Professional tools are discussed in some detail in chapter 7, so I won't repeat the discussion here.) There may be special cases where other choices are acceptable, but a qualified book designer certainly shouldn't be working with a consumer-level program that's more appropriate for simple church bulletins.

The related question is *"Can you work with files from (name of your word processing program)?"* Obviously, if the designer can't work with your files, it's going to be a very frustrating project. Most experienced book designers work with files from a variety of sources and have no problems with those created in more broadly used applications. Fortunately, most word processors can save in a format that is easily translated by other programs (RTF or rich text format).

Don't worry if your designer works on a Macintosh if you work on a Windows computer. The designer is probably experienced with making transfers between the platforms, and Macintoshes have been very cross-platform compatible for many years.

QUESTION: *Will you show me sample pages before the project is really under way?* Again, the answer has to be yes. Whenever I start a new project, I give my client sample pages of my initial design and get approval before proceeding. Even after careful conversation and looking at examples from other projects, there's nothing like seeing the actual pages to understand the impact of the decisions you've made. In many cases I may make up two or three sets of sample pages with slight variations in style. The client may say, "I like the heads in sample one and the body type in sample three." Now, *that* narrows things down for me. I would hope that most book designers will have similar practices.

QUESTION: *How many sets of proofs will I see?* While different designers will have different policies, in most cases you should see two full proofs and followup pages for the handful of corrections (called "correx" in the business) that escape the earlier proofs. You will probably be billed a modest amount for additional full proofs provided.

QUESTION: *How do you charge for changes I make?* When I managed a typesetting service that worked with major publishers, "Author's Alterations" (AAs) and "Editor's Alterations" (EAs) were an important profit center for our service. Frankly, we would usually just about break even on the basic typesetting charges—all the profit came from billing for the changes. "Printer's Errors" (PEs), the term applied to typesetting errors, were not charged, but they rarely amounted to more than a small fraction of the total number of changes.

These days, the counting and charging "by the error" is usually too time consuming—and trusting non-professional proofers to properly mark AA, EA or PE by each change is unrealistic. Many designers will charge for the time it takes to input corrections with some "free" allowance (to account for PEs). Personally, I often waive a correction charge if the changes are "reasonable"—perhaps 30 pages with changes in a 300 page book. Beyond that, an hourly charge kicks in for AA/EA corrections.

So, minimize the corrections by making sure that the manuscript is as clean as possible before submitting it to the designer/typesetter. After the manuscript is returned from your editor, proof it on screen. Spell-check it—for whatever that's worth. Then print it out and proof it again. Have different people, who are unfamiliar with the material, do the proofing. Consider hiring a proofer—at this stage a person with good secretarial skills may be sufficient as technical typesetting issues aren't an issue with the raw manuscript.

QUESTION: *How will the book be submitted to the printer?* These days the answer should (almost universally) be: "I will create an Acrobat PDF and send the file electronically or via a courier service (as required by the printer)." Some projects may be sent via the Internet, but most frequently, the printer wants a proof, printed single side, to compare to the typeset file. So, to keep the whole project together, the proof and a disc go into a package and are sent via an overnight courier service who offers tracking and delivery confirmation. Usually, by the time the project reaches this stage, it's desirable to not delay by using ground service or other less reliable shipping methods. The designer will usually pass on the cost of the courier service to you.

QUESTION: *How long will it take you to turn the project around?* You should be able to go over the schedule with a commitment (on your part) that you will deliver the finished manuscript on a particular date—so the question may be phrased as "If I give you the finished manuscript on (insert date), when can I see sample pages and then the first proof?" You may also wish to state that you wish to get the project to the printer by a certain date. At this point, the assumption is that you will be able to promptly correct and return proofs you receive.

The answer can be a bit complex. Usually you should see sample pages within a week or two from the time the designer receives the files. A first proof should often be available within three to four weeks. An overall schedule should run two or possibly three months under normal production (non-rush) circumstances. In practice, when the client receives the first proof, many times it's like it fell into a black hole! If you keep a proof for more than a week, you can expect that the schedule will be extended—at the least by the time you hold the proof, but possibly longer. I try to "turn" proofs back to clients forty-eight hours after they arrive—but I (and most designers) normally have several projects ongoing at any one time. I have a policy of "first in, first out"—unless a quick look at a manuscript shows (1) a remarkably small number of corrections (fewer than ten) or (2) a remarkably large number of corrections (several on every page). In the first case, I might turn around such a proof very quickly as there's little for me to do and I can get it out of the way. In the second case, I know I'll be tied up with fixing corrections for some time, so I'll send it to the bottom of the stack until I can clear out the other work. (Surely this might be an incentive to take the advice to make sure the original manuscript is as clean as possible!) Ultimately, most books can make it through the production cycle in two to three months, assuming no extreme delays to the proof while in the author's hands. In practice, authors tend not to return proofs as promptly as they expect—so, many projects take three to six months in the production cycle before going to the printer. (Keep in mind that the large publishers frequently take two years from acceptance of an unedited manuscript until a printed book appears.) Just remember, the designer isn't responsible for delays caused by the author and generally can't "make up" time when the project is delayed outside the designer's control.

QUESTION: *Will you give me an itemized estimate that will tell me what you will do for the basic charges and what are billed as extras? Will you give me a written agreement covering your services?* Designers may be more-or-less business-like. The brain organization that makes a designer creative may work against them in the business detail area. However, it's reasonable to expect that a designer can give you a basic run down of the usual expenses. Covers are usually billed at a "flat" rate—that is, so much for the cover that includes preparation of two or three "concepts" and the rendering of a final version once a concept has been approved. Additional concepts or significant changes to a design once the final version rendering is underway will generate additional charges. The key is to stay at the concept stage (if you don't like any that are offered) rather than "approving" one and moving forward with a design you don't like. Here's where a termination fee is important. How much will it cost if I don't like your cover concepts and want to "pull the plug?" Hopefully, a careful review of the designer's portfolio will keep you from having to go through a change in designers, but it happens and you need to plan for that possibility. The range of charges from competent designers ranges from $600 to $3,000 per cover. Under $600, the designer is likely to be unfamiliar with designing a cover and may not be able to complete a "press ready" file. Over $3,000 and you're probably paying more for the designer's reputation than for the design itself. While the prestige of "important" designers may be appropriate to corporate patrons, it's a rare small publisher who should be paying for reputation. The median charge is about $1,500 or so. Any commissioned art or stock photography will probably incur an extra charge as will courier fees.

An interior designer/typographer will commonly charge by the page. Page rates will vary based on the complexity of the design and typesetting. A simple all-text novel with a 5½ x 8½ trim size might cost as little as $4.00 per page, while a complex book with many photos, multiple columns, and color work might cost $20 or even $30 per page. Shorter works may have a higher page charge or the designer may have a minimum. Expect to pay extra for scans (charged "each" plus an hourly charge for clean up and editing in Photoshop). Commissioned or clip art or stock photography, and courier charges are extra—along with the charges for corrections. Due to the wide variety of material, it's hard to give a typical charge—but consider that the U.S. average hourly charge for graphic design is slightly over $60 per hour and the rates (for both interior and exterior) will reflect that reality.

A designer should be willing to give you a written contract. The contract should contain a rights clause that spells out exactly what copyright interest that the designer might have and how it's to be assigned to the client. A publisher should expect an assignment that carries few restrictions or residual charges. It is fair and reasonable to give credit to the cover and interior designers (usually put on the copyright page). The designer may also reserve a right to use of the materials in their portfolio and for other promotional purposes. (I had one client object to this term—she was concerned that I was claiming a "right" to reproduce her book and sell it in competition with her. All I can say is that it's hard enough for the publisher to sell books—I can't imagine that a designer is "setting up" to compete with the publisher! However, I need a right to reproduce sample pages for other potential clients or to transmit portions electronically as part of my promotional needs.) As with any time a legal document is involved, if you have questions, ask a suitably qualified attorney to review the contract. Usually, when publishing is involved, an "intellectual property" attorney is appropriate.

QUESTION: *Can you give me contact information for references?* Now, here's the exciting part: call or email the references! While any reasonably bright designer will only provide references who are assumed to be willing to give a "good review," a conversation (rather than an email) may gain you information that will make a difference in your selection. Remember, you're evaluating the "get along and work together" aspect of the designer's personality as well as the price, responsiveness, and service quality. While some clients may prefer a designer to "guide" them, other clients may resent a designer who "imposes" ideas on them. It's only through conversations with past clients that you can get some ideas about how easy a designer may be to work with. Be aware that you may need to encourage a reference to clarify if his or her comments are ambiguous.

Be aware that efforts to save money by going outside the book/cover design community or hiring someone close to you (a relative or friend who says they can do the job) is a risky endeavor. In the long run, you're likely to get more headaches, more trouble, and spend more money using these "money saving" choices. (I've generated some good fees reworking books and covers produced by these low cost alternatives. 'Nough said?)

Sources to find designers include contacting local publisher groups similar to the [San Francisco] Bay Area Independent Publishers Association or getting newsletters from the Publishers Marketing Association (PMA) or the Small Publishers Association of North America (SPAN). Web site information for these organizations is in the resources section at the back of this book.

Preparing a manuscript for the designer

This is one of those times where *less is more*. Whenever designers get together, they usually end up sharing "horror stories" about files delivered by clients with too much formatting, hard returns, and forced page breaks. Usually, the author has created all these items while trying to get an idea of what the book might look like when it is finished. This is understandable, but not helpful. If you want to experiment, use a copy of the manuscript file(s) and try some of the formatting suggested in chapter 8.

As for what you give to your book designer, as indicated, a "less is more" approach is desired. Use minimal formatting. Set the body text in any convenient typeface at a comfortable reading size. Use the default margins of your word processing program. Do not insert forced page breaks, section breaks, or extra returns.

Insert one return at the end of each paragraph. If you wish, you may hit two returns at the end of each paragraph to assist you during writing. If you are comfortable with the features of your word processor, it's better to avoid the two returns, and use the first line indent (¼ to ½ inch) to identify paragraphs. If you can be consistent, create a body text paragraph style and apply it to all body text paragraphs—but do not apply the style to any other elements (such as chapter titles, or headings). Input the text without trying any fancy typographic effects. Use italics and bold as appropriate. Do not underline text (use italics or bold). If you have used the "show revisions" feature (that marks deleted text with strike-through), be sure to "accept revisions" before sending the file to your designer.

If your document has tables, use the table feature of your word processor. While

PageMaker does not import tables, both XPress and InDesign handle them well. (If you know that PageMaker is the target program, input the table material with a single tab between each column, even if the default tab settings don't properly align the table material.)

If your material should be indexed, and you know how, use the index tagging feature of your word processor to mark the index words. These tags will carry over into both InDesign and PageMaker. (XPress requires use of a third party Xtension.)

If your manuscript has photos or illustrations, do not import them or create them in the word processing document. Illustrations created in a word processor are usually not of sufficient quality for formal publication and do not import into the page layout program. To make matters worse, photos or imported illustrations are exceedingly difficult to extract from most word processing files. Instead, place notes to the designer in the text to indicate where graphic elements should appear. Please, create a systematic method of marking and identifying the image files, especially if there are more than a few.

Marking suggestion: the first photo to appear in chapter one might have a file name of PIC01_01.TIF; the second PIC 01_02.TIF, and so on. Use the greater than/less than symbol (or any other symbol that does not regularly appear in the text) to enclose notes and tags for the designer. For photos, tables and other elements, put them on a line by themselves like this:
<Please insert photo PIC01_02 here.>

The designer will remove such notes as the files are worked. Using an otherwise unused symbol (such as the <> symbols) will allow a global search to be made to ensure that all such notes have been removed from the final document.

Likewise, if there is special formatting, such as a lengthy quotation that should be given special indenting, use the <> symbols to note the special instructions. Use the coding method of HTML to mark the beginning: <indented> and at the end: </indented> of special formatting.

If you have a complex structure with heads, subheads and sub-subheads, you may need to provide some guidance to enable the designer to understand the structure. If you can use the word processor's style sheet consistently, you may apply appropriate styles to indicate the status of the various heads. A somewhat safer approach is to use the tagging method described above to tag each head. <h1> would be the highest level heading, <h2> the next and so on. Put these tags at the beginning of the line with the head. (No need to use the </h1> convention to mark the end of a head, as it would be assumed to end with the next return.) Symbols for chapter number <cn> and chapter title <ct> can also be used in a similar manner, if it isn't otherwise obvious. Again, the designer will remove these tags as the manuscript is turned into a book, but it will eliminate guessing as to the status of any particular element.

Finally, proofread your work. Then, proofread it again. Have a friend or professional proofreader check the manuscript. The cleaner a manuscript is, ultimately, the less the project will cost to typeset.

Give the designer the word processing file along with a version saved as rich text format (RTF). Most designers will open the word processing file in the word processing program to check for common situations (see "file clean up" at the beginning of chapter 9.) A relatively clean file will be appreciated and potentially eliminate errors. There will, however, always be some small errors that slip through, so a wise designer will always preprocess word processing files with their usual procedure before importing it into the page layout program.

Cross platform issues

It is a fact of life that most designers work in a Macintosh environment (Macintosh has an 80% to 90% share of the design market depending on how it is measured) and nearly everybody else works on Windows computers. Apple Computer has done its best to make the Macintosh compatible with Windows computers. However, due to design choices made in the early history of both the Macintosh and MS-DOS computers, there are differences in the encoding positions of certain characters in the font used on each system. This difference only affects those characters beyond the basic alphabet and numbers. (For the technologically inclined, in the 256 character encoding scheme, the differences occur after the first 128 characters, i.e. those with the "high bit" on.) As a result, there are common problems translating documents between the platforms. The most troublesome characters are "curly" quotes, both double and single, em- and en-dashes, apostrophes, and other non-alphabetic symbols.

Most of the time a Microsoft Word document created under Windows and opened in the Macintosh version of Word will automatically be translated to the correct font. If your designer can accept files in MS-Word format, that's usually the ideal way to avoid some of the font issues.

Files transferred in RTF (rich text format) or other formats not specific to a particular program will usually exhibit the font problems. Most designers are alert to the problem and have devised an appropriate series of search and replace routines (or written macros) to remedy the problem. However, when checking proofs, you should be alert to the cross-platform font issues and watch for systematic errors related to this problem.

Anything else?

At this point, preferences tend to become quite personal to each designer. So, before you do your final proofread, contact your designer and find out if there's any specific tips that you should follow for your project. Good communication always helps to smooth the process of getting from manuscript to finished book.

Glossary

THIS GLOSSARY GOES BEYOND merely listing definitions of words used in the text by also listing many words you'll encounter while going about the tasks of publishing your book. It includes terms used by those involved with computers, design, printing, publishing, photography, typography, and related general industry terms.

10BaseT, 100BaseT, 1000BaseT (networking) A type of wiring often used on Ethernet networks. Why the weird name? Well, it stands for 10 megabits, 100 megabits, or 1000 megabits (or 1 gigabit) per second (the data transmission rate), baseband (just data, no carrier), and twisted-pair wire (as opposed to coaxial cable). It's a medium-performance relatively inexpensive wiring system that's the most popular way to do Ethernet. 100BaseT is the usual choice, these days, for small LANS. 1000BaseT has recently become economical for small LANs, but is not yet in very wide use.

24-bit color (display) Color system in which every pixel has a 24-bit number attached to it describing exactly what color it should display. The color value is calculated with 8 bits each for red, green, and blue. 24 bits can describe just over 16.7 million different colors.

AA (editorial) Abbreviation for authors alterations (usually when added to a typeset proof) See also EA. See correx.

AAP (publishing) Association of American Publishers. Mostly the larger publishers are members of this organization.

ABA (publishing) American Booksellers Association. Trade association for non-chain bookstores in the U.S. Major trade show is BookExpoAmerica (BEA). Regional affiliates hold much smaller trade shows, such as the Northern California Independent Booksellers Association (NCIBA).

abrupt serif (typography) A serif which breaks suddenly from the stem at an angle.

absorption 1. (paper) The property which causes paper to take up liquids or vapors in contact with it. This is one of the ways that ink dries after being transferred onto a sheet of paper. 2. (optics) The partial suppression of light through a transparent or translucent material. This is the basis of the workings of optical filters and how ink pigments are combined to create colors.

accordion fold (bindery) A term used to describe two or more parallel folds which open like an accordion.

acid free paper (paper) Paper made without acidic content. Printing papers made with wood pulp are more naturally acidic. Wood pulp was not widely used for book paper before 1930. Later, it was discovered that such papers would self-destruct after about 50 years, causing consternation in research libraries. See alkaline paper.

Acrobat (software) Adobe software that embodies and supports PDF. See PDF.

active window (interface) The front-most window on the desktop; the window where the next action will take place. An active window's title bar is highlighted.

A/D converter (industry) Analog to digital converter. A device that converts analog signals or data to digital. See digitizer. See Scanner.

additive primaries (color reproduction) Red, green, and blue are the primary additive colors. When equal amounts of light of these colors are combined, they produce the sensation of white light. (The light isn't truly white—true white light has the complete spectrum of colors.) This is the technique used for creating images on color monitors (and televisions).

addendum (publishing) Supplemental material that is added to a book.

adhesive binding (binding) A case binding method where adhesive (glue) is used to attach the book block to the spine and covers. See notch binding, smyth sewn.

adnate serif (typography) A serif which flows smoothly to or from the stem, also known as a bracketed serif.

Adobe Font Metrics Files (typography) AFM files come with many PostScript fonts. These files contain information that describes the dimensions of the font for applications that require the information in this format. The only programs (at one time) that require font information in this format are FrameMaker and Interleaf. (Both are layout programs similar to PageMaker.) It is unlikely that most people will ever need these files so they may be kept in archival storage and removed from the active hard drive.

advance (publishing) Compensation paid to an author once the author's book is contracted but before it is published. Typically, one-half of the advance is paid upon signing of the contract, the remaining half upon delivery of the final manuscript. Advances are paid against future earnings (royalties), which means the author doesn't receive royalty payments until the advance has been "earned out."

advance copy (printing) A copy (or copies) of a book sent out to the publisher by the printer in advance of the main shipment. May be used as an approval copy.

advance reading copy (artwork) Another name for bound galley. See bound galley.

AFM Files (typography) See Adobe Font Metrics Files.

against the grain (bindery) This is the term used to describe folding or feeding paper at right angles to the grain direction of the paper. Also called cross grain.

Agate (typography) A measure equal to $\frac{1}{14}$ of an inch. Agate type is a traditional name for 5½ point type (which is about $\frac{1}{13}$ of an inch in height). An agate line is one column wide by ¹⁄₁₄ inch, traditionally used in publications for selling advertising space. Fortunately, Agate measurements are no longer frequently used in the United States.

AIM (industry) Abbreviation for Automatic Identification Manufacturers Association. This organization has branches in most countries or regions. It is supported by manufacturers and suppliers of automatic ID equipment (barcodes) and services.

airbrush (graphics, artwork) A small paint "spray gun" shaped like a pencil used to apply watercolor pigments. Used to correct or obtain tone or graduated tone effects. Paint can be applied in very thin, translucent layers, building up a color over the base image. This is also a "tool" used in most photo retouching programs (like Adobe Photoshop) that allows the application of color over the base image. In the print shop, an airbrush is used with pumice-like abrasives to remove spots or other unwanted areas from metal printing plates.

ALA (publishing) American Library Association. Holds an annual trade show where books are displayed. Related regional associations also hold smaller shows.

alert box (interface) A message box that appears when the computer needs to tell you something important. It usually brings bad news and an alert tone.

algorithm (programming) Any specific procedure for solving a problem in a finite number of steps. Named after al-Khuwarizmi, a ninth-century Arab mathematician. Programmers use this term to refer to pieces of code.

aliasing (interface) 1. The stair stepping appearance of diagonal lines and curves on a low resolution display. 2. A metallic distortion heard when digitized sounds are sampled at too low a rate, forcing the computer to fill in the gaps. In both cases, anti-aliasing resolves the problem by smoothing things out, sacrificing some detail in order to look or sound better.

alignment (typesetting) The positioning of type so that each character rests on an imaginary line, called the baseline. In traditional typography, type is measured from baseline to baseline. Some page layout programs have options using other measurement systems, but the baseline system is usually the best choice for best accuracy.

alkaline paper (paper) Paper made with a synthetic alkaline size and an alkaline filler like calcium carbonate which gives the paper over four times the life (about 200 years) as compared to acid-sized papers (40 to 50 years). Sometimes called archival paper. Almost all book papers are "acid free" now. Always specify "acid free" paper for printing your book.

alpha channel (display) When you create 24-bit images in a 32-bit graphics world, you have eight extra bits to play with. Software developers use these bits in a variety of ways; to carry transparency information, to create masks, and the like.

Alphanumeric (industry) A character set including numbers and letters.

AM (halftones, screening) (Amplitude Modulation) This is the "normal" halftone screen, where the dots vary in size and have equal spacing from center to center. See FM (Frequency Modulation) screening. Also, see halftones.

ampersand (typography) A mark derived from the Latin *et*, meaning "and," used in place of "and" in titles of businesses. The creation of the ampersand invites typographic play and personal expression in type design.

analog (interface) A flow of information where things change smoothly and have an infinite number of values and imitate (are analogous to) the real-life source. For example, on an analog phonograph record, the wiggles in the groove get bigger as the music gets louder. Contrasted (and often converted to) digital, which divides up differences into pieces for storage on a computer.

analog color proof (color reproduction) Off-press color proof made from separation films. Sometimes called a color-match print. There are also many newer techniques for creating color proofs from digital image files.

analphabetic (typography) A typographical character used with the alphabet but lacking a place in the alphabetical order. Examples: the acute accent, the umlaut, the circumflex, and the asterisk.

ANSI (industry) (American National Standards Institute) A group that publishes many computing standards, which are usually upheld by the industry. The U.S. representative in the ISO (International Standards Organization). Often used as shorthand for a particular standard, as in "It's ANSI-compliant," although these days you're more likely to see or hear "complies with ISO 9000" (an international quality standard).

anti-aliasing (interface) Software techniques used to smooth out jagged-appearing edges of curved lines, such as the blur tool in Photoshop. See aliasing.

anti-offset or set-off spray (printing) Fine powdered starch dry-sprayed onto the paper as it passes from the press to prevent wet ink from transferring from the top of one sheet to the bottom of the next sheet. The spray powder should be used very sparingly to avoid contaminating the ink on the press, especially if the sheets must pass through the press more than once. Work received with excess spray powder may reflect poor production practices by that particular press operator.

antiqua (typography) German for Roman: Post Antiqua would be called Post Roman if translated to English.

antique finish (paper) A term describing the surface, usually on book or cover papers, that has a rough, natural finish.

aperture (photography) The lens opening or lens stop expressed as an "f-stop" such as f/16. This represents the ratio of the lens opening diameter to the distance from lens to the film-plane or digital sensor. The inverse-square law governs the amount of light passing through the lens. F stops represent a reduction of ½ from each lower number to the next higher number. Typically lenses will have some or all of these settings: f/2, f/2.8, f/4, f/5.6, f/8, f/11, f/16, f/22, and f/32. Changing the lens setting between adjacent f-stops will either reduce (larger f-stop number) or increase (smaller f-stop number) the amount of light falling on the film or digital sensor.

apex (typography) The peak of the triangle of an uppercase A.

apochromatic (photography) Color corrected lenses which focus all colors of the visible spectrum in the same plane. Most modern photographic lenses are (nearly) apochromatic. Lenses that do not focus all colors in the same plane are said to exhibit chromatic aberration.

applet (computer programming) An applet is a small application. The term was coined in connection with the programming language Java but current usage extends the term to any small application, particularly those that may be called upon to operate within a larger application. Also called a widget.

application (interface) A collection of tools that is programmed to perform a specific set of tasks; also known as a program. Contrasts with a document, a piece of software that is acted upon by an application.

application font (system software) The font an application will use unless you specify another. Usually this is Geneva (Mac) or Times (Windows).

aqueous coating (printing) A shiny (glossy) water based coating applied over a printed sheet for protection or appearance. Gloss coatings often give the impression of greater depth and richness to printed color.

arm (typography) A horizontal stroke that is free on one or both ends. The uppercase E has three arms. The cross-bar of the uppercase T is also an arm.

ARPA (industry) (Advanced Research Project Agency) An arm of the U.S. Department of Defense that funds technology research that funded the ARPANET, which evolved into the Internet. Now called DARPA for Defense Advanced Research Project Agency.

arrow (interface) The default shape of the cursor used for selecting objects, etc.

art (graphics, printing) All illustration and copy used in preparing a job for printing. Often produced as a digital file.

artifact (artwork) A visible defect in an electronic image.

ascender (typography) The part of the lowercase letters b, d, f, h, k, l, and t, that extends above the height of the lowercase x.

ASCII (industry) Acronym for American Standard Code for Information Interchange This particular mapping of the letters of the Roman alphabet and Arabic number system to number codes is understood by nearly all computers (except IBM mainframes, which use EBCDIC). Documents containing only text and numbers are sometimes called ASCII files (or text files). ASCII is the U.S. adaptation of an international standard. A 7-bit binary code is used. ASCII is universally supported in computer data transfer.

automatic processor (photography) A machine to automatically develop, fix, wash, and dry exposed photographic film. Also used for photographic papers used in preparing output from an imagesetter (Linotronic). In the print shop, a machine used to develop, rinse, gum, and dry metal printing plates.

A/W (artwork) Abbreviation for artwork.

axis (typography) The real or imaginary straight line on which a letterform rotates.

backbone (bindery) The back of a bound book connecting the two covers, also called spine.

background (system software) When an application runs in the background it is operating behind the scenes while you use another application. This is a common feature of modern computer operating systems.

backing up (printing) Printing the reverse side of a sheet already printed on one side.

backing up (computers) Creating a second copy of files or contents of complete hard drives onto another media. This is done to prevent loss of files should the primary storage device fail. For piece of mind, back up early and often!

backlist (publishing) Books from previous seasons that are still in print. In children's publishing, a book can remain on a publisher's backlist for many years, particularly if they are award-winners or regarded as "classics." (Children's backlist books often outsell new titles, and for that reason many bookstores carry more backlist than "frontlist" titles.)

back matter (publishing) Printed material that appears in the back of a book, after the main body of text. Typical inclusions are glossaries, footnotes, indexes, author and illustrator biographies, etc.

back slant (typography) Type that slants to the left, the opposite of italic or oblique.

backup (storage) 1. (verb) Actually spelled "back up." The very necessary process of copying important software and documents onto some other medium (floppy disks, magnetic tape, etc.) to guard against their loss should anything happen to the original. 2. (noun) The media containing copies of important software and documents.

back-track mottle (printing) Non-uniform trapping of previously laid down ink film(s) onto an offset blanket caused by improper ink setting or inappropriate ink tack sequencing in multicolor printing. (This is a printing flaw sometimes noticeable in color printing.)

bad break (typography, composition) Starting a page or ending a paragraph with a single word (a widow). Also, when software gives a poor result when selecting the hyphenation point in a word.

bad copy (typesetting) Any original text that is illegible to the typesetter.

ball terminal (typography) A circular form at the end of the arm in letters such as a, c, f, j, r, and y. Examples of faces which use ball terminals are Bodoni and Clarendon.

banding (artwork) Refers to visible steps in shades in a gradient. Usually caused by a lack of resolution in the output device.

bandwidth (networking) The amount of data that can be transmitted over a channel (data link), measured in bits per second. (For example, a T1 line has a full bandwidth of 1.544 megabits per second. Cable-based networks often handle 3 to 4 megabits per second. DSL typically handles 384 kilobits to 1.5 megabits per second (depends on your distance from the switching facility). Note: do not confuse megabits with megabytes—8 bits equals 1 byte.)

banner (typography) A large title or headline that runs across the full width of a page.

bar (typography) The horizontal stroke in the A, H, e, t, and similar letters.

bar width reduction (barcodes) The thinning of bars within a bar code which compensates for print (dot) gain.

baseline (typography) The imaginary line on which letterforms rest. (Round letters like "e" and "o" normally dent it, pointed letters like "v" and "w" normally pierce it, and letters with foot serifs like "h" and "l" usually rest precisely upon it.)

basic size (paper) In inches: book papers: 25 x 38; cover papers: 20 x 26; bristols: 22½ x 28½ or 22½ x 35; index 25½ x 30½. See also parent sheets.

basis weight (paper manufacture) The weight in pounds of a ream (500 sheets) of paper cut to a given standard size for the grade (the parent sheet); e.g. 500 sheets 25 x 38 inches of 60-lb book paper weigh 60 pounds.

beak terminal (typography) A sharp spur, found particularly on the f, and also often on a, c, j, r, and y in many 20th century Romans. (Examples: Perpetua, Pontifex, Ignatius.)

bearer bars (printing) The protective printed lines or box around a bar code which help to support the printing plate and reduce the squashing effect on the bar code image.

beta (industry) The testing stage of a piece of software or hardware in which problems or bugs are discovered and (we hope) corrected. Usually, beta versions of a product are given to people called beta-testers who report problems to the designers. After the internal alpha-test, before a paying customer's gamma-test.

bézier curve (graphics) (pronounced "BEZ ee yay") A curve described by mathematical equations. The computer presents these curves as being composed of anchor points (where the curve starts, stops or changes direction) and control points, which you use to alter the deflection of the curve. This is the basis for drawing curves in programs like Adobe Illustrator or MacroMedia Freehand. PostScript objects and some scalable fonts are based upon Bézier curves.

bicameral (typography) A bicameral alphabet has two alphabets joined. The Latin alphabet, which you are reading, is an example; it has an upper and lowercase. Unicameral alphabets (the Arabic and Hebrew alphabets) only have one case.

binary system (programming) A number system composed entirely of zeros and ones. This is the base-2 system, where the equation 10 + 10 = 100 is the equivalent of our (base-10) 2 + 2 = 4.

binding (printing) The various methods used to secure the pages of a book to each other and to the cover. Usually hard (or case-) bound, perfect bound, saddle-stitched or a mechanical method, such as Wire-O, spiral, or plastic comb.

BISAC (publishing) A list of categories used to organize a bookstore developed by the Book Industry Systems Advisory Committee. The BISAC category should appear at the top left on the back cover. This assists the bookstore clerk to file a book in the right place. The associated code number is used by distributors to notify bookstore buyers interested in stocking specific categories of books.

bit (programming) The basic unit of digital information; it is a contraction of BInary digiT. A very small piece of information equal to a value of 0 or 1, off or on. Lots and lots of these together instruct your computer on how to work and compose all your data. 8 bits equals a byte.

bit depth (display) 1. A characteristic/setting of a display card that determines how many colors you can show at once on a screen. RGB 24-bit color means a pixel depth of 8 bits per color or 256 levels per color. In combination, 24 bit color creates millions of unique colors. 2. The number of bits of tonal range capability in an output device.

bitmap (computer graphics) The electronic representation of an image or page, indicating the position and value of every possible spot. A black & white bitmap will have either a 1 (bit on, black) or 0 (bit off, white) for each position on the page or screen. A color or grayscale image uses additional values to indicate how dark or light the shade of gray or what color value to assign for each position on the page. These positions are often called pixels which is a combination of picture and element (pix=picture and el=element).

bitmapped graphics (interface) A type of graphic composed of lots of pixels. Bitmapped graphics can be edited dot by dot, at low resolution (screen resolution, 72 dots per inch or DPI) in a simple "paint" program or at incredibly high resolution in Photoshop. Each dot can contain a lot of information (up to 32 bits worth) —color, grayness, transparency, etc.—but it's still a dot. Contrasted to object graphics, made up of lines and components. Because video screens are made of dots called pixels, the computer's video driver deals exclusively in bitmaps and rasterizes (converts) object graphics into bitmaps for display. A bitmapped image has a fixed resolution which is determined when it is created, so it can only print at whichever resolution is lower, its own or the printer's.

black printer (printing, color reproduction) The black plate, made to increase contrast of dark tones and make them neutral.

black-and-white (printing) Originals or reproductions in a single color, as distinguished from multi-color. Term is often shortened to B&W. This is also the format of many classic movies before Turner Broadcasting used computerized methods to "colorize" them. For the very ancient, this was how television was originally viewed, although the images were actually grayscale.

blackletter (typography) A general name for a wide variety of letterforms that stem from northern Europe. Blackletters are generally tall, narrow, and pointed. In architecture, comparable to the gothic style.

blanket (offset printing) A rubber-surfaced fabric which is clamped around a cylinder (the "blanket cylinder"), to which the image is transferred from the plate, and from which it is transferred to the paper. The image on the plate is (normally) "right" reading while the image on the blanket is reversed.

bleed (printing) An extra amount of printed image which extends beyond the trim edge of the sheet or page. Bleeds are usually between ⅛ to ¼ inch—this can vary by the type of job and press used. After trimming, the image appears to run off the edge of the sheet. Always verify the amount of bleed required with your printer. If you are using a "quick printer," who is printing on exact-size paper stock, often a bleed of ⅟₃₂ inch can accommodate printing to the edge of the sheet. (Note: This technique causes ink to build up on the blanket. The printer will have to stop the press from time to time to wipe off the excess ink. This technique should only be used on relatively short runs. Also, it is only possible to bleed to three sides of the sheet with this technique, as the press requires between ¼ to ⅜ inch on one edge of the sheet for the gripper.)

blind embossing (printing) A design which is stamped without foil or ink giving a bas-relief effect. The blind embossing can be done either before or after the sheet is printed. If the design involves matching a printed image in registration with the blind embossing, then it is often better to do the blind embossing first to allow the printer to bring the ink in registration with the stamped design.

blind image (lithography) An image on a printing plate that has lost its ink receptivity and fails to print.

blowup (photography, printing) 1. A photographic enlargement. 2. What a customer (sometimes) does when a print or design job does not meet his or her expectations. 3. The title of an incomprehensible film by a famous Italian director that I saw in my youth.

blueline (printing) See blueprint.

blueprint (printing) A print made from stripped-up negatives or positives, used as a proof before final printing. An ultraviolet light-sensitive paper is (usually) used to create the print. It is important to evaluate this proof carefully, since this is the final version before the actual printing plates are made. Check that all elements are present, that screens and half-tones are correct, and that there are no instances of error due to damaged or improperly exposed negatives. Also, ensure that all pages are present and in the right order. This blueprint proof is also often referred to as a "blueline." Sometimes it may be called a "Dylux." which is the name of one brand of the ultraviolet light-sensitive paper used to make blueprints. With the coming of the digital era, bluelines may not be possible for jobs that are imaged directly to the printing plate or direct to the press (DTP—Direct To Plate or Direct To Press). If possible, it is desirable to have a proof made using the same RIP (Raster Image Processor) as will be used for the DTP output. This will ensure that errors induced by the RIP are identified before the job runs.

body 1. (printing ink) A term used in connection with the viscosity or consistency of an ink (e.g. an ink with too much body is stiff). 2. (publishing) The main text of a book.

body size (typography) The height of the face of the type. Originally, this meant the height of the face of the metal block on which each individual letter was cast. In digital type, it is the height of its imaginary equivalent, the rectangle defining the space owned by a given letter (different from the dimension of the letter itself).

body type (typography) The type used for the main part of text of a printed piece, as distinguished from the heading(s).

boilerplate (publishing) Refers to publishers' standard contracts prior to any changes by an author or agent. Most publishers have a variety of boilerplate contracts to meet different needs. Boilerplates are always weighted in favor of the publisher and should be regarded by authors only as a starting point for hammering out agreeable terms.

bold face type (typography) The name given to type that is heavier than the text type from which it is derived. (The basic weight of a typeface is often called plain or Roman.) Many modern typefaces are sold in families including Roman (or plain), bold, italic, and bold-italic. Some faces have been designed with additional weights of light, (plain), medium, demi-bold, (bold), heavy, and ultra-bold (with or without italic variations in these additional weights). While these weights are relative within a particular typeface, there is not necessarily a relationship between differing faces (e.g. Souvenir demi-bold may appear as heavy (dark) as Times bold).

bond paper (paper) A grade of writing or printing paper where strength, durability, and permanence are essential requirements; used for letterheads, business forms, copy paper, etc. The basic size is 17 x 22. The most commonly used basis weights are 20 lb. and 24 lb.

book packager (publishing) Book packagers—also known as book producers or book developers—create new titles from concept to bound book for publishers, as opposed to publishers using in-house staff and existing authors to do the same. Book packagers are often receptive to new authors, though payment is frequently in flat fee and less generous than with publishers.

book paper (paper) A general term for coated and uncoated papers used throughout the printing industry. The basic size is 25 x 38. Book papers come in a huge variety of colors and qualities. Most printers have substantial collections of sample sheets to help you select a book paper suitable for your job. Uncoated book papers are approximately equivalent to the following bond weights: 50 lb. book is similar to 20 lb. bond and 60 lb. book is similar to 24 lb. bond.

book-plus (publishing) A term occasionally used to describe packages that combine both a book and components, such as a book on magic tricks packaged together with devices mentioned in the publication. Book packagers often prepare these specialized products for publishers.

book type (typography) Same as body type—the type used for the main part of the text, as distinguished from the headings or display type.

Boolean algebra (programming) A branch of mathematics that concerns itself with binary logic where every answer is either true or false. Boolean operators include "and," "or," "not" and combinations such as "and not." Computers love this stuff. You probably have encountered it when searching a database. Named for its inventor, George Boole, a nineteenth-century English mathematician.

boot (interface) To start up your computer; that is, turn on the power. This odd-sounding bit of jargon dates back to the very early days of computing. Early computers had no ROM (permanent memory). When you first turned on the power, the machine was truly a blank slate. To start up one of those old machines, you had to first enter a short "loader" program (in binary) by flipping switches on the control panel of the machine; this program was just sufficient to let the machine use its paper tape or punched card reader to load in a longer, full-fledged loader program, which in turn could be used to load the program you wanted to run! Loading a loader in order to load the loader that would load your program reminded operators of the old phrase "Pulling yourself up by your own bootstraps"; hence the short loader came to be known as the boot-strap loader, and the whole process "booting up."

border (typesetting) A decorative line surrounding a typeset area.

bound galley (artwork) Originally, a bound version of a book made from the raw type-setting before final pagination. Since digital production, a bound sample copy of a book in an early stage of production, usually for advance review. Also called advanced reading copy. See galley proof.

bowl (typography) The generally round or elliptical forms which are the basic body shape of letters such as (uppercase) C, G, O, P, and (lowercase) b, c, e, o, and p. Similar to the space known as an "eye."

box (typography) A section of text marked off by rules (or otherwise set off) and presented separately from the body text. Sometimes called a sidebar—most common in magazines rather than books. This book, however, does have a few such boxes.

break for color (composition, artwork) To separate the parts to be printed in different colors. See also Color Separation.

brightness (paper, photography) In connection with paper, brightness refers to the re-flectance or brilliance of the paper. In photography, this refers to the amount of light reflected from the copy.

brochure (printing) A pamphlet. Brochures are often created from a single sheet and folded into (approximately) thirds. This is referred to as a "three-panel" bro-chure as it has three "panels" on each side of the sheet suitable for placing copy. Brochures can also be made up of a number of pages and bound together. (Of-ten they are saddle-stitched.) A brochure or pamphlet is no more than 16 pages. If the work is larger (but less than 96 pages), it is called a booklet; 96 pages or more, and the work is called a book.

broken line (typesetting) A rule with short gaps (a dashed line).

broken type (typesetting) Type that is improperly formed. This is uncommon in the digital era. In the days of metal type, the cast letters might become damaged during handling or worn from use on a press. With offset, broken type was more often caused by flaws introduced when making negatives, due to stripping errors or contamination of the developer leaving "gunk" on the negative.

bronzing (printing) Printing with a sizing (adhesive) ink, then applying bronze pow-der while the sheet is still wet to produce a metallic lustre.

buffer (programming) A generic term for an area of memory used to store informa-tion while it is being collected and before it gets passed on to its final destina-tion.

bug (programming) Unexpected behavior usually caused by a mistake in program-ming but sometimes by a hardware malfunction. According to the late com-puter pioneer Grace Murray Hopper, the first computer bug was just that — a moth that became stuck in one of the relays of the Mark II, a very early (1940's vintage) electromechanical computer at Harvard, causing it to malfunction. Sometimes (humorously) referred to as an "undocumented feature."

bulk (paper) The degree of thickness in paper. Papers with the same basis weight can have a side variation in their bulk depending on the surface finish and other manufacturing techniques. Vellum finish papers generally are thicker than smooth finished papers. The term is also used in book printing to indicate the number of pages per inch for a given paper type.

bullet   (typography) A large dot or other symbol used at a beginning of a block of text for emphasis or for organization. When several items are involved, called a bullet list. See also numbered list.

burn   1. (plate making) The term commonly used to describe exposing a printing plate. 2. (hardware) A term used to describe making a CD or DVD. The laser in the CD/DVD writer hardware heats a portion of the encoding media, causing it to change its reflective characteristics thereby storing the digital data.

bus   (hardware) Circuitry that transfers information between the parts of a computer. Bus also refers to connections between computers on a network or allowing you to plug in other peripherals, such as a SCSI bus, PCI bus or IDE bus.

button (interface) An area of the screen which sends a command to an application when you click on it. A button with a double-thick or darker border or colored interior is the default button and hitting the return or enter key on the keyboard selects it.

byte   (computers) A unit of digital information corresponding to a single character. One byte is made up of 8 bits. It may be a computer code or represent a (human) readable character. A byte usually corresponds to a single character in Western alphabets.

C1S   (paper) Coated one side—refers to paper most frequently used for soft cover perfect-bound books. The coating, used on the outside of the book allows superior reproduction of the cover art. The uncoated side adheres to the binding glue more securely than a coated paper would. Most commonly 10 or 12 point weight is used. The 10 and 12 refer to the caliper (thickness) of the paper, 10 points being .010 inch. The 10 point weight is generally sufficient, although a book expecting heavier use or in a larger trim size might be more durable with 12 or even 15 point cover paper.

CAD   (industry) (Computer Aided Design) Using computers to perform many of the engineering and architectural tasks traditionally done by draftsmen. This term does not apply to graphic arts. CAD technicians (formerly draftsmen) are given rather different training and education than graphic artists.

CAD/CAM (computers) Acronym for Computer Assisted Design/Computer Assisted Manufacturing.

CAE   (industry) (Computer Aided Engineering) Vertical-market software to solve engineering questions, such as stress on beams and seismic strengthening.

calender rolls (paper making) A set of polished steel rolls at the end of a paper machine. The paper is passed between the rolls to increase the smoothness and gloss of its surface. A super-calendered paper has a very smooth surface finish and has been passed through a large number of calender rolls.

caliper   (paper) The thickness of paper, usually expressed in thousandths of an inch. The term refers to the tool used to perform the measurement. See also bulk.

call out (editorial) 1. a brief instruction to the compositor placed in the body of a manuscript flagged with some special symbol. Call outs should be removed during typesetting so they do not appear in the final work. 2. A phrase or short sentence taken from the body text set in larger type in a separate location to draw attention to the material. Call outs are usually used only in magazines, but I've seen them used occasionally in books. See pull quote.

camera-ready (artwork) Copy which is ready for prepress photography. This also refers to files that are ready for digital output to an imagesetter or to another device that prepares the final printing product.

cap height (typography) The distance from baseline to cap line of an alphabet, which is the approximate height of the uppercase letters. It is often less, but sometimes greater, than the height of the ascending lowercase letters.

caps and small caps (typography) Two sizes of capital letters made in one size of type, commonly used in most Roman typefaces. Small caps use the form of the capital letters, but are sized at the x-height of the particular typeface. Most page layout programs can generate faux small caps by decreasing the size of the standard capital letters; however, these computer-generated small caps are generally lighter than true small caps that are part of "expert" or "typographers" sets available for some typefaces.

caption (typography) Text set below or otherwise in an association with a photo, illustration, or other element that provides specific information or identification.

capture (hardware) Obtaining video or sound from the outside world and digitizing it for further processing. Performed by hardware built in to the audio-video circuits or available on third-party cards. Frame Capture is digitizing a still frame from a video source. Uncompressed video capture uses up about 1 megabyte of hard drive space per second. See MPEG and JPEG for descriptions of common compression technologies.

card (hardware) A fiberglass board composed of circuitry and chips which do something specific when placed inside a computer. You can get video cards, accelerator cards, clock cards, printer cards or whole computers on a card. Cards plug into slots inside most computers. There are ISA, EISA, PCI and other kinds of slot interfaces commonly used.

carding (typography) A method of vertical justification that places extra white space equally between all lines on a page.

caret (interface) 1. Generic name for any symbol indicating the place in a block of text where new text will be inserted. The little wedge symbol, ^, by typing Shift-6, used by many programs during find-and-replace operations. For example, in MS-Word you specify a tab character by entering ^T in the Find window. In many programming languages the caret is used to indicate exponents of numbers, for example, 2^8 means 2 raised to the eighth power, which equals 256.

carpal tunnel syndrome (interface) An occupational disease caused by doing the same movements with your hands over and over. The symptoms include shooting pains, tingling, or numbness in the wrist, hand or finger joints. Prevention is the best way to deal with this: keep your wrists straight while you type and/or use wrist rests, lower the keyboard, vary your tasks, take breaks, sit up straight. CTS can be permanently disabling, so if you suspect anything, see your doctor now. In extreme cases, surgery may be necessary, in lesser ones, the lifelong wearing of a wrist brace. See also repetitive stress syndrome.

case 1. (bookbinding) The case is the covers of a hard bound book. Another term for hard bound book is case bound book. 2. (interface, typography) Whether a letter is capitalized (uppercase) or not (lowercase). Many word processors have commands that let you change the case of a word or group of words. Some searching and sorting functions may be made "case sensitive," that is, capital letters sort before small letters and a small letter does not equal its capitalized self. Most search functions are case-insensitive unless told to be otherwise. 3. (typography) Type is divided into upper- and lower-case sizes. Uppercase is the capital letters, while lowercase refers to the non-capital letters. The term developed from the habit of early typesetters to arrange their type in a (somewhat) standardized manner, with the capitals stored in the upper portion of the type case and the small letters stored in the lower portion of the type case.

casewrap (printing) Same as lithowrap. See lithowrap.

cast coated (paper) Coated paper dried under pressure against a polished cylinder to produce a high-gloss enamel finish. Cast coated paper stocks are frequently used as the cover material for soft bound (soft cover) books.

cast off (typography) The term used to describe the method of estimating the space copy will use or estimating the length of a book. See characters per inch.

CCD (computers) Acronym for Charged Couple Device. This chip is the heart of most low and mid-priced scanners and many digital cameras.

CD-ROM 1. (multimedia) Acronym for compact disc–read only memory. Optical media which can store about 700 megabytes of computer data, or about 70 minutes of audio. The information on the disk cannot be changed, although it can be copied and read. 2. A mass storage device based on the consumer CD technology. The computer devices are more robust in their construction and can operate at much higher speeds than the music-oriented CD players. Other CD-related technologies include CD-RW for compact disc read-write (a re-writable CD) and CD-R for compact disc-record which is a device that will create a CD-ROM. The CD-R discs are usable in the consumer CD players (if music is transcribed). A newer technology, DVD (which started out to mean Digital Video Disc and now is called a Digital Versatile Disk or perhaps doesn't mean anything) uses multiple layered disks that can hold over 4 gigabytes of data on a single side and double sided or multi-layered DVDs can hold nearly 9 gigabytes of data. Recordable and re-writable versions of the DVD are available.

cedilla (typography) The accent (Ç), used primarily in French, to soften the letter C.

cell (interface) A single block in a spreadsheet capable of holding data or a formula. Each cell is identified by its row and column numbers. In many ways analogous to a field in a database.

chalking (printing) A term that refers to the improper drying of ink. Pigment dusts off because the solvent has been absorbed too rapidly into the paper. It is also caused by improper use of "ink dryer" additives mixed into the ink. This is an error that is within the control of the printer and, if found, should result in a rerun of the job or a reduction in the charges for the work performed.

chap book (publishing) A short book, often home-made, that authors use to share their writing. Frequently used by authors of poetry.

chapter books (publishing) A loose category of books for children ages 9-12, typified by R.L. Stine's "Goosebumps" series. Chapter books may be the first books of length that children read when they grow too old for picture books, though others are more comfortable working their way into chapter books by tackling "easy readers" first. Chapter books often carry one black and white illustration per chapter.

character generation (typography, video) The production of typographic images using master font data. Several software approaches to character generation are in use. However, the most common is the use of Adobe's generic font replacement used with Adobe Type Manager and Adobe Acrobat. Similar technology is used with Adobe Multiple Master fonts. Other companies have offered similar technologies, but at the present are not in wide use. The fonts can be displayed on screen or output to various imaging devices and printers. Generic font replacement maintains the line and page endings (font metrics) but does not reflect the true appearance of the type. In Video, this refers to the creation of fonts for display on screen over the video image. Used for titles or sub-titles.

character count (typography) The number of characters in a document, including letters, numbers, spaces, symbols, and punctuation. Getting a character count is the first stage in preparing a cast off. See characters per inch.

characters per inch or characters per pica (typography) This is a measurement of the width of a particular typeface based on the average number of characters that can be set in a unit of a line. Using this factor, along with the line length, number of lines per page, and other factors (such as number of chapters, number of subheads, and number of photos, illustrations, or tables) a cast off can be prepared to estimate the length of a book.

chase (typography) Rectangular frame used to lock lines of metal type into position in letterpress use.

check box (interface) A special button that adds or removes an option.

checkdigit (industry) The last digit in an EAN barcode or ISBN. Used to verify correct reading of the barcode through a mathematical calculation. See also checksum.

checksum (programming) A simple kind of error-checking that adds up the bits in a piece of information, divides by some number, and checks the remainder.

chemical pulp (paper manufacture) Chemical treatment of groundwood chips to remove impurities such as lignin, resins, and gums. The two most common types are sulfite and sulfate. Mixed with water, sulfuric acid is formed. See alkaline paper.

chemistry (photography and platemaking) A term used to describe the processing solutions used in photography and platemaking, e.g., a printer might say, "It's time to clean the processor and replace the chemistry."

chip (hardware) That truly amazing and remarkably tiny piece of silicon that has an entire electronic circuit embedded in its surface. The chip is the basis of the computer industry.

chokes and spreads (artwork) Refers to the overlap of overprinting images to avoid color or white fringes (or borders) around image detail. Called trapping in digital imaging systems. Traditionally chokes and spreads were prepared optically (by the print shop) during the negative and platemaking process. As digital production methods have become almost universal, trapping has moved into software. However, when the graphic artist is responsible for trapping, care must be exercised to create proper traps for the particular press and printing process being used (talk to the printer). Newer imagesetters perform trapping in the RIP software, removing this responsibility from the graphic artist. The RIP software creating the trapping should be calibrated to the press and printing process in use.

cicero (typography) A typographic measurement system used in Europe. Slightly larger than the pica (0.1776 inches). The sub-unit of the cicero is the didot point. There are 12 didot points to each cicero.

CIE Color Spaces (artwork) Three dimensional color mapping systems, CIELAB, CIEL*a*b and CIELUV which are used to plot the three color attributes X, Y, and Z. The CIEL*a*b color space (labeled as Lab Color) may be used as an editing mode in Photoshop. It can be uniquely useful, as the luminosity is separated from the color attributes, unlike the CMYK or RGB color spaces.

CIP (editorial) Abbreviation for Cataloguing in Publication. CIP is data provided for the card catalog in a library. It may be provided by the Library of Congress (called LCIP) or by the publisher (through a commercial service or willing librarian—called PCIP for Publisher's Cataloging in Publication.)

CISC (hardware) (Complex Instruction Set Computer) - Processor architecture that is capable of handling lengthy, complex instructions (code) which often require more than one clock cycle to complete. Program file length is kept to a minimum; the trade-off is that this type of processing runs slower than comparable RISC (Reduced Instruction Set Computer) processing. Motorola's 680x0 chips, and Intel's Pentium chips are popular CISC devices. Until the introduction of the PowerPC chip, all desktop computers used CISC architecture.

classical type (typography) Letterforms having vertical axis, adnate serifs, style teardrop terminals and moderate aperture. Originated in the 18th century.

click (interface) The action of moving the mouse cursor (usually a little arrow on the screen) on top of something on the screen and pressing and releasing the mouse button. Clicking will do one of three things 1) it will select the item clicked on, if that item is selectable (like an icon on the desktop); 2) it will perform some action if the item clicked on was a button; 3) it will do nothing or deselect anything previously selected if the item clicked on was neither a button nor selectable. Some items that do nothing when clicked on will do something if you click and move or drag the mouse without releasing the mouse button.

client (networking) One-half of a two-part software system. The other half is the server software. Your computer is the client and you log into the server and communicate with it. This relationship is used on the Internet.

clipboard (system software) The place in memory where the things you Cut or Copy are stored. The clipboard usually can contain only one selection at a time. This information is erased whenever you turn the computer off. The kinds of information you can transfer between applications using the Clipboard includes plain text and pictures from the screen. Inside an application, however, the clipboard can hold almost any type of information.

close up (editorial) To remove space between paragraphs, letters, or words, usually indicated by a proofing mark.

CLUT (display) (Color Look-Up Table) A table containing all the colors that a specific image uses, mapping color codes to particular color values. CLUTs usually contain 256 colors (8 bits) but can hold many, many more if the video card can handle it. Often the CLUT (called a palette by some applications) can be customized for the needs of the particular image on display. In graphics programs, the currently active window uses its CLUT for the whole screen, which can make the images in other windows look temporarily discolored (with 8-bit or less color). The Macintosh OS and Windows OS do not use identical CLUTs, however, there are 216 colors in common between the platforms. If you create an Internet web site, you should be careful to use the common CLUT or color palette.

CMYK (printing) Acronym for Cyan, Magenta, Yellow, and blacK; the subtractive primary colors. These are the standard ink colors used for process color reproduction. Also called four-color printing. CMYK printing has certain limitations and recently multi-color inks have been developed to improve color gamut. One system, using six colors, is called Hexachrome. See Hexachrome. See also Gamut.

coated paper (paper) Paper having a surface coating which produces a smooth finish. Surfaces come in matte, semi-gloss, and high-gloss finishes. (Different manufacturers will have various names to describe these various finishes. You will need to see samples and make comparisons between brands to determine exactly the relationship between similarly named products.)

coating (platemaking, printing) In platemaking, refers to the light-sensitive mixture applied to a printing plate. In printing refers to an emulsion, varnish, or lacquer applied over a printed surface to protect it.

codabar (barcodes) A bar code type which may encode 0-9 and six special characters. Four different start and stop patterns are used to signify different uses.

code (programming) The programming content of any application. The ROM chip in your computer contains bite-sized segments of code which applications may call upon. Programming languages, such as C or Pascal, have to be boiled down by a compiler program into a binary code called machine language in order to be run on the computer. Others, such as BASIC, are interpreted and the binary instructions are sent to the machine by the interpreter.

code 128 (barcodes) A full alphanumeric bar code type which can encode the entire ASCII character set.

code 39 (barcodes) An alphanumeric bar code type which can encode capital letters, numbers and seven special characters.

cold color (printing) A printed piece where the colors exhibit a bluish cast.

cold type (typesetting) Type set using photo-composition equipment. This term was used to distinguish the new generation of equipment from the earlier "hot type" machines (linotype) that used melted lead to cast type a line at a time.

collate (binding) The gathering of sheets or signatures to make ordered sets.

colophon (editorial) A Colophon is a listing, usually on the last page of a book, providing design and production credits. See page 53.

color balance (printing) The correct combination of cyan, magenta, yellow, and black to reproduce a photo without a color cast, to produce a neutral gray, or to reproduce the colors in the original scene or object. Color perception is highly subjective, so achieving proper color balance also a subjective process although many tools are used to try to make objective color measurements in arriving at a "correct" color balance. Intermediate steps in creating a particular image can have a substantial impact. For example, Kodachrome (slide) film generates images with enhanced reds and skin tones. A tomato photographed with Kodachrome film and put through the creative process may result in an image in a final printed piece looking rather more red than the original tomato.

color bars (printing) A series of colors printed in the trim area to allow measurement of ink densities in process color printing.

color correction (artwork, printing) Any method such as masking, dot-etching, re-etching and scanning, used to improve color rendition. Today, almost all color correction is done using software, like Photoshop, to modify scanned images.

color filter (photography) A sheet of dyed glass, gelatin, or plastic used in photography to absorb certain colors while transmitting others. Color filters are used to translate color negative images onto color photographic paper positives. Varying the "filter pack" can remove unwanted color casts. In color separation, red, blue, and green filters are used, although today almost all color separation is done using digital scanners.

color keys (printing) Off-press overlay color proofs using 3M Color Key® materials.

color management (prepress) A system of hardware, software, and calibration procedures designed to ensure color accuracy throughout the production process. It sometimes works. Best if systems are calibrated to a particular press. See ICC. See ColorSync.

color proofs (printing) Preliminary color images of a final printed piece made with various digital or analog methods. As the graphics industry moves toward fully digital production processes, creating color proofs has become subject to debate as several competing methods are used to create the proofs. Reasonably accurate, moderate cost proofs are now frequently made with high-end Epson inkjet printers. See also off-press proofs, progressive proofs.

color separation (photography, printing) With spot color jobs, color separation is the process whereby printing plates are prepared for each spot color used in the project. With process color jobs, it is the process of separating the color originals into the primary printing color components in negative or positive form.

With digital processes, color separation is done in software. The former traditional methods require the use of analog scanners or photographic methods using primary color filters.

colorimeter (printing, graphics) An instrument for measuring color as seen by the human eye.

ColorSync (system software) Apple's color management system that provides device-independent color consistency. Software is used to create profiles of the various display and output devices so that image files may be adjusted during output to best reflect the intended colors. If incorrectly used, can make a real mess of the colors. Similar color management software is available for Windows computers.

column inch (typography) A measure, usually used in magazines or newspapers, to indicate the space taken up by an article or advertisement. The width of a column by the length (in inches).

comb binding (bindery) See plastic comb binding.

Command key (Macintosh interface) A combination of keys pressed while holding down the Command key, such as Command-S to save a file. Command keys most often perform the same function as menu commands. Actually the term is "Command-key equivalent," but most people say "Command key" for brevity. The Command key is sometimes referred to as the "splat" key in reference to the symbol (⌘) used to signify the key having an appearance similar to a squashed bug. Windows computers mostly use the control key for equivalent commands.

command-line user interface (interface) The way in which users controlled most computers until the Macintosh came along. A command-driven operating system, such as DOS or (early) UNIX, is strictly text based. It prompts you for input with something like ? or C : or > and you type something like DIR, which means "give me the directory of things on this hard drive." It is also the basic interface for telecommunications, where a terminal is the lowest common denominator. It's called "clooey" (CLUI) as opposed to "gooey" which stands for graphical user interface (GUI).

commercial register (printing) A standard where the misregister in color printing is within one row of (halftone) dots. This term may appear in printing contracts. Ask your printer to show you examples of work that just meets this standard to get an idea of the least acceptable job under this standard.

common impression cylinder press (printing) A press with a number of printing units around a large impression cylinder. This can be as simple as a "T-head" press in a quick print shop or a very complex press used for flexography, letterpress, or lithography. Some digital imaging offset presses are designed with a common impression cylinder.

comparison read (proofing) Act of reading two pieces of copy to determine if there are any differences. One of the most accurate proofreading methods is to have one person read out loud from one copy while another person follows along with the other copy. Usually, this technique is reserved for critical material, due to the time and expense involved. (We regularly did this when I supervised tariff publishing at Southern Pacific. An error in a freight tariff had the potential to cost the company a significant sum due, in part, to regulatory constraints that required twenty days notice before an increased rate could be put into effect.)

compiler (programming) A program for creating programs, which takes a series of commands (written in a computer language like C or Pascal) and converts it into the binary numbers (called object code) that a processor can understand. Why is a translation necessary? Well, if you had to write 1101 0011 0001 0011 1110 every time you wanted to add two numbers together, you'd be pretty unhappy. A compiler lets you write "add 1 to 2"; then it does the hard part.

composition (typography) Another term for typesetting. Before the era of photographic or digital type, typesetting was done in a "composing room." Newspapers may still use this term.

comprehensive or comp (artwork) A layout showing the exact position of all elements used as a reference when preparing the mechanical. Comps are often used to show a client the expected appearance of a final design at an early stage. A comprehensive may be quite simple or elaborate. See also dummy.

computer, analog (hardware) A computer that solves a mathematical problem by using analogs, like voltage or density, of the variables in the problem. (Many older 35 mm consumer cameras used simple analog computers as a light meter to set the exposure. A slide rule is an example of an analog calculator, where the calculations are derived by the relative positions of the scales marked on the ruler.)

computer, digital (hardware) A computer that processes information in discrete, digital form. When used with natural processes (such as maintaining a particular temperature), an analog to digital converter is often used to translate the data into a digital format. Digital computers are the most common type of computer in use today.

computer science (programming) A collective term for the mathematical and theoretical underpinnings of computers and computing.

computerized composition (typography) An all-inclusive term for the use of computers to automatically perform the functions of hyphenation, justification, and page formatting. This term was often used in connection with cold-type composition equipment used prior to the arrival of desktop publishing. Today, page layout software (Quark Xpress, Adobe InDesign, etc.) performs this function on general purpose desktop computers, and the term is now rarely used.

concept books (publishing) A type of picture book for preschool children (sometimes as young as six months) in which a basic concept is introduced. Among the most often created concept books are those that deal with the alphabet, shapes and sizes, numbers, and colors. However, concepts of socialization—such as sharing—are increasingly gaining favor with publishers. Many concept books carry only illustration, or art with only a few words per page.

condensed type (typography) A narrow or slender typeface often modeled on a wider "normal" typeface.

connector (networking) The physical device that electronically joins two pieces of hardware and exchanges information.

contact print (photography) A photographic print made from a negative or positive in contact with sensitized paper, film, or printing plate.

contact screen (photography, prepress) A halftone screen on film having a dot structure of graded density, used in vacuum contact with the original photograph (or other artwork) to produce halftones.

continuous tone (photography) A photographic image which contains gradient tones from black to white or with the full range of colors present.

contone (photography, printing) An abbreviation of continuous tone.

contract (publishing) A legally binding agreement in which an author or illustrator sells to a publisher some or all rights to a creative piece of work. Contracts spell out what rights are being surrendered, for how long, under what circumstances, and for what compensation. In publishing, contracts are also referred to as "Publishing Agreements."

contract proof or print (printing) Usually the final accepted proof of a project that establishes the standard against which the final printed project will be judged.

contrast (photography) The tonal gradation between the highlights, middle tones, and shadows in an original or reproduction. Too much contrast often causes a degradation of detail in highlights and shadows while too little contrast results in a flat, muddy image with no true blacks or whites.

coprocessor (hardware) A separate microprocessor, sometimes on its own board, that works alongside your regular processor to do a particular task. There are numeric coprocessors which speed up calculations (especially when rendering three-dimensional images), video coprocessors which take over the screen display duties. See also multi-processor.

copy (interface) 1. To place what has been selected into the clipboard. The selection can be text, pictures or almost anything. 2. To duplicate a file, especially between volumes or drives. 3. (printing) Any material to be used in the production of printing. Copy can be the manuscript, pictures, or other artwork.

copy editing (publishing) Checking a manuscript for spelling, grammar, and content errors. See also substantive editing and proofreading.

copy fitting (typesetting) A mathematical procedure to determine the area required to typeset a given amount of copy in a specified typeface. In the desktop publishing era, it's more likely to "flow" (import) the text into a sample document, then adjust the factors (type face, size, and leading) to determine the copy fit. See also cast off.

copy holder (proofreading) The one who reads (out loud) the material when comparison reading is used to proofread a document.

copy preparation (printing) Directions for, and checking of, desired size and other details for illustrations, and the arrangement into proper position of various parts of the page to be photographed or electronically processed for reproduction. With the ever more digital workflow systems evolving for use in the publishing/printing industry, the responsibility of copy preparation has moved "upstream" from the printer to the originator of the work to be printed.

copy protection (storage) Security added to software so users cannot simply or easily make copies. Designed to prevent illegal distribution of software, it also inhibits easy use of hard drives and thwarts backups.

copyfitting (composition) Copyfitting is the calculation of how much space a given amount of copy will take up in a given size and typeface. Also, the adjusting of the type size (or other factors such as leading or tracking) to make it fit in a given amount of space.

copyright (publishing) Legal protection for original works of authorship that are fixed in some tangible way—as in a manuscript. When you sign a contract, you essentially agree to relinquish some or all rights for an agreed-upon price and length of time. Copyright law can be complex, but in most instances new works are protected for your lifetime plus seventy years.

corona wire (electrostatic imaging) An electronic component in a copier or laser printer that passes an electric charge to materials passing nearby. There is a corona wire that applies an electrostatic charge to the imagining drum or belt and corona wire that applies an opposite electrostatic charge to the paper. These electrostatic charges cause the toner particles to first adhere to the imaging drum or belt, then to transfer to the paper.

counter (typography) The full or partially enclosed space within a character, such as the letter "e."

counter (typography) The white space enclosed by a letterform, whether wholly enclosed (as in "d" or "o") or partially (as in "c" or "m").

correx (typesetting) Corrections to a typeset document, e.g., there are a lot of correx marked in this proof copy.

cover letter (publishing) A brief letter that accompanies a submission to a publisher. See query letter.

cover paper (paper) A term applied to a variety of papers used for the covers of catalogs, brochures, booklets, books and similar pieces. Cover papers are often made to match a variety of writing and printing papers. Business cards are generally made from cover paper stock. Also called cover stock. The most common cover paper for soft cover perfect bound books is called "c1s" which is an abbreviation of the term coated one side. See also c1s.

CPI (industry) 1. Characters per inch. 2. Consumer Price Index. A financial calculation used to measure inflation.

CP/M system (software) Acronym for Control Program/Monitor. An older operating system, commonly used on microcomputers in the 1970's. An ancestor of DOS. Written by Gary Kildall of Digital Research. There are many myths in the history of desktop computers describing how IBM and Kildall failed to come to terms to adapt CP/M for the IBM-PC, allowing Microsoft and Bill Gates the opportunity.

CPU (hardware) (Central Processing Unit) The chip that does all the work. Also commonly refers to the computer component, such as the circuit board or box, in which it is housed, e.g., I have seven monitors but only five CPUs in my office.

creep (bindery) 1. The distance margins shift when paper is folded and/or inserted during finishing. The amount of creep will vary depending on both the number and thickness of the sheets and must be compensated for during layout and imposition. Creep has a large effect on saddle-stitched booklets, but is also a factor in the folds and signatures of offset-printed books. See shingling. 2. The acronym of the "committee to re-elect the president" during the campaign for the second term of the Nixon administration. Now, *that's* just creepy.

crop (artwork) To eliminate portions of the copy, usually on a photograph or plate, indicated on the original by crop marks. The cropping tool is used in page layout software to perform this function in a digital environment.

crop marks (artwork) Fine lines marked on the mechanical showing where the edge of the paper should be. Additional marks, ⅛ inch away may show the extent of the bleed area. See trim marks.

cross direction (paper) The direction across the paper grain. Paper is weaker and more sensitive to changes in relative humidity in the cross direction than in the grain direction. When scored or folded, some papers are more likely to split when the score or fold is parallel to the cross direction.

cross-platform (system software) Software or files written to be able to run or used on multiple platforms, for instance Macintosh and Windows computers, without changes. Current Macintoshes are designed to be highly compatible with cross-platform transfers. Additional utilities that perform translations between programs and platforms can be used to enhance compatibility.

cross marks (printing) see register marks.

CRT (computers) Acronym for Cathode Ray Tube or the video display.

CTP (printing) Acronym for computer-to-plate or computer-to-press. Used in the digital workflow where the digital file is imaged directly onto a printing plate. This eliminates the need for making negatives, stripping, and then making plates. Some presses (e.g. the Heidleburg GTO-DI) image the blank plate after it is mounted on the press. By imaging plates on the press, the work is more likely to be in near-perfect register and make-ready work is substantially reduced. The downside is that the press is not available for work while the plates are being imaged and additional plates cannot be prepared while the press is in operation.

cursive (typography) Refers to type that looks like handwriting.

curl (paper) Refers to the distortion of a sheet due to differences in structure or coatings from one side to the other, or to absorption of moisture on an offset press. Also refers to the distortion that occurs in a copy machine or digital (toner-based) press. The curl that develops in an offset press is due to the moisture present while the sheet is being printed. In contrast, the curl that develops in a copy machine is due to the drying effect of heat applied to set the toner.

cursor (interface) A part of the screen display that moves when you move the mouse. The shape of the cursor varies depending on what application you are using or what the computer is doing. A wrist-watch cursor, sand-clock, or spinning beach ball means you must wait while the computer does its work. In programs that allow text-editing, the flashing bar that indicates your place in the text is called the "insertion point" or "I-Beam cursor."

cut-off (printing) Refers to the cut or print length of a sheet prepared on a web press.

Web presses often have a limited range of adjustment to the cut-off, causing excess waste for projects with incompatible page sizes.

cutscore (die-cutting) A sharp-edged knife, usually several thousands of an inch lower than the cutting rules in a die, made to cut part way into the paper or board to create a scored line for folding.

cyan (color theory, photography) Hue of a subtractive primary color and one of the 4-color process inks. It reflects blue and green light and absorbs red light.

cylinder gap (printing) In a printing press, the gap or space in the cylinders where the mechanism for plate or blanket clamps and grippers (sheetfed press) are located.

daisy-chain (computer) To connect together several devices in a line with a single cable path that runs through each device. Firewire devices may be daisy-chained, but USB devices must directly connect to the port or to a hub that multiplies the available connections.

dampening system (lithography) The mechanism on a press for transferring dampening solution to the plate during printing. Dampening systems come in a variety of arrangements, but are divided into integrated and conventional systems. An integrated dampening system places the fountain rollers in contact with the ink rollers transferring the ink and water solution to the plate at the same time. In a conventional dampening system, the fountain rollers transfer the water solution directly onto the printing plate (separate from the ink rollers). Integrated dampening systems are common on small duplicator presses because it is easier for the press operator to manage the ink/water solution balance; however the disadvantage is that the ink has a tendency to become emulsified with the dampening solution, causing unacceptable print quality. When ink emulsification occurs, the press must be shut down and cleaned.

dash (typography) A short line, longer than a hyphen (-), used in typography. They come in two lengths, em-dash (—) and en-dash (–).

database (applications) A document designed to store lots of information in predefined categories, and which allows this information to be arranged and manipulated easily. The term "database" can refer to the application which creates and manipulates these documents or to the document itself. A database contains records, and each record is made up of fields.

daughterboard (hardware) A circuit board that attaches to the motherboard (aka logic board) and gets its power from same. A less sexist term is subassembly.

DCS (prepress) Acronym for Desktop Color Separation. A data file used to assist making a color separation. Using DCS, five files are generated: the four color files; cyan, yellow, magenta, and black; and a composite color preview file. Photoshop can create DCS files. This format has been superseded by PDF.

DDES (industry) Acronym for Digital Data Exchange Specifications.

dead copy (typesetting) The original manuscript used to compare against a typeset version.

debossing (printing) A process where depressions are created in a sheet of paper. See embossing.

debug (programming) The process of fixing errors in programming, removing the bugs. Developers spend a large amount of time and resources testing their products (some more than others!). Special programs called debuggers are employed to track down gross errors, and people called beta-testers put the software through its paces.

descender (typography) The part of a lowercase letter which extends below the main body, as in the g, j, p, q, and y.

decimal tab (interface) A tab character used to make columns of numbers (like dollars-and-cents values) line up vertically on their decimal points at the tab's location.

deckle (papermaking) The width of the wet sheet as it comes off the wire (a conveyer and strainer) of a paper machine.

deckle edge (paper) The untrimmed feathery edges of paper formed where the pulp flows against the edge of the wire frame.

default (interface) The parameters your program uses in the absence of instructions to do otherwise. For example, when you open a new document in your word processor, the typeface is usually Times-Roman 12 and the tab stops are whatever the program's designers decided. You may reset some or all of the defaults in most applications by choosing "Preferences" from one of the menus.

delimiter (interface) A character used to separate items of information. Tabs, commas, and returns are the most popular delimiters, but some applications allow you to specify whatever you want. Delimiters are most important when importing and exporting data from one application to another. For example, if you save an Excel spreadsheet as a text-only file and import it into a word processor, the data from each cell will have a tab character after it, and each row's data will be followed by a carriage return. Tab-and-return delimiters are also commonly used in database import/export. One thing to remember: a delimiter character should not appear elsewhere in the data.

densitometer (photography, printing) A photoelectric instrument which measures the density of photographic images, or of colors. In printing: a reflection densitometer is used to measure and control the density of color inks on the substrate (paper).

density (photography) The degree of darkness (light absorption or opacity) of a photographic image.

descender (typography) The part of the letters g, j, p, q, y and sometimes (uppercase) J that extends below the baseline.

desensitizer (lithographic platemaking, photography) A chemical treatment to make non-image areas of a plate repellent to ink. Photography: An agent for decreasing color sensitivity of photographic emulsion to facilitate development under comparatively bright light.

desktop (interface) 1. The area of your screen behind all your windows. 2. The adjective added to certain types of computing work (desktop publishing, etc.) which can be accomplished by a personal computer.

developer (photography, lithographic platemaking) The chemical agent and process used to render photographic images visible after exposure to light. Plate-making: the material (solvent) used to remove the unexposed coating.

device independent (prepress, computer) A characteristic of software or file format that allows different output devices to output a file at the best quality that each device is capable of producing. PostScript, EPS, and PDF are generally device independent file formats.

dialog box (interface) A message box that appears on your screen when the computer needs you to give it further instructions or information. Dialog boxes usually have one or more buttons that allow you to respond to the message displayed.

diaresis (typography) The accent used to separate the pronunciation of two consecutive vowels, as in coördinating. Similar to the umlaut.

didot point (typography) The sub-unit of the cicero (0.1776 inches) is the didot point (0.0148 inch). There are 12 didot points to each cicero. (The didot measurement system is not commonly used in the U.S.)

die-cutting (printing) The process of using sharp steel rules to cut special shapes for labels, boxes, and containers, from printed sheets, Die-cutting can be done on either flatbed or rotary presses. Rotary die-cutting is usually done inline with the printing. Die-cutting is often done at the same time as embossing.

die-stamping (printing) An intaglio process for the production of letterheads, business cards, etc., printing from lettering or other designs engraved into copper or steel. This is a relatively expensive process that has been substantially replaced by Thermography. See Thermography.

digital (industry) Information represented as discrete numeric values, such as 0 and 1. The opposite of analog. In digitization, which is what a scanner does, the flow of information is sampled at regular intervals and converted into numeric values.

digital asset management (industry) Acronym DAM often used. Systems and procedures used to catalog digital media (text, images, video, audio) and some physical media to enable efficient storage, retrieval and use. Extensis' Portfolio is one program often used as the cataloguing part of a DAM system.

digital color proof (printing) An off-press color proof produced from digital data without the need for separation films. Many techniques are being used to produce digital color proofs: dye-sublimation, ink-jet, and electrostatic. Dye-sublimation prints are considered the "best" from a quality standpoint (they appear much like a continuous tone image). A criticism of digital color proofs is that (most) methods do not show the actual half-tone dots as will be used in the final image—moiré patterns not visible in the "proof" may occur in the final image.

digital ink (printing) See toner.

digital photography (photography) Cameras using CCD or CMOS sensors to capture images electronically instead of silver-based photographic film. While early (and low-end consumer) cameras produce low quality images unsuitable for print reproduction, current mid-level, prosumer, and professional level digital cameras produced excellent images. Indeed, it appears that professional digital cameras with sensors exceeding 16 million pixels provide higher resolution than photographs made with high quality films. While the data encoding capacity of film suggests that theoretical parity might not have been reached until 50 million pixel sensors were used, in practice the lack of "grain" in digital sensors causes the measurable resolution to be matched or exceeded at much lower sensor pixel density. Well photographed images of 3 to 4 million pixels and up are generally suitable for print reproduction. Somewhat higher resolution (perhaps 8 million pixels and up) is more appropriate for edge to edge cover images.

digital plates (printing) Printing plates that are exposed by laser or other energy sources (heat) in a digital platesetter. (A platesetter is similar to an imagesetter, except it exposes metal plates rather than film or photographic paper. Many imagesetters can expose polyester-based plate material in addition to film or paper. The polyester plates produced by digital means are also considered digital plates.)

digital printing (printing) Printing by plateless imaging systems that are imaged by digital data from prepress systems. A common system is the Xerox DocuTech. Full color digital printing is produced on systems from Canon, Hewlett-Packard, Xerox, and various other companies. Xerox, Océ, and IBM also make stand-alone high-speed printers normally used with mainframe computers but adapted to digital text and graphic printing uses.

digitized typesetting (typography) The creation of typographic characters and symbols by the arrangement of black-and-white spots called pixels (picture elements) or pels. This term is generally applied to the generation of typographic production equipment available before the widespread use of desktop publishing software.

digitizer (hardware) A computer peripheral device that converts an analog signal (images or sound) into a digital signal. A scanner is one type of digitizer.

digitizing tablet (hardware) A device generally used with a special pen or stylus that allows drawing on a surface that is directly input as "pen strokes" into a drawing program. Often, an original drawing can be secured to the digitizing tablet so that the image may be traced with the stylus. Digitizing tablets can be substituted for the mouse on most computer systems.

dimensional stability (printing) Ability to maintain size, resistance of paper or film to dimensional change with change in moisture content, relative humidity, or heat. Good dimensional stability in films, plates, and paper is necessary for high quality production of four-color process printing.

dingbat (typography) 1. A traditional printer's term for ornamental or picture characters like stars, bullets, little boxes, hearts, diamonds, tiny flowers and snowflakes. The Zapf Dingbats font (designed by famous German typographer Hermann Zapf) is built into most PostScript laser printers. See also pi fonts. 2. The nickname given by Archie Bunker to his wife, Edith, in the "Archie Bunker" television show popular during the 1970s.

DIP (hardware) (Dual Inline Package) A kind of chip that has two rows of connectors. DIP SIMMs are RAM SIMMs using DIP chips, which are taller than normal SIMMs and may not fit some computers. Also refers to a type of switch (DIP

Switch) found on some computer boards used to configure a device.

direct screen halftone (color separation) A halftone negative made by direct exposure from the original on an enlarger or by contact through a halftone screen.

directory (interface) 1. The list of the contents of a folder or disk, ordered by name or icon or date, etc. 2. The first few sectors on every disk, containing information about what files are on the disk and the size, location, and type of these files. The list of where the parts of all files are located on a disk is called a file allocation table. When a disk is damaged the problem is often due to a garbled directory.

directory dialog box (interface) The dialog box the computer puts up when you want to Open or Save an application or document. You use it to navigate to the correct folder, then select the desired file.

disc (storage) The correct spelling of "disk" when referring to compact discs or DVDs.

disk (storage) A round piece of plastic with magnetic stuff in it where you store information you want to keep around after you turn the computer off. Comes in hard and floppy varieties.

dismount (storage) Causing the computer to give up all claim to a volume (a floppy, hard drive, cartridge, etc.) and remove it from the desktop. A removable cartridge drive, CD, or DVD must be dismounted before it can be removed from the drive. (Mac)

display type (composition, typography) Type set larger than the text. Also, type that is designed for use in larger sizes only. Many "display type" faces are created for use in advertising. These highly decorative faces are only useful in relatively large sizes and when only a few words are set. They should not be used for extensive passages of text.

distributor (publishing) An intermediary in the bookselling trade. Distributors usually require an exclusive contract and a discount of 65 to 70 percent off list. The presence of a sales force distinguishes a distributor from a wholesaler.

dithering 1. (printing) Mixing colors you have to provide the illusion of more colors than are available by sacrificing resolution. A process commonly used in printing to reduce the number of colors needed for a particular print job. Also used with black and white images to give the illusion of gray when seen from a distance. 2. (display) A technique of filling the gap between two pixels with another pixel having an average value of the two to minimize the difference or add detail to smooth the result. This is similar to anti-aliasing, but applied over a large area. For example, if your computer is set to display 256 colors on screen, software will often create an image using dithered colors of a file created with a larger number of colors.

division (publishing) Usually refers to a branch of a company that is not separately incorporated. Larger publishing houses typically contain several divisions. A division may also be an imprint. See also imprint.

doctor blade (printing) In gravure printing, a knife-edge blade pressed against the engraved printing cylinder which wipes away the excess ink from the non-printing area. In laser printers and electrostatic copiers, it is a blade, often made of plastic, that wipes away left-over toner to prepare the drum (or imaging belt) for the next image. A defective doctor blade can leave streaks on the output.

document (system software) A file on the computer containing information to be acted upon (as contrasted with an application which does the acting).

DOS (system software) (Disk Operating System) The software instructions that tell a computer how to operate. The most common flavor is Windows, in various versions (made by Microsoft). Others are UNIX, Linix, and Mac OS X.

dot (prepress, printing) An individual element in a halftone. Common usage does not clearly differentiate between dots and spots. The fineness of a screen is measured in lines per inch (lpi). A dot is made up of (potentially) many spots (in digital halftones); that is, the marking engine (laser) in the output device places groups of spots to create halftone dots of various sizes. The physical number of spots per inch possible for the output device ultimately limits the maximum shades available at any particular line per inch ruling. For example, a 300 "dpi" (actually spots per inch) laser printer can only produce a half tone at a very rough 55 lpi, while a 600 dpi laser printer may manage about 85 lpi. Most laser printers now use technology that allows some variation in spot size to increase practical output to 85 lpi or more. In AM screening, dots vary in size while in FM screening, the dots are all the same size.

dot etching (photography) Chemically reducing halftone dots to vary the amount of color to be printed. Dot etching on negatives increases color; dot etching on positives reduces color.

dot gain (printing) A defect in which dots print larger than they should, causing darker tones or stronger colors. Dot gain varies by the press used (both type of press and size) and by the paper used. Highly absorbent paper will experience more dot gain than hard, coated papers. Fine screen halftones will produce very poorly on soft, absorbent papers. The lpi value of the halftone screen used should be selected in consideration of the paper being used. The printer can usually compensate for dot gain caused by the type and size of the press during creation of the printing plates. You should discuss dot gain with your printer before final art is released to ensure that dot gain has been considered in your project.

dot matrix printer (output) A printer that uses dots to create the text and image on the page. This term is most commonly used to refer to impact dot matrix printers which use small metal pins striking the ribbon and paper to create text and images. These have been nearly eliminated from the market by inkjet and laser printers, however some are still used where multi-part forms are printed.

dots per inch (dpi) (graphics) A measure of the resolution of a laser printer, scanner or other digital device. (In reality, this actually refers to spots per inch, see dot.) Do not confuse dots per inch (dpi) with lines per inch (lpi) which is the term used in connection with the frequency of a halftone screen. To maximize quality, an image should be scanned at a dpi resolution that is approximately twice the lpi frequency of the halftone screen. For example, a scan at 300 dpi is required for a halftone screen of 150 lpi. If the scanned image is enlarged or reduced, then the scan frequency should be adjusted to reflect the enlargement or reduction.

Double storey (typography) Seen in the lowercase "g" with the closed tail and lower-case upright finial "a."

double-click (interface) To click the mouse twice quickly in succession without moving it. This usually tells the computer to open or launch whatever was double-clicked. It also enacts the default option in dialog boxes.

download (communications) Transferring a file from a remote computer to your computer, using a terminal program and a transfer protocol.

downloadable font (output) Descriptions of fonts that are not built into a printer but are kept on a hard disk (connected to either the printer or the computer) and sent to the printer as needed. This process takes time and makes printing slower.

dpi (output) (dots per inch) A measure of resolution on a printer or scanner. The more dots per inch the device is capable of producing or reading, the more detailed and smooth the resulting output. Early laser printers printed at 300 dpi, while most laser printers today print at 600 dpi or even 1200 dpi; the standard computer screen is between 72 and 96 dpi. Typeset-quality print is generally considered anything above 1000 dpi. Most image/plate-setters output at 1200, 1270, 2400 and/or 2540 dpi. (1200 dpi laser output often falls short of "typeset quality" due to the toner particle size and the scattering of toner caused by the electrostatic process used to create the image.)

drag (interface) With the cursor over an object, press and hold the mouse button pressed. Then move the mouse and watch as the object moves with the cursor.

DRAM (hardware) (Dynamic Random Access Memory) The most common kind of RAM, the one people mean when they just say "RAM." Most computers now use Static or pseudo-Static RAM, which is abbreviated as SDRAM.

draw-down (inkmaking) Term used to describe an ink mixer's method to roughly determine an ink's color shade. A small blob of ink is placed on paper and drawn down with the edge of an ink (putty) knife to get a thin film of ink.

drier (inkmaking) A chemical additive used to hasten drying of ink. Interestingly, most ink driers must be used most sparingly, as an excess will cause the ink to dry more slowly or not at all! The most common drier requires only one or two drops in an ink fountain. Beginning press operators easily overdose the drier then compound the problem by adding even more drier when the ink does not react as expected. When this happens, the press operator must completely clean the ink fountain and dispose of the contaminated ink. The press rollers may need partial or complete cleaning depending on the severity of the contamination.

drive (storage) A device that moves a storage medium and reads it or writes to it. Drives either spin disks or transport tapes. There are floppy drives into which you insert a floppy disk; there are hard drives that are sealed inside the computer or the external box they come in. There are also removable media drives, including those that have a large slot into which you insert a hard disk cartridge, tape drives, DAT drives, CD or DVD drives and optical drives.

driver (system software) A piece of software which controls communication between your computer and some peripheral device. The most common drivers are those needed for printing. Scanners, CD-ROM drives and removable media drives. Many drivers are built into the operating system (particularly with Mac OS X) allowing many devices to "plug and play."

drop cap (typography) A large initial capital in a paragraph that extends through several lines. If the sentence with the drop cap is within quotation marks, the leading mark is not used.

drop folio (typography) A folio (page number) dropped to the foot of the page when the folios on other pages are carried at the top. Drop folios are often used on chapter opening pages.

drop-out (printing, prepress) Portions of originals that do not reproduce, especially colored lines or background areas. Drop-out is often done (with a graphic arts camera) on purpose to delete unwanted material on an original. "Non-repro blue" pencils or preprinted forms with "non-repro blue" ink were used to mark paste-up sheets to assist in alignment of artwork. These techniques are mostly in the past with the transition to digital work flows.

drum scanner (prepress) Device that uses photo multiplier tubes (PMT) and produces color separations of higher resolution and dynamic range than those produced with CCD-based scanners. The advantage of PMT-based scanners over high-end CCD scanners has become quite small and it's often hard to justify the high price of scans from drum scanners. Note, a drum scanner only works with flexible originals.

dry reading (proofreading) Reading typeset copy without reference to dead copy.

DSL (communications) Abbreviation for digital subscriber line. DSL gives a broadband connection to the Internet at speeds that decrease as the subscriber is further from a telephone switching center. It is generally unavailable beyond three miles (as the wires travel) from a switching center. Unless located quite close to a switching center, better broadband performance is usually available with a cable connection. Connections often prove difficult, interfering with both computer and telephone communication. Several of my friends have suffered through "DSL hell."

dummy (prepress, printing) 1. A preliminary layout showing position of illustrations and text as they are to appear in the final publication. 2. A set of blank pages made up in advance to show the size, shape, form and general style of a piece of printing. Dummy books are usually provided by children's book printers to allow finalizing artwork to the actual measurements taken from the dummy.

duotone (artwork) The creation of a two color halftone reproduction from a contone original. Usually done in 'finer' printing projects that are not budgeted for full color (CMYK) printing. One very effective example matched black ink with metallic silver ink to create stunning duotone images in the book. Duotones can be created in Photoshop, but use of third party or other custom duotone profiles are generally recommended for best results. (This isn't for a beginner.)

duplex paper (paper) Paper made with a different color or finish on each side. Often manufactured by gluing dissimilar sheets together. Most often available only in cover weights.

duplex printing (hardware) A printer that automatically prints on both sides of the page. A handy feature that can save paper and ease preparation of sample books. Most large copier/printers have duplex capability. This feature is now being seen more frequently on smaller workgroup and individual-use laser printers. See perfecting press.

DVD (computers) A storage technology, similar to a CD-ROM that holds more data in about the same space. DVD, which started out to mean Digital Video Disc and now is called a Digital Versatile Disk (or perhaps doesn't mean anything), uses media that can hold over four gigabytes of data on a single side. Double sided, multi-layered DVDs can hold nearly nine gigabytes of data. Recordable and re-writable versions of the DVD are readily available. DVD drives with write capability for double layer discs have recently come onto the market. Most DVD writers can also write CDs. There is a format war between groups of manufacturers supporting "–R/RW" and "+R/RW" discs. Fortunately, many of the latest DVD writers now support both the + (plus) and – (dash) disc varieties, avoiding the consumer dilemma of choosing one format over the other.

DVORAK (interface) An alternative arrangement of the keys on a keyboard. DVOR-AK (named after the person who invented it) is nonstandard, but allows more rapid typing than the standard QWERTY layout, some say.

Dylux™ (printing) A brand-name of an ultraviolet light-sensitive paper used to make blueprints used as a final proof before printing plates are made. See blueprint.

dynamic range (prepress) Density difference between highlights and shadows of scanned images. See gamma.

EA (typesetting) Editor's alteration. A change to a document initiated by the editor.

EAN (barcodes) Acronym for European Article Number. The international standard for coding retail goods. The type of barcode used by ISBNs.

early readers (publishing) Sometimes also called "beginning chapter books," early readers are books targeted at kids making the transition from picture books to lengthier chapter books. Though many publishers don't provide guidelines for early readers, most such books are aimed at children ages 8-11. The books typically run about 64 pages when published and still feature generous use of illustration, often in black and white.

edit (editorial) To check for facts, word usage, spelling, grammar, punctuation, and consistency of style. See chapter 2.

editorial board (publishing) A decision-making body within a publishing house that votes whether or not to move forward with a project. Editorial boards are typically composed of an acquisitions editor, sales director, marketing manager, publisher and finance individual, all of whom debate the various editorial qualities, marketing prospects, and costs associated with a potential book. Some boards are smaller, but rare is the publishing house that doesn't include some structure designed to determine the feasibility of any potential book before an offer is made to an author. (In other words, it isn't enough for an individual acquisitions editor to like a manuscript for that manuscript to become a book.) In a smaller publisher, the "editorial board" may consist of only one or two people, but the same editorial, marketing, and financial considerations will apply to the potential project.

EDG (digital imaging) Acronym for Electronic Dot Generation. A method of producing halftones electronically on scanners and other prepress systems.

EEPROM (hardware) (Electronically Erasable Programmable Read Only Memory.) Allows certain "permanent" programs that help operate a computer or other device to be upgraded "in the field" by running a special program (that replaces the original programming) instead of swapping chips. See also firmware.

email (networking) (electronic mail) Text messages sent from computer to computer over a network or over phone lines. The term is often spelled e-mail, but seems to be evolving to eliminate the hyphen.

ear (typography) A small stroke that projects from the upper right side of the bowl of the lowercase Roman g.

Egyptian type (typography) Letterforms having square serifs and almost uniform style thickness of strokes. Has nothing specific to do with Egypt.

electronic printing (digital printing) Any technology that reproduces pages without the use of traditional ink, water, chemistry, or plates. Also known as plateless printing.

electrophotography (digital printing) Image transfer systems used in copiers and laser printers to produce images using electrostatic forces and toners.

electrostatic plates (digital printing) Plates for high-speed laser printing using zinc oxide or organic photoconductors.

ELF (display) (Extremely Low Frequency.) A type of radiation given off by almost any electric device, but potentially dangerous amounts come from power lines, heavy machinery and some video monitors. The best prevention is to sit at arm's length from your monitor and stay away from the back or sides of one—that's where the radiation is highest. There are some studies that indicate that earlier concerns about ELF radiation are overblown, however recently manufactured CRT displays comply with low-ELF safety standards. ELF is not an issue with LCD displays.

elliptical dot (halftone) Elongated dots which give improved gradation of tones particularly in middle tones. Also called chain dots.

em (typography) A unit of measure exactly as wide and tall as the point size being set. As such, the em is relative to the type face and type size being used. Name comes from the practice in early type manufacture to cast the letter M on a square body. May also be called an em quad.

embossed finish (paper) Paper that has a raised or depressed texture resembling wood, cloth, leather or some other pattern. Often used with hard cover books.

embossing (printing) The impressing of an image in relief to achieve a raised surface. Often overprinted, if left unprinted, called blind embossing. Can easily be combined with foil stamping and die cutting. Debossing is the same process used to create a depressed surface.

em dash (output, typography) A dash (—) as wide as the character "M" in a given font used in text to separate a parenthetical note as an alternate to parenthesis.

em space (typography) A distance equal to the type size — 12 points in a 12 point typeface, 11 points in an 11 point typeface and so on. Also known as a "mutt."

emulation mode (system software) Software running on hardware for which it is not primarily written, such as DOS software running on Macintoshes. Not all features may work properly, and speed is degraded.

en (typography) One half the em width. May also be called an en quad.

enamel (paper) Another term for coated paper or for a coating on a paper.

en-dash (output) A dash (–) half as wide as the em-dash in a given font.

en space (typography) Half an em. Also known as a "nut."

end note (editorial) A note appearing at the end of a chapter or in a section at the end of a book providing supplemental information about a reference marker placed earlier in the chapter or book. Usually limited to source reference, but sometimes notes include supplemental information that did not belong in the main text.

end sheets (binding) In case bound books, end sheets are used to glue the book block to the cover and to hide the edges of the fabric facing material that wraps around the edges. End sheets may be plain, colored paper, or printed.

English finish (paper) A grade of book paper with a smoother finish. Probably a reference to Bond, James Bond, a smooth Englishman.

Enter key (interface) Key that functions the same as the Return key.

EPS (output) Abbreviation for Encapsulated PostScript. A graphic file format composed of two parts: a simple bitmapped image that the computer reads and displays on the screen and a complex PostScript code that a PostScript printer reads. EPS files are called "device-independent" or "resolution independent." This means they will print at whatever resolution the printer happens to be.

etch (printing) 1. An acidified gum solution used to desensitize the non-printing areas of a printing plate. 2. An acid solution added to the fountain solution to help keep non-printing areas of a plate free of ink.

Ethernet (networking) A hardware protocol for two or more nodes (devices) on a LAN. Regular Ethernet uses a transmission of 10 megabits per second baseband transmission (or 10Base-T). See also 100Base-T and 1000Base-T.

expanded type (typography) A typeface made wider than its normal version.

exposure (photography, platemaking) The step where light or other radiant energy is used to produce an image on a photosensitive coating or sensor.

extender (typography) Descenders and ascender; i.e., the parts of the letterform that extend below the baseline (p, q) or extend above the x-height (b, d).

eye (typography) The enclosed part of the lowercase e.

face (typography) The full range of type of the same design. See typeface.

family (typography) A range of typeface designs that are all variations of one basic style. For example most typefaces are available in regular, italic, bold, and bold-italic. Some, more elaborate families, also have light, semi-bold, black, condensed, compressed, and/or extended variants that may also have italic variants as well. All of these related variations are part of a typeface family.

fanout (printing) Distortion of image on the press due to waviness in the paper caused by absorption of moisture at the edges of the sheet, particularly across the grain.

feeder (printing) That portion of the press that separates the sheets and feeds them into position for printing.

felt side (paper) The top side of the sheet as it passes through the paper machine. Usually the smoother side of the sheet.

field (interface) A discreet piece of information in a database, such as the ZIP Code in an address book. A database contains records, and each record is made up of fields. In a program such as FileMaker, you can format the display of each field's information and specify the type of data, for example to make sure there are the right number of digits and no letters in a phone number. Very similar to a cell in a spreadsheet.

file (system software) A self-contained single set of digital information on disk or in memory. A file is either an application or a document, anything from a brief letter to a QuickTime movie to a gargantuan application like Excel.

file format (system software) The particular structure that a document (spreadsheet, graphics, text, etc.) is saved in. Text is a standard file format for words and numbers, and many applications can read text documents. Text file formats include ASCII (text only) and RTF (Rich Text Format), in addition to the particular format for your word processor. TIFF is a standard file format for bitmapped graphics, and EPS is a standard file format for object-oriented graphics. Many page layout programs can readily use PDF files for graphics as well.

file server (networking) A computer that saves and retrieves files on a network. Often, one computer and its hard drive are dedicated to the job of being a file server. File server software controls who gets to read and use what, and how many people get to do it at the same time. Many home networks use "peer to peer" file sharing without using a server.

file sharing (networking) Using files on another computer via a network. It can be as simple as two computers sharing a printer, or as complex a network as you can create.

filling in (or filling up) (printing) A condition where ink fills the space between halftone dots or plugs (fills in) the type.

film lamination (binding) A plastic with adhesive that is applied over a cover. Adds to the thickness and durability of the cover. Care must be taken to order "lay flat" laminate. Laminate is usually more durable than aqueous or UV coatings.

filter (system software) A small piece of code (module) used by an application to convert a document created by another application into a file format that it can read or modify the content of an image. Common use: Photoshop filters.

firmware (hardware) Software that lives in chips, such as your computer's read-only memory (ROM), and can't be changed. Some computers and other devices use Electronically Erasable Programmable ROM (EEPROM) instead of ROM, so firmware upgrades can be installed by downloading and running a program.

fixing (photography, platemaking) Chemical action (with "fixer") to convert unexposed silver halide (in an image) to soluble salts that can be rinsed away to make the image stable and insensitive to further light exposure.

FKEY (interface) A key along the top of the keyboard that initiates some function in software. See function key.

flat   1. (prepress) The assembly of negatives on goldenrod paper ready for platemaking. See stripping. 2. (photography) A lack of contrast in a photograph.

flatbed scanner (hardware) A device that captures digital images designed similar to a copy machine with a glass plate (the platen glass) upon which the artwork is placed face down for scanning. Some flatbed scanners also have the ability to scan transparencies using a light that shines from the top.

flat fee (publishing) One-time compensation that generally provides a lump sum for an author or illustrator's work in exchange for all rights. This is in contrast to a contract that includes an advance and royalties. While similar to work-for-hire, the flat fee arrangement does not remove the author's credit from a project.

floating point (programming) Mathematics in which an essentially unlimited number of digits after the decimal may be used. Mathematically speaking these are "real" numbers (as opposed to integers). Some graphics programs, particularly those that work with 3-D images extensively use floating point math.

folder   (interface) A place to hold documents, applications, aliases and other folders on a Macintosh disk. Folders have icons on the desktop and hold virtually an unlimited number of files except at the root (top) level of a disk, where it is possible to encounter a limit. It's best to have a (relatively) few top-level folders (directories) and store files and folders below the top level. You move items into a folder by dragging the item's icon onto the folder. The easiest way to open a folder is by double-clicking on its icon. Windows uses a similar arrangement. Folders are the equivalent of "directories" in MS-DOS and UNIX.

flush cover (printing) A cover that has been trimmed to the same size as the inside pages. Almost all soft-cover perfect-bound books have a flush cover.

flush left (or right) (typography) Type set to line up on the left (or right). Most of this book is set flush left *and* right, or full justified.

flush paragraph (typography) A paragraph with no indentation. Paragraphs starting with a drop cap (at the beginning of a chapter) or following a heading are generally set flush.

flying paster (or splicer) (printing) A device on a web press that automatically glues a new roll of paper onto an expiring roll without stopping the press. As the process leaves a length of "doubled" paper, it would be unusual to encounter this on a book press. It seems to be common with presses used for newspapers, at least where I live.

FM screening (printing) Frequency modulation screening, a digital screening process that varies the frequency of equal-sized dots to create a halftone image. At one time considered a "breakthrough" technology for more attractive color images with smoother appearance. In practice, it requires significantly higher precision in printing and press operator skill. See stochastic screening.

focal length (photography) For a camera lens, all parallel light rays entering will be focused (on the opposite side) to a point referred to as the principal focal point. The distance from the optical center of the lens to that point is the principal focal length (f) of the lens. For a thick (or multi-element) lens made from spherical surfaces, the focal distance will differ for different rays, and this discrepancy is called spherical aberration. The focal length for different wavelengths (colors) of light will also differ slightly, and this is called chromatic aberration. The principal focal length of a lens is determined by the index of refraction of the glass, the radii of curvature of the surfaces, and the medium in which the lens resides. In 35 mm photography a "normal" lens has about a 50 mm (2 inch) focal length.

fog (photography) Image density in unwanted areas, often caused by unintended light striking the film (or sensor). With very sensitive films (those with high ISO sensitivity ratings, e.g. above 500), repeated exposures to x-rays during airline security checks may cause the film to fog.

folded and gathered (bindery) Abbreviated "F&G." The stacks of unbound books after gathering has been completed. See gather. For books printed in color, F&Gs are often provided as advance copies rather than bound galleys.

folio (typography) The page number. A drop folio is a page number at the bottom of the page. This book has drop folios on the opening pages of the chapters.

font 1. (output) A digitized set of characters in a single typeface design. A digital font can be any size unlike old-fashioned typography where each size is considered a font (and all sizes a typeface). Plain (or Roman), Bold, Italic, and Bold/Italic are the traditional styles within a font family. Fonts come in three varieties, PostScript, TrueType, and OpenType. 2. (typography) A set of characters including uppercase, lowercase, numerals, and punctuation. In the world of metal type, this means a given alphabet, with all its accessory characters, in a given size. All sizes of a particular design are a typeface. In the world of digital type, it is the character set itself or the digital information encoding it. The term "font" is often used when "typeface" is more correct. Confusion of the terms font and typeface first came about when non-typographer programmers started writing software that made significant use of typefaces —which they (incorrectly) called fonts.

footnote (editorial) A note appearing at the bottom of a page referring to a reference mark somewhere on that page. The typographer's dilemma occurs when a reference occurs near the bottom of a page so that when room is made for the footnote, the reference point moves to the next page. As a result, I usually prefer to use end notes. Sometimes, supplemental comments, not specifically relevant to the main text are placed in notes (either foot or end). In some cases, it's appropriate to use end notes for source references, but to use footnotes for expository material on the page where the reference occurs. In that case, use superscript numbers for the end notes, but use asterisks for footnotes. See end note.

form (printing) 1. (offset) The assembly of pages and other elements for printing on a single sheet. 2. (letterpress) Type and other material locked in a chase (frame) for printing.

format (artwork) (noun) The size, style, type page, margins, printing requirements, etc. of a printed page. (verb) To prepare artwork in the size, style, etc.

form rollers (printing) The rollers, either inking or dampening that directly contact the printing plate on a printing press.

fountain solution (printing-offset) A solution of water, natural or synthetic gum, and other additives used to dampen the plate and keep non-printing areas from accepting ink. The evaporation of the solution also cools the press, reducing heat build up (a problem with "waterless printing"). A primary additive is isopropyl alcohol, which is the main source of volatile organic compounds in offset printing. (Use of petroleum-based or vegetable oil-based inks have little, if any, impact on VOCs in printing.) Considerable research is directed at removing VOCs from fountain solutions. Use of acid-free papers has also created problems with fountain solution becoming too alkaline during press operation—requiring testing and use of acidic additives to keep the solution neutral. Proper use and adjustment of fountain solution is one of the most complex areas of press operation and is one of the more difficult challenges for a press operator. (Most print-quality deficiencies can be traced back to fountain solution adjustment problems.)

FPO (artwork) Acronym for "For Position Only." Sometimes artwork is prepared with proxy images in a document to be replaced at a later time with the final image.

free sheet (paper) Paper free of wood pulp. (The irony of the name is that such paper is far from "free" but usually is rather more expensive.)

freeware (industry) Software that some nice person creates for themselves and then puts out into the world at no cost to others. Usually available at web sites run by computer magazines or at others like www.VersionTracker.com.

French spacing (typography—metal era) Text composed *without* a second space after a period in a sentence to create tight spacing. This is considered normal spacing with digital fonts.

front end system (prepress) The workstation(s) with the software used to prepare images and text. The term is usually reserved for proprietary systems rather than desk top publishing systems.

frontlist (publishing) Books published in the current season, and featured in the publisher's current catalog.

front matter (publishing) Printed material that appears in the front of a book, before the main body of text. Typically includes the title page, copyright page, dedication, table of contents and preface. May be numbered with lowercase Roman numerals (in which case the body of the book begins numbering in Arabic numbers from page 1) or may be numbered consecutively with Arabic numbers throughout the whole book.

f-stop (photography) Fixed stops for setting lens aperture. See aperture.

FTP (communications) (File Transfer Protocol) The Internet's protocol for moving files from one computer to another.

function key (interface) A key at the top of a keyboard that carries out some operation when pressed. Most function keys are application-specific. Big in the pre-Windows world, but discouraged in Macintosh software where the graphical interface is supposed to make them unnecessary. See FKEY.

furniture (typography) Small blocks of wood of various sizes, rectangular in shape, used in conjunction with a quoin to lock hand-set or machine-set lines of metal type within a chase (frame).

fuser (electrostatic printing) The component in a copier or laser printer that applies heat and pressure to the toner and paper, causing the resins in the toner to melt and bond to the paper.

fuser roller (electrostatic printing) The heated roller inside the fuser unit that passes over the paper and toner, performing the fusing function. See fuser.

galleys or galley proof (artwork) Before software was developed that allowed books to be formatted "direct to page," typesetting was output in long sheets called galleys, later the galleys would be cut up in page-sized blocks and "pasted up" (with headers and folios) as pages. Before the considerable paste-up work was performed, the typeset galley proofs would be provided to proofreaders and editors for checking. Authors are generally also given an opportunity to review galleys for errors or significant changes. How much an author may change the typeset copy is often spelled out in contracts. This is also the source of the term "bound galley."

gamma (artwork) A measure of contrast in photographic images.

gathering (bindery) The assembly of folded signatures into the proper sequence.

gauge (typesetting) A metal or plastic ruler calibrated in picas on one edge and inches on the other. Transparent rulers may have a small section marked off in points. Used to measure type. I have also found a simple depth gauge (available in most hardware stores) to be quite useful. Typically about 6 inches long, made of metal with a sliding pocket clip with a perpendicular metal strip that assists in saving a measurement. Very useful for checking that type is parallel to the edge of a page or for measuring a stack of paper to determine a spine width.

GCR (prepress) Acronym for gray component replacement. In creating a color separation, black ink is used to replace portions of cyan, magenta, and yellow ink in colored areas as well as in neutral areas. GCR separations tend to reproduce dark, saturated colors and maintain gray balance better on press than other techniques. See UCR.

gear streaks (printing) Parallel streaks appearing across the printed sheet at the same interval as gear teeth on the cylinder. Obviously something you hope you'll never see.

geek (programming) A name given to socially inept experts, particularly in the computer field. Once a pejorative (among high school students) it is now more of a term of adulation. The current "richest man in the world," Bill Gates, was often considered a geek. Now he is the ultimate example of the geek's revenge.

generation (printing, copying) Each succeeding stage in reproduction from the original copy. One advantage of direct-to-plate printing is that the plate is the first generation and the printed books are the second generation. With the traditional original, negative, plate approach, the printed copy is (at least) third generation. There is always some slight degradation in the image with each additional generation in the process.

genlock (display) The ability to synchronize two video signals—for example, a video image with a computer-generated title on the same screen.

genre (publishing) Genre generally denotes nothing more than a category of a book, as in self-help, mystery, romance, western, sci-fi, historical, etc.

ghost writer (publishing) A writer who prepares material that is published in someone else's name; often performed under a work-for-hire agrement or a flat-fee contract. In some cases (particularly with better-known writers) some form of credit is given, for example a book by "famous celebrity" as told to "Mary Smith."

GIF (graphics) A Compuserve graphic format with 1 to 8 bits per pixel. It is compressed and is sometimes used for images on the Internet.

gigabyte (storage) A unit of measure, technically 1024 megabytes, which is roughly one billion bytes—1,073,741,824 to be precise. Impress your friends with this useful knowledge.

glossary (self-reference) 1. A list of terms used in a particular discipline and their meanings. The better ones are cross-referenced.

goldenrod paper (offset) A specially coated masking paper of yellow or orange color used by strippers to assemble and position negatives for exposing printing plates.

gothic (typography) Another name for type designed without serifs, that is, sans serif. Sometimes confused with highly ornate early type designs known (in the U.S.) as blackletter. It is also the term used to designate blackletter type in the U.K.

grabber (interface) The little hand icon with which you can move your document around in its window. First seen in the original MacPaint, the grabber has become a standard tool in most graphics programs.

grain (paper) The fibers in paper tend to align with the direction that the sheet moves through a paper-making machine. Try tearing a sheet of paper, and you'll quickly discover that you can make a relatively straight tear with the grain, but it is most difficult across the grain. Presses usually operate best if the paper moves through the rollers oriented parallel to the direction of travel. The paper grain should be parallel to the spine in books to reduce likelihood of distorted spines or curling.

graphical user interface (interface) The interface used on the Macintosh, developed at Xerox PARC and more recently copied by Windows. This type of interface uses pictorial representations of real-world things on the screen and gives you a mouse or other pointing device to interact with them. Nickname: "gooey" (GUI).

graphics tablet (interface) A flat panel used with a special pen (called a stylus) to move the cursor around the computer screen, especially in graphics programs. More advanced models are sensitive to the amount of pressure you're applying and alter the line weight accordingly. See digitizing tablet.

gray balance (artwork) The dot values (intensity) of cyan, magenta, and yellow that produces a neutral gray.

grayscale (display) 1. The number of bits per pixel determines the number of values of gray that can be differentiated; for example, 8 bits permits a maximum 256 shades of gray. A TIFF file can be saved as a grayscale document. Grayscale images are typically used for photos in printed books. Some reproduction methods reduce the number of shades that can be observed. Most printed halftones may only reproduce 80 or 90 shades of gray. 2. (artwork) A strip of standard gray tones ranging from white to black, placed next to an original copy during prepress photography to measure the tonal range and gamma (contrast) obtained.

greek (output) Nonsense text used to simulate the finished appearance of a document without distracting you with its content. This term also means the gray lines that stand in for text when a document has been reduced on the monitor.

greeking (typography) The use of gray bars or "dummy" characters to represent text that is too small to be legible when displayed on the screen. Also, in graphic design, the use of dummy text in a layout so that the design of the document will be emphasized rather than its content. See lorem ipsum.

gripper edge (printing) 1. The leading edge of a sheet as it passes through a press. 2. The edge of a printing plate that is attached to the front clamp on the plate cylinder.

gripper margin (printing) The unprintable area on the sheet where the grippers clamp, usually ½ inch or less.

grippers (printing) On a sheetfed press, metal fingers that clamp the sheet and control its movement through the press. It is the action of the grippers that give printing presses superiority over laser printers and copy machines for front-to-back registration. (Sheetfed laser printers and copy machines only use rollers and belts to move paper through the device.)

grotesk (typography) Another way to describe letters without serifs. This is the German spelling and is often applied to typefaces that were designed in Germany.

groundwood pulp (paper) A mechanically-prepared wood pulp used in the manufacture of newsprint and publication papers.

GUI (interface) Acronym for Graphical User Interface. Pronounced "gooey." See graphical user interface.

gum arabic (printing) Used with offset plates to protect them from oxidation or other degradation while being stored.

gumming 1. (platemaking) The process of applying a thin coating of gum arabic to a lithographic printing plate. 2. (eating) What a toothless person does with food.

gutter (artwork) The blank space or inner margin from printing area to binding.

hairline (typography) 1. A thin stroke usually common to (modern) serif typefaces. 2. A thin rule, usually a quarter point. Some computer programs create a line one pixel wide. This is acceptable on the screen and on some laser printers, but usually results in a line that will not reproduce clearly from an image or platesetter. It is best to specify a .25 point line, when you wish to use a "hairline."

hairline register (printing) Register (between two or more colors) of less than ½ row of halftone dots.

halation (photography) A blurred effect, resembling a halo, usually occurring in highlight areas or around bright objects. It's caused by light reflecting off the base layer of photographic film.

halftone (output) An image that is composed of solid black dots of various sizes at equal spacing, which creates an illusion of various shades of gray. (Also called amplitude modulation screening.) This is how continuous-tone images, such as photographs, have traditionally been reproduced by printing presses, which are not capable of generating true grays. See also FM or Frequency modulation screening.

hanging bullets (typesetting) An arrangement where all bullets are set flush left and all lines of copy that follow are indented.

hanging indentation (typesetting) The first line of text is set to the full width and the following lines are indented. This glossary uses hanging indentation. Also called "hanging indents."

hanging punctuation (typography) In justified text, hyphens, periods and other punctuation marks at the ends of lines are positioned partially or completely outside the main text block margins to improve the appearance of the typography. InDesign is the only page layout program that automatically generates hanging punctuation. It is used in this book. Also called "optical margin alignment."

hard copy (output) The printed version of what you have in the computer.

hardcover (publishing) Books bound with a hard, cloth-over-cardboard cover and covered with a paper dust jacket. Also called case bound.

hard dot (printing) Halftone dot with little or no fringe and prints with little or no dot gain. See soft dot.

hard drive (storage) The machinery that permanently encloses a disk together with the disk itself. The disk is a round piece of hard plastic or metal coated with a magnetically sensitive surface. Most hard drives have more than one disk inside, stacked on a single shaft, with a read/write head for each surface. The machinery spins the disk and reads its contents very quickly. Hard drives come in 2.5″, 3.5″ and 5.25″ diameters. Those most commonly used can hold from 40 gigabytes up to 1000 gigabytes or more. The maximum size available has been doubling about every 24 months. Same as hard disk.

hard proof (artwork) A proof output on paper or other substrate as distinguished from a *soft proof,* which is an image on a video display.

hard return (output) A carriage return that will always end the current line regardless of where it is on the page. A hard return starts a new paragraph.

hard space (output) Sometimes called a non-breaking space. A space that looks like a space to you but not to the computer. The sentence will not be broken by word wrapping at a hard space. On the Macintosh, type Option-Space to insert a hard space into a document.

hardware (hardware) The parts of the computer or any peripherals that you can bang on. The monitor, keyboard, modem, scanner—all the things you can touch are hardware. As opposed to software.

head crash (storage) What happens when the read/write head of a hard drive contacts the disk surface, or even when a piece of dirt gets in between them. Very bad. You can't replace your divots on a hard disk. The only protection is to backup frequently.

headline (editorial) The title of an article. Also called a head.

head margin (artwork) The white space above the first line on a page.

header (typography) A line of type above the main text block on a page. See running header.

He/Ne (hardware) Helium-neon red laser with a wave length of 632 nm. The most common laser used in platesetters. It's proven more difficult (expensive) to produce shorter wave-length (blue) lasers, but the shorter wave length allows more precise and higher resolution spot placement.

hexadecimal (system programming) A number system consisting of sixteen integers from 0 to 9 and A through F. This system, aka base-16, dovetails beautifully with the binary system since a pair of hexadecimal numbers can represent discreet values from 0 to 255 in base-10, the same as two bytes can. Hence 01 hex equals 00000001 binary and 1 in base-10, while 61 hex equals 01100001 in base-2 and 97 in base-10, and is used to represent the letter a. Hex is much easier for us humans to read than a mosaic of zeros and ones.

hickeys (offset printing) Spots or imperfections in the printing due to dirt or debris on the press, usually paper particles or dried blobs of ink. Has nothing to do with what went on in the backseat of a teenager's car. In the press room, other, less delicate terms are also used.

hierarchical menu (interface) A sub-menu that pops up when you select a specific menu item. Hierarchical menus are used to provide many choices within a category without bringing up a dialog box. Menu items that lead to sub-menus are designated by a black triangle, on either side of the menu, pointing sideways.

high contrast (photography) A reproduction with high gamma in which the difference in darkness (density) between neighboring areas is greater than the original. Also refers to films commonly used in graphic arts that are designed to expose with maximum contrast, rather than capture intermediate levels of gray.

high-level language (programming) Programming languages that are closer to natural speech such as BASIC. They are easier to program but often result in bulky, slower programs.

highlight (interface) 1. To make something visually distinct from its background by changing or inverting its colors. Objects that are selected or chosen are highlighted. 2. (artwork) The lights or whitest parts in a photograph represented in a halftone reproduction by the smallest dots or the absence of dots. Where significant portions of a photograph's highlights have an absence of dots, it is said that the highlights are *blown out*.

hi-lo books (publishing) Books that have a high-interest level but low reading level. Hi-Lo books are often used to encourage reluctant readers from middle grades and up.

hint (output, typography) Techniques developed along with outline fonts that let any printer accurately render serifs, stems and any other character components at any type size or orientation. Hints are important for on-screen and low resolution output (on a laser printer), but are not important for output above 1000 dpi.

holdout (printing) A property of coated paper with low ink absorption which allows ink to set on the surface with high gloss. Papers with too much holdout may cause problems with set-off.

hot metal (typesetting) Refers to an earlier generation of typesetting machines that used melted lead to cast type as each project was produced. The most common machines were the Linotype and the Monotype.

hot spot (interface) The one pixel in the cursor that actually shows the mouse's location on screen, and the part of the cursor that must be aimed accurately. On the arrow cursor, the hot spot is the tip of the arrow.

HSV (color) Acronym for Hue, Saturation, and Value (or luminance). A color space used in some graphics programs. Similar in some respects to LAB color.

HTML (industry) Acronym for HyperText Markup Language. This is how material on the Internet is usually encoded for display by a Web browser. HTML is a subset of SGML and is being superseded by XML.

hub (networking) A connecting point for network cables. Most hubs handle the topography issue for the network and have electronics to ensure that the signals are being transmitted and received reliably. Network topography refers to the arrangement of computers on the network. When computers are connected sequentially they are said to be "daisy chained;" while computers connected through a hub are usually in a "star" topography.

hue (color) The main attribute of a color that distinguishes it from other colors.

humanist type (typography) Letterforms which originate from the humanists of the style Italian Renaissance. There are two kinds of humanist letterforms: Roman (based on Carolingian script) and italic (first appearing in the 15th century). Humanist letterforms show the clear trace of a broad-nib pen held by a right-handed scribe.

hydrophilic (printing) Water receptive. Corresponds to the non-printing areas on a lithographic printing plate.

hydrophobic (printing) Water repellent. Corresponds to the printing areas on a lithographic printing plate.

hydrophobia (medicine) A longer name for the disease, rabies, based on a symptom where rabid animals refuse to drink water.

hypermedia (industry) Simply the extension of the hypertext idea to cover non-textual information, like pictures and sound (not much of an intellectual leap).

hypertext (industry) A concept articulated by Ted Nelson in the 1960's, in which any piece of textual information on a computer can be connected to any other. Users can jump from one piece of information to other related pieces quickly and thereby learn things in a nonlinear, semi-random sequence. Users should also be able to create these links between pieces of information. This is the basis of the extensive linking that occurs on the Internet.

I-beam (interface) The cursor shape that looks like the edge-view of an I-beam. When you want to enter text, position the I-beam cursor and click. The insertion point (a flashing vertical bar) appears where you clicked, and you're ready to go.

ICC (industry) 1. Acronym for International Color Consortium. The ICC was formed in 1993 for the purpose of creating and promoting the standardization of an open, cross-platform, vendor-neutral system for color management. See *www.color.org* for more information. 2. The Interstate Commerce Commission, formed in the 1890s to regulate surface transportation services in the United States. It was rendered moot by deregulation of transportation and was dissolved in about 1990 (its few remaining functions were absorbed by the Department of Transportation). The author became interested in publishing and printing as Tariff Publishing Officer for Southern Pacific Railroad and he had frequent contact with the Federal ICC.

icon (interface) A small picture representing something. On the Macintosh, icons represent documents, applications, devices and sometimes processes. For example, a disk icon represents a physical disk but the Trash Can represents the action of throwing something away. Windows has copied this system.

image assembly (prepress) See stripping.

image processor (output) An application that takes images and edits them. You can edit your brother's ex-wife out of a family snapshot, or add yourself to the cantina scene in Star Wars. As this technology becomes undetectable advertisements are more enticing than ever. It lends a new twist to Groucho's joke, "Who are you gonna believe, me or your own eyes?" Photoshop is the professional standard for this software. Consumer-level versions are often crippled for professional use by not saving files in the CMYK color space.

imagesetter (output) A very high resolution laser-based printer that prints on photographic paper. A related device is the platesetter. These devices image directly to metal or plastic printing plates. Image- and platesetters can create output at up to 3,600 dpi, but most text is output at 1200 or 1270 dpi and images are often output at 2400 or 2540 dpi. The data passes through a raster image processor (RIP) that converts PostScript data to the position data necessary to create the output image. These machines differ from a traditional typesetting output device primarily in the format of the data used to generate the output.

import (applications) Bringing a file created by one application into a document created by another. To successfully import something you need to know if the application doing the import is capable of handling the file format of the incoming document. Many programs use special filters to convert formats, and there are file conversion programs such as DeBabelizer. You can almost always import a plain-text file, and tab-and-return delimited databases go to and fro quite easily. It gets more complicated with graphics.

imposition (prepress) The positioning of pages on a signature so that after printing, folding, and cutting, all pages will be right-side-up and in the right sequence. One of the reasons that stripped negatives are rarely compatible between various printers is that they use different imposition strategies to accommodate their unique blends of processes and equipment.

impression cylinder (printing) The cylinder on an offset printing press against which the paper picks up the impression from the inked rubber blanket (blanket cylinder).

imprint (publishing) The name of a publisher's specific line of books that has its own, distinct characteristics. HarperSan Francisco, for instance, is the New Age and religious book imprint of HarperCollins.

incremental backup (storage) A type of backup where only those things which have been added or changed since the last backup are copied.

indent (output) White space between the margins on a page and a body of text.

Normally used only as a first-line indent, either inside the margins like you learned in typing class or outside as a hanging indent. Most indentation is accomplished with margin settings since word processors give each paragraph its own ruler. Not really a character (no ASCII value), but it's almost one.

indicia (artwork) 1. Identifying marks or indications. 2. A box with information describing the mailing status of postage paid printed documents. Most commonly used with junk mail, but may also be used with newsletters and other bulk mailed items.

initial caps (typesetting) Describes the situation where the first letter of each word is capitalized. Most programs (that have a "change case" feature) call this the "title case."

initialize (storage) The process of erasing a disk's directory so that new data can be written over the old. Since initialization doesn't remove the data, it may be recovered with a disk utility if the disk is initialized by mistake.

ink fountain (printing) The device on a press that stores and supplies ink to the inking rollers. The device also has a means to control the rate at which ink is passed on to the rollers along their length. The press operator (or software) must adjust these controls to allow just the right amount of ink into the system so that the image is neither starved of ink nor the plate overwhelmed with ink.

inkjet printer (output) A type of printer whose print head squirts droplets of ink through tiny nozzles onto plain paper. This amazing technology made very low cost 360 to 1440 dpi printing available to most users. With most technologies in common use, the ink can smear, even after it's dry (if it gets wet). Also, the ink tends to wick through the paper fibers with untreated or uncoated papers, causing the output quality to be less satisfactory than the resolution would suggest. With suitable paper, some printers produce very attractive photographic quality prints from digital images.

ink mist (printing) The tendency of some low-tack inks to form filaments (or threads) during press operation.

input device (input) Any device through which you get information into or control the computer. This term includes keyboards, mice, data sensors, etc., but usually excludes things like disks and modems.

insert (industry) A printed piece prepared for insertion into a book, magazine or other printed piece.

insertion point (interface) The place where what you type will next appear on screen, identified by a flashing vertical bar. You can move the insertion point by clicking somewhere else in the text; you can get rid of it by clicking outside the text area.

installer (system software) Software that automates the process of putting an application or even a new operating system onto your hard drive.

institutional sales (publishing) Refers to books primarily sold to schools and libraries. Both trade and mass market books can have institutional sales. Children's book publishers typically rely on institutions for a large portion of their sales.

interface (interface) Broadly, the way any two things communicate with each other. In this glossary interface is short for "user interface," which is the way a computer appears to the user, and the rules by which they communicate. The Macintosh and Windows has a Graphical User Interface, nicknamed "gooey" (GUI) because it makes use of pictures and images to convey meaning. Most other (older) computers use a Command-Line User Interface, called "clooey" (CLUI) which uses specific typed commands to convey instructions.

interlace (display) A kludgy trick to work around the limitations of the NTSC video system, which is the original standard in the United States. NTSC consists of 30 frames per second at 525 lines per screen resolution. That is low resolution compared to Europe or Japan which use up to 1,000 lines of resolution in the video signal. In order to "fill out" the image and minimize flicker, one-half of the image is drawn every $\frac{1}{60}$ second, first the 262.5 odd numbered lines, then the evens. Couch potatoes are waiting until HDTV comes along. Computer displays are non-interlaced—a major reason why they cost more—and many computer CRT displays run at 66 to 80 frames per second.

interleaved 2 of 5 (barcode) A numeric only bar code symbology. Characters are encoded in pairs with the bars representing the first character and the spaces representing the second.

Internet (communications) "The Internet"—The international telecommunications network formed by thousands of networks connecting millions of academic, industrial, government and personal computers that exchange messages.

interpreter (programming) A program like BASIC that takes a program written in a high-level language and translates it "on the fly" to machine code that the processor can carry out. Interpreters are very convenient to work with since you can interrupt the process at any point to make changes or debug. However, since the translation takes substantial time, your program runs more slowly under an interpreter.

invisible file (system software) A document or application which is on a disk but whose icon (or name) is not shown. It is not counted as an item, and cannot be selected in any standard way. Such files can still be accessed by applications and will be seen by utilities. Any file can be made invisible.

IR   (industry) Abbreviation for Infrared Radiation above 700 nm.

IRT   (transit) Abbreviation for Interborough Rapid Transit, the first subway in the U.S. When attending BEA in NY, you take the IRT from LGA. (The IRT is now merged into the MTA.)

ISBN   (industry) Acronym for International Standard Book Number. Represented by the ISBN/EAN bar code on books. A numbering system used by the publishing industry to identify each book. ISBNs have recently been expanded from 10 to 13 digits.

ISDN   (communications) Acronym for Integrated Service Digital Network. A worldwide standard for digital telecommunications. ISDN features two channels running at 64 kilobytes per second plus a third low-bandwidth channel. ISDN is still available in some locations, but has been generally displaced by DSL.

ISO   (industry) Abbreviation for International Standards Organization. The really big organization headquartered in Geneva that slowly works out telecommunications, industrial, and scientific standards for the whole world.

ISO 9000 (industry) A set of standards for quality management systems that is accepted around the world. Currently more than 90 countries have adopted ISO 9000 as national standards. Involves substantial record keeping and bureaucratic nonsense so that a claim that "the standards are met" may be made. Includes process analyses techniques using the label "continual improvement" (unless it was changed to "continuous improvement" last week). Yet another management fad that is more of a substitute for good management than a systems improvement. The best management will generate excellence with or without these distractions.

ISSN   (publishing) Acronym for International Standard Serial Number. Represented on magazines and periodical publications by the ISSN/EAN bar code.

italic   (typography) A class of letterforms more cursive than Roman but less cursive than script. It was originally designed to replicate handwriting. Italic typefaces have been designed to correspond with related Roman typefaces. Now, italic is used to provide emphasis or otherwise set-off portions of text. In typography, italic is used instead of underline.

jaggies (graphics) The stepped effect of bit-mapped type and graphics caused when square pixels represent diagonal or curved lines. Especially noticeable when a low resolution image is enlarged significantly.

Java   (computer programming language) Java is a cross-platform computing environment that can work with existing computers using network connections. An important aspect of Java is that it eliminates demands on central servers and moves processing to the "client" computers. Java "applets" and applications can be dynamically downloaded and run. Java applications can run on any "Java enabled" web browser like Netscape Navagator™ or Microsoft's Internet Explorer. Java has been widely accepted in the computer industry and is supported under Windows; Macintosh OS; IBM OS/2; UNIX (from many vendors); Novell NetWare 4.0; and IBM's mainframe operating system, MVS. Microsoft attempted to co-op Java by creating "extensions" to the language that are only availabe on Windows computers. This has been the subject of lawsuits between Sun Computers (originator of the Java language) and Microsoft. See also applet.

JDF   (industry) Abbreviation for Job Definition Format. A data exchange standard that will act as an electronic job ticket that contains control data from print buying through estimating, customer service, prepress, press, finishing, and dispatch. JDF contains production information rather than content data. At this time, there's a lot of talk about JDF in the printing industry, but implementation is in the very early stages.

jog   (bindery) To align sheets of paper into a compact stack. When I opened my print shop, a "jogger" was part of the equipment package. I quickly learned (after demonstrations from experienced press operators) to hand-jog fairly large stacks of paper. The jogger became an expensive "work order holder" near the press. I doubt that it was used more than once a year.

JPEG   (output) Acronym for Joint Photographic Experts Group. A graphic file format used for compressing large bitmapped graphic images. QuickTime uses JPEG to compress images to as little as ½₀ their original size. However, there is a trade-off: JPEG is called "lossy" because when you decompress an image, you don't get back the exact original. Repeated cycles of compression (saving) and decompression can significantly degrade an image, and should be used with care. JPEG is the default storage format for most digital cameras. Once a file from a digital camera is opened in an editing program, it should be saved with a non-lossy compression method until all processing/editing is complete.

justify (typography) To space out lines to line up uniformly on the left and right. Most of the text of this book is justified. Scientifically devised tests have shown that reading comprehension improves with well-justified text. Likewise, poorly justified text can retard comprehension.

kern (typography) 1. (verb) To adjust the spacing between two characters. Specifically, to alter the fit of certain letter combinations so that the limb of one projects over or under the body or limb of another. Typefaces are designed with specific space around each letter, but some combinations, such as WA look awkward unless brought closer together. Kerning can be done in any amount, but small amounts are best. 2. (noun) Part of a letter that extends into the space of another.

kerning table (typography) Special instructions for adjusting the spacing between specific pairs of letters in a typeface. Better quality typefaces have from several dozen to hundreds of "kerning pairs" defined. Use of optical justification with InDesign reduces the importance of kerning tables. That is, a typeface with a poor or completely absent kerning table can be made usable in InDesign with optical kerning turned on.

kerning (typography) The act of adjusting the space between two characters, usually to make them closer together.

key (typography) (verb) To type copy into a document. Also, to code copy into a dummy by means of symbols, usually letters. See lorem ipsum.

keyboard (interface) The input device that reminds me of my old "loyal" Royal typewriter. The time-honored key layout is called QWERTY, but some pioneering types use the more efficient DVORAK layout. Other developments include the curved Maltron keyboard, and a keyboard which separates into two parts so you can hold your hands more naturally. Advocates claim these last two help avoid the dreaded carpal tunnel syndrome.

keyline (artwork) 1. An outline drawing of finished art to indicate the exact shape, position and size for such elements as halftones, sketches, etc. 2. A narrow line, usually ½ to 1 point wide, drawn around a halftone to give it a more defined edge. Once quite difficult to execute using analog methods (a very steady hand with a razor blade being required), they are now quite simple to create with page layout programs.

kilobyte (storage) Abbreviated as K, kb, or KB. 1024 bytes, exactly 2 to the 10th power. Kilo- is Latin for thousand. The most common measurement for computer file length.

kiss impression (printing) A very light impression, just enough to produce an image on paper.

kludge (programming) 1. Awkward, makeshift and non-intuitive. Used to describe applications or methods of getting things done . 2. A work-around that effectively solves a problem, usually in an innovative way. Sometimes used as: "It was a kludgy way to handle the problem."

kraft   (paper) A paper or board containing unbleached wood pulp, brown in color, made by the sulphate process. Sometimes used to describe a brown color. Corrugated 'cardboard" cartons are usually made from kraft papers.

lachrymal (typography) See teardrop terminal.

lacquer (printing) A clear resin/solvent coating, usually glossy, applied to a printed sheet for protection or appearance. May be limited to give a "spot" effect. See UV coating, aqueous coating.

laminate (bindery) A thin plastic film applied over the printing. Protects a book cover, and adds shine (gloss laminate) or creates a dull finish (matte laminate). A plastic laminate will eliminate the danger of ink cracking on the corners of the spine. While an extra cost, it pays for itself by protecting the cover from scuffing or from moisture transferred from people's hands.

LAN   (networking) Acronym for Local Area Network. Computers joined together with hardware and software to form a Local Area Network (LAN), allowing hard drives, printers, and other devices to be shared. LANs typically have 50 or fewer computers hooked together in the same building, but very large installations do exist. A LAN may operate on a peer to peer arrangement (usually limited to smaller networks with fewer than 10 computers) or it may have servers running special software to support the linked computers with various services.

laser   (industry) Acronym (now considered a word) for Light Amplification by Stimulated Emission of Radiation. The laser is an intense light beam with very narrow bandwidth used in digital imaging devices to produce images by pulses against a light-sensitive material. Also used in pointing devices, CD and DVD players, and numerous other applications.

laser printer (output) A great leap forward in printing technology that allows us to produce high-quality graphics that look like they were done by professional printers. The output of a laser printer is a page whose image is composed of extremely small, evenly spaced dots of ink at resolution of 600 dpi (600 dpi is most common, 1200 dpi is now available). The laser beam electrically "etches" the image upon a statically charged drum which then picks up charged powdered ink. The ink is then handed off to the paper and made permanent by a hot roller. First developed by Xerox, laser printing technology is based on photocopier technology.

launch (interface) To run, or start, an application.

layer   (interface) A system built into image drawing, manipulation, and page layout programs (such as Illustrator, Photoshop and InDesign) in which layers can be established to hold individual elements without affecting items on other layers (except to 'hide' lower items where a higher layer overlays those below). These layers can be turned on or off as needed. For example, in a newsletter project, advertisers are listed on a layer in rectangles "reserving" their space. The ads are positioned on another layer. That way, proofs can be sent out for editorial review without the large file size if the actual ads were included. An advertiser index can easily be made by indexing the advertiser names (on the "list" layer).

When the newsletter is prepared for final output, the "list" layer is made invisible and the actual ad layer is present.

layout (artwork) 1. The drawing or sketch of a planned printed piece. 2. The act of producing a printed piece using layout software. 3. (platemaking) A sheet showing the settings for a step-and-repeat machine.

LCD (display) (Liquid Crystal Display) A relatively flat screen whose circuitry can cause individual pixels to display a color. The screen itself is transparent and requires some sort of back light for adequate visibility. They're also impossible to repair (except to replace the backlight), so if more than a few pixels are bad the factory has to dispose of the screen. The most common type of LCD is called "active matrix" and has a transistor for each pixel. First used only on laptop computers, LCD displays are now being used for desktop computers and for high definition TV sets.

leader (typography) A small character such as a period, hyphen or underline that leads your eyes from one column of information to another, such as in a table of contents. Many applications let you attach leaders to tab stops.

leading (typography) (pronounced led-ing, rhymes with sledding.) An additional space added between the bottom of the descenders and the top of the ascenders in subsequent lines of type. Usually it is expressed as a baseline to baseline measure, e.g., 10 point type with 12 point leading, therefore an additional 2 points of space is added between the rows of type. In the metal-type era, this was usually done with strips of lead.

lead in (typesetting) The first few words in a block of copy set in a different or contrasting typeface. This book uses a lead in of small caps following the drop cap at the beginning of each chapter.

ledger (paper) 1. A kind of very durable paper used for accounting work. 2. Paper that measures 17 x 11 inches oriented in the landscape direction as distinct from tabloid paper, 11 x 17 inches oriented in the portrait direction.

leg (typography) The lower diagonal stroke of the letter k.

letterspacing (typography) The placement or adjustment of space between letters in a word. When using negative letterspacing, exercise care to avoid actually having the letters touch one another. Very wide letterspacing is sometimes used for artistic effect.

ligature (typography) Two or more letters tied into a single character to perfectly design their spatial interaction and eliminate unpleasant touching. Most common with fi and fl.

light face (typography) A lighter variant of the standard typeface weight.

line copy (artwork) Any copy suitable for reproduction without using a half-tone screen. Text, pen & ink drawings, and simple line drawings are usually suitable line copy.

line for line (typesetting) Instructions to a typesetter to set the copy with the exact same line lengths as they appear in the original copy.

line height (typesetting) The measure of a line of type plus the leading.

line length (typesetting) The maximum line length of the copy, usually measured in picas but often expressed in inches in the desktop publishing era.

line spacing (typography) The distance from the baseline of one line of text to the baseline of the next, measured in points (¹⁄₇₂ of an inch). Some word processing programs might offer "single line spacing" or " double line spacing" to reflect the practice used with typewriters. This term is more often used by those whose background is computer programming or word processing rather than typography or graphics.

link (typography) The stroke that connects the bowl and the loop of a lowercase Roman g.

linoleum block (artwork) A style of illustration carved in relief into a block of wood covered with a sheet of linoleum. Originally used with letter press equipment. Clip art designs in EPS with the linoleum-block-look are readily available. See also wood cut.

lists (publishing) When publishing professionals talk about a "list," they are referring to the books designated for publication for any given selling season. Most often, publishers offer new lists twice every year—spring and fall.

list box (interface) A window or dialog box that contains a list of things (or even a series of icons or pictures) that you can select from.

literary agent (publishing) Experienced book industry people who represent authors and illustrators for a percentage of their clients' profit. Agents typically claim 10%-20% of advances and royalties, but are generally more experienced at contract negotiations and have market knowledge and contacts that most authors do not. Some agents charge reading fees to review a prospective client's manuscript. Those fees can be very high, and do not guarantee placement of the manuscript with a publisher. In some cases, literary agents earn more from reading fees than from commissions—try to avoid such agents!

Literary Marketplace (publishing) An annual industry guide that lists publishers, editors, marketing executives, agents and more, along with their addresses and other pertinent information. The guide is usually available in public libraries, most often in the reserve section.

lithowrap Hard cover book where the cover is printed with the design. This is usually used only with children's picture books or text books. See page 173.

local area network See LAN.

localization (programming) Adapting software to another country, culture, or language by rewriting the menu names and menu items, default settings, and anything else necessary to make the application understandable and appropriate.

lock (storage) To protect a file or disk (or other storage medium) from being changed, written to, or erased. Protection can be achieved physically, by sliding the write-protect tab on a floppy disk to where you can see through the hole, or in (Macintosh) software, by clicking the checkbox named "Locked" in the file's Get Info window.

logotype (or logo) (artwork) The name of a company or product in a special design or symbol used as a trademark in advertising. The Apple Computer logo is a silhouette of an apple with a "bite" (byte?) missing from one edge. (⌘).

loop (typography) The lower portion of the lowercase g (in most serif typefaces).

loose leaf (binding) Leaves punched to go into a loose leaf binder. May be a way to distribute material subject to frequent change. Most of the freight tariffs at Southern Pacific Railroad were issued as loose leaf. Only those pages that changed needed to be distributed, a tracking method was used so that the reader could be aware of what the most current pages should be.

lorem ipsum (artwork) The name given to filler text used as an example in a document before the actual text is ready. Usually, it is derived from part of a passage from Cicero, specifically *De finibus bonorum et malorum*, a treatise on the theory of ethics written in 45 B.C.—some samples have been designed to replicate the word lengths and structures of English to give a reasonable idea of the actual text appearance. Other samples have been massaged to include some humor.

lossless (storage) Compression without data loss. The Stuffit or Winzip programs compress files without data loss.

lossy (storage) An adjective to describe a compression scheme, typically for a bitmapped image like a captured frame or photograph or a video sequence, that loses a little bit of data every time a file is compressed. Lossy compression algorithms like JPEG squeeze files up to ten times tighter than lossless compression like Compactor.

lowercase (typography) Noncapital letters such as a, b, c, etc. Derived from the practice of placing these letters in the bottom (lower) case of a pair of typecases.

lpi (output) Acronym for Lines Per Inch. The number of lines in a halftone image. Halftone images are composed of solid dots of various sizes at equal spacing, which creates an illusion of various shades between black and white (gray tones). This is how continuous-tone images, such as photographs, have traditionally been reproduced by printing presses, which are not capable of generating true grays. Resolution when defined by the number of lines (of dots) per inch, does not readily translate into dots per inch on an electronic output device such as a laser printer, because these are digital devices with dots all the same size. In fact, sophisticated algorithms are required to translate from lines per inch (halftone) to dots per inch (electronic). PostScript is an example of such an algorithm. When preparing an image, the rule of thumb is to make the dots per inch resolution about twice the lines per inch value used in printing. This provides sufficient data to make a reasonably good rendering during conversion to the halftone by the output device.

M 1. (storage) Abbreviation for Mega, commonly used to mean one million. In actual computer terms, it stands for 1,048,576, and is used to indicate an amount of storage as in megabyte. 2. (industry) Abbreviation for 1000. (From the Roman numeral for 1000). "We printed 24M sheets for the order."

machine coated (paper) Paper which is coated on one or both sides on the paper making machine.

machine direction (paper) Same as grain direction in paper. See grain.

machine language (programming) Computer instructions at the lowest level—the language of the processor itself. Serious 0's and 1's. This code is almost always the end result of a program written in a mid- or high-level language, then turned into machine language by an interpreter, or the duet of a compiler or assembler and then a linker.

macro (interface) A series of commands, mouse movements, and keystrokes that are recorded and played back (actually, reenacted). Macros are used to automate complex or repetitive tasks. They are created typically by having the computer record a series of actions as you go through them (this is called a "watch me" macro) or by writing out instructions in a special programming language. The languages used for making macros are usually not as powerful as traditional programming languages, and so they are often called "scripting" languages, and the macros are called "scripts." Microsoft products (Excel and Word) have fairly extensive macro capability, but are also notorious for harboring virus-like macros. Macros in Photoshop (called "actions") and in InDesign can be quite useful for automating repetitive tasks.

magenta (color) Hue of a subtractive primary and one of the four color process inks. It reflects or transmits blue or red light and absorbs green light. (If a CMYK image looks too green, you add some magenta to compensate.)

magenta screen (prepress) A type of contact screen (magenta in color) used in making halftones with analog techniques.

magnetic storage (storage) Any disk, tape, strip, drum, or core that is used to store digital information. Such storage is susceptible to damage by stray magnetic radiation. For example, placing a BART ticket near a magnet (as a magnetic latch on a purse) may erase the information stored in the magnetic strip.

magneto-optical drive (storage) A combination of optical and magnetic storage technology. The disk itself is magnetic, but a laser beam is used to encode the surface. These drives have much greater capacity because the light beam can be focused much tighter than a magnetic field.

magnification (barcodes) The method of describing bar code width and height in some symbologies. 100% magnification is nominal size. Most books print EANs at the minimum allowable 80% size.

mail merge (output) A document combined with a database, usually for the purpose of making form letters. For example, a letter would have "holders" in place of name, address and title, and a database containing that information could be merged with that letter to create a series of "customized" letters.

majuscule (typography) Archaic term for an uppercase letter, see also minuscule.

makeover (printing) Repeating a process, usually due to an error. For example, a plate may be remade, resulting in a makeover plate. Or worse, a complete job may have such serious production errors that it must be printed again. A key management goal (in printing) is to reduce the number of makeovers in a shop.

makeready (printing) All the work done to set up a press for printing, such as mounting plates and adjusting registration. Short run printers focus on reducing makeready time and effort as that is a major disruption to productivity. Short run book printers may run 10 or more jobs per day on a press; fast makeready is imperative as no revenue is generated during makeready. (In this sense, each signature requires its own makeready, so a book with 5 signatures requires 5 makereadies before the printing is completed.)

manuscript (publishing) An original work of an author submitted for publication. Normally tendered in electronic form as word processing files. Abbreviated MS.

marking engine (hardware) The device within an image- or platesetter that actually generates the image on the substrate.

malware (software) A computer program written to cause harm to the computer it is running on. See virus.

margins (typography) The non-printed area around the edges of the page.

markup (typesetting) Specifications for typesetting noted on the original copy.

marquee (interface) A selection area indicated by a rectangle of dashed lines that move around the boundary, featured in the Finder and lots of graphics programs. The result looks kind of like marching ants, or like the chaser lights on a theater marquee. Hence the name.

mask (output) The act of covering up a portion of an image either to avoid seeing it or to protect it from some process you intend to perform on the unmasked portion. This term is also used as a noun for the thing itself, such as an opaque material used to protect selected areas of a printing plate during exposure.

mass market books (publishing) Paperback books that are smaller (and cheaper) than trade paperback books. In addition to bookstore placement, these so-called "rack sized" books are often distributed through drugstores, airports, supermarkets and the like. The cheap paper used in their manufacture is the source of the term "pulp fiction."

mass market publishers (publishing) Publishers who concentrate on high-volume releases of paperback books that typically fit current popular market needs. Titles tend to be short-lived compared to hardcover books, but often are printed in far greater quantities.

master (printing) A plate for a duplicating machine. ("Presses" in local retail print shops are usually classified as duplicators or duplicator-presses.) See paper masters.

matte finish (paper) Dull paper without gloss or shine. Also used to describe plastic laminate that has a dull finish without gloss or shine.

matrix (typography) The copper block onto which the steel die for a letter was stamped. The matrix served as the mold for the face of a type or for a printing plate.

meanline (typography) An imaginary line that establishes the height of the body of lowercase letters.

measure (typography) The width of a line of type, usually expressed in picas.

mechanical (prepress, artwork) A term for a camera-ready paste up of artwork, including type and images all on one piece of art board.

mechanical binding (bindery) A method using fasteners to bind a book. Wire-O, spiral, plastic comb, and Velobind are methods of mechanical bindings.

mechanical pulp (paper) Groundwood pulp made by mechanically grinding logs, woodchips, or lumber mill waste. It is mostly used for newsprint or as a base for low grade printing papers.

megabyte (computer industry) Abbreviated as MB, mb, Meg, or M. A byte is a small unit of data used by the computer. For example, one byte is required to store one letter of the English alphabet. (Some eastern alphabets require two bytes.) When accumulations of bytes are discussed they are referred to as kilobytes (1024 bytes) and megabytes (1024 kilobytes or 1,048,576 bytes or 2 raised to the 20th power). Computers count in these "odd" numbers due to their binary nature. Bytes are made up of 8 BInary digiTs (bits) and since the computer counts by "twos" everything works out to a power of 2 (1024 is 2^{10}). Megabytes are convenient for measuring the amount of RAM and size of Hard Disks. See also Gigabyte.

megahertz (industry) Abbreviated as MHz. Frequency of one million cycles per second. Measures bandwidth or electronic analog signals. Not long ago, was the typical, though imprecise, scale used to "rate" the speed at which computer processor chips operated; however, the most common computer chip speeds are now rated in gigahertz. These speed ratings are valid for comparing chips of the same design, but may be misleading when comparing chips of different designs.

memory (hardware) Where information is stored. It comes in two varieties, volatile and nonvolatile. Volatile memory lives in chips and is dependent on a continuous electronic current and can be easily lost (RAM is volatile memory). Nonvolatile memory does not depend upon current (e.g. a floppy disk). Memory is measured in groups of bytes (kilobytes, megabytes, gigabytes, etc.).

menu (interface) A list of commands which appears on screen so you can select commands from your mouse. Pull-down menus which are accessible at the top of the screen or window, as well as pop-up menus in some dialog boxes. Choosing a menu item results in immediate action unless the item ends with three dots in which case a dialog box listing more options will appear.

menu bar (interface) The horizontal strip usually visible at the top of the screen which contains the titles of menus.

metric system (industry) A decimal measurement system adopted by most countries for liquid (volume), solid (weight), and distance measurements. While extensively used in the U.S. in manufacturing and engineering, it has been resisted in the consumer marketplace, probably because it was first conceived in France.

microcode (hardware) Machine-language instructions that tell the CPU how to handle complex instructions such as those that take more than one clock cycle and/or use a lot of different parts of the chip. A high-overhead feature of CISC architecture, virtually eliminated by RISC chips.

micron (industry) One thousandth of a millimeter. Measurement often associated with the positioning of barcodes.

middle reader (publishing) A general description of books intended for children ages 9-11.

middletones (photography) The tones between shadows and highlights in a photograph or halftone.

midlist (publishing) Titles on a publisher's list that are not expected to be big sellers, but might introduce a new author or find audiences in niche markets. Midlist books are often mainstream books, as opposed to books that neatly fit into genres

In recent years, large publishers have concentrated on potential "blockbuster" titles, and have decreased the number of midlist books published.

minuscule (typography) Archaic term for a lowercase letter, see also majuscule.

MIPS (hardware) Acronym for Millions of Instructions Per Second. A common, if somewhat misleading, measure of a computer's processing power.

mode (interface) A state in which your choices are limited. Good software design tries to avoid modes and let you choose any command at any time. Dialog boxes are a willful intrusion of modality because they occur when specific information is needed before the program can continue.

modem (communications) Term is derived from MODulator-DEModulator. A device that enables computers to communicate over telephone or cable television lines by converting digital signals into frequency modulated tones or signals and vice versa.

model (programming) A graphic or mathematical representation of an idea, whether a scientific problem or a three-dimensional object. The earliest kinds of modeling were used for weather forecasting (and bomb blasts). In fact, these studies fostered the emergence of chaos theory. Graphic modeling is used to create architectural views and fantastic visual effects for Hollywood movies.

modern type style (typography) Letterforms with flat serifs, abrupt and exaggerated strokes, and vertical shading. Originated by Francois Didot in the late 18th century, this style represented a casting away of the decorative baggage of the rococo era. An unfortunate term, as the passage of time has made "modern" historic and not at all contemporary.

modifier key (interface) Any key that has no character associated with it, but which changes the behavior of the keys. The modifier keys on the Macintosh are Caps Lock, Command, Control, Option, and Shift. When using modifier keys, you hold those keys down and then tap the character key that it modifies. For instance, the keyboard shortcut for Paste is (usually) Command-V. A few applications also recognize the Num Lock key as a modifier. Under Windows, the Control key usually replaces the Command key for most circumstances.

monotype (typography) A typeface design with uniform stroke thickness, examples include Courier and Kaufmann. See monoweight. Monotype was also the brand name of a "hot metal" typesetting machine that competed with the Linotype.

monoweight (typography) A typeface with no thick/thin transition, the width of the strokes making up the letter are all the same. See monotype.

moiré (artwork) The undesirable pattern caused by incorrect screen angles of overprinting colors in CMYK halftones. Similar patterns produced when scanning (or re-screening) a halftone image. Sometimes taking a photograph of a halftone (with a regular camera), may sufficiently eliminate the effect. Some scanners have "descreening" features in their software.

molleton (printing) A thick cotton fabric similar to flannel used on the dampening rollers of a press.

monitor (computers) A video screen on a workstation. See CRT.

montage (artwork) Several photographs (or other elements) combined to create a composite illustration. Montages are now relatively easy to prepare in Photoshop, and are thus somewhat overused. Still, when well crafted, can make an attractive cover design.

motherboard (hardware) A piece of fiberglass onto which the most important chips fit, including the CPU which runs your computer. Also called a "logic board."

mottle (printing) A spotty or uneven appearance in solid areas of ink. Frequently a problem with covers printed digitally (toner simply doesn't "lay down" as well as inks). A solution is to add "noise" to the solid areas in Photoshop. In offset printing, other techniques are often required. When printing solid colors of an individual ink (a PMS color), double striking (using two plates) the sheet will often help, especially if one plate is "screened" at about 50%.

mount (storage) In the old days, "mounting" meant taking a reel of computer tape and mounting it on the tape drive's spindle so the computer could use it. Nowadays it means making any volume (a floppy disk, removable cartridge, CD, DVD, hard drive or even a disk partition) available to the computer. Usually this happens without your having to do anything specific.

mouse (interface) A little hand-held rolling device that moves a cursor around a computer screen and comes with one or more buttons. Moving the mouse cursor and clicking the button is one way of controlling a computer. The latest mouse designs use optical scanners rather than mechanical rollers (eliminating cleaning problems) and some are wireless, using a radio signal to transmit the movements to the computer.

MS and MSS (publishing) Abbreviation for manuscript or manuscripts.

mullen tester (paper) A device for testing the bursting strength of paper.

multi-tasking (programming) The ability of a sophisticated computer to calculate separate problems at the same time. Multi-tasking has been used for many years on large computers to handle multiple users. On a non-multi-tasking computer, when the computer has to figure something out the user has to wait. The latest versions of Windows and Mac OS are multi-tasking operating systems.

multimedia (multimedia) Any combination of graphics, video, animations, text and sound. A leftover word from the 1960's (you might enjoy reading "The Electric Kool-Aid Acid Test") that was all the rage in the '90s.

Mylar® (printing) A polyester film especially suited to stripping positives because of its strength and dimensional stability. (Film positives are rarely used in the U.S., but are quite common overseas.)

nanometer (industry) One billionth of a meter. A unit in which wavelengths of light and other radiant energy is expressed. Also used when measuring the width of circuits on computer chips. A human hair is about 80 nanometers in diameter.

nanosecond (hardware) One billionth of a second, or the time it takes light to travel one foot. In this realm, the distance between chips on a circuit board becomes an issue. The speed that RAM chips are able to accept read and write operations is measured in nanoseconds. With speedier processors you need faster RAM chips to keep everything flowing. Faster RAM chips are being introduced along with faster CPU chips with each new generation of computers.

native mode (system software) Software specifically written for the hardware it is running on, thereby taking full advantage of the capabilities of that hardware and running at maximum speed.

negative (photography) Film containing an image in which the values of the original are reversed so that dark areas appear light and vice versa. Negatives are frequently used as an intermediate step in the process in getting from original copy to the exposed printing plate.

negative letterspacing (typography) A type specification in which the space between characters is reduced beyond the default setting either by kerning or tracking.

negative leading (typography) A type specification in which there is less space from baseline to baseline than the size of the type itself (for example, 40-point type with 38-point leading).

net royalty (publishing) A royalty paid to an author based on the money the publisher actually receives from each book sale after discounts are given to book stores or buyers. This royalty calculation method tends to be favored by small publishers. It also eliminates the need to contractually adjust royalties for deep discount sales, making the publishing contract easier to administer.

network (hardware) Computers connected together in order to exchange information and resources. Usually this is accomplished with cables but sometimes with modems and telephone lines. See LAN.

newsprint (paper) Paper made mostly from groundwood pulp (with some chemical pulp) used for printing newspapers. Usually, the lignin liquor (a component of wood) is not completely removed, causing the paper to rapidly turn yellow and become brittle in the presence of strong sunlight.

node (networking) A device that is attached to and is capable of communicating with a network. Includes computers, workstations, printers, servers, bridges, gateways, and routers.

noise 1. (communications) Any unwanted data mixed in with the good stuff you're after. In telecommunications it is static in the phone line or satellite link. In graphics it can be odd bits of dirt on your image or colored specs in a digital photo. 2. (marketing) The general level of messages being sent to a potential buyer. To get your message through, you must penetrate the noise to get the attention of a prospect.

nominal size (barcodes) The standard size for a bar code of a given symbology.

nonbreaking space (typography) In typesetting, a special space character placed between two words to keep the words from being separated by a line break. See also hard space.

non-impact printer (hardware) An electronic device, such as a copier, laser, or inkjet printer that creates an image without mechanically impressing the substrate.

nonreproducing or nonrepro (artwork) A special pen or pencil that leaves marks invisible to graphic arts films, usually a light blue color. No longer used in the all-digital era.

notch binding (binding) A method of adhesive binding where groves are cut perpendicular to the spine to hold additional glue. Provides superior spine strength compared to simple adhesive binding. See smyth sewn.

NSA (industry) Acronym for National Security Agency. The NSA is in this glossary only because they possess the largest concentration of computing power in the U.S., and quite possibly in the world. Just thought you might want to know.

NTSC (display) (National Television System Committee.) This name is most commonly used as an adjective for a kind of video format, notably the standard for broadcast video in the U.S. NTSC IV video has two interlaced fields per frame and 30 frames per second. Wags say that it stands for "Never Twice the Same Color." Other standards are PAL (Britain) and SECAM (France). Some day HDTV will replace this half-century-old standard, but the programs will probably not be any better.

oblique (typography) A text style created by slanting a Roman font to simulate italics. Commonly used with sans-serif fonts.

oblong (bindery) A book or catalog bound on the shorter dimension. Also called landscape orientation.

OCR (input) Abbreviation for Optical Character Recognition. A process whereby hardware and software look at printed or typed words and interpret them as words rather than as pictures. OCR technology has improved and someday may actually be convenient. OCR software works with a scanner and (in theory) converts printed pages into word processing files. It's still a toss-up whether it's more efficient to scan/OCR/correct a document or to simply retype it. When conditions are ideal, it can be a big time saver.

OCR-A and OCR-B (industry) Character sets or fonts both of which may be machine or human read. Used for the human readable information in bar coding.

OEM (industry) Acronym for Original Equipment Manufacturer. The company that actually makes the hardware that other companies buy and resell under their own labels. You rarely need to know the OEM, as computers contain many components built by others (optical drives, hard drives, power supplies, LCD panels), but they are usually covered by the computer makers' warranties and service agreements.

offline (communications) The state of not being connected to another computer. Used to describe your computer when it is performing actions that do not depend upon being connected to another machine. A printer can be offline if it's turned off but still connected.

off loading (computer) Relieving the central processing unit (CPU) of intensive computing tasks, most commonly done with sophisticated video display cards that handle graphics calculations and display tasks outside of the CPU.

off-press proofs (prepress) Proofs made by photomechanical or digital means in less time and at lower cost than press proofs. These are the usual way proofs are made in North America. Press proofs are more common with color printing done overseas (Asia mostly), but digital proofing methods are starting to become more common with overseas printers as well.

offset (printing) 1. More correctly offset lithography. The most common commercial printing process in which the ink is transferred from the plate to a rubber blanket cylinder before being transferred to the paper. 2. Sometimes (incorrectly) used to indicate set-off, where damp ink is transferred to the back of the sheet next above in a stack. This is an undesirable situation. It can be controlled by use of spray powder and/or removing the paper from the output section of the (sheetfed) press before the stack of printed pages has become too heavy.

old style (typography) Dating from the 1490s, old style letter forms have the weight stress of the rounded forms at an angle. The serifs are bracketed by a tapered curved line. The top serifs on the lowercase letters are at an angle.

oleophilic (printing) Oil receptive. See also hydrophobic/hydrophilic.

oleophobic (printing) Oil repellent. See also hydrophobic/hydrophilic.

online (communications) Connected to another computer, usually the Internet. Any device that is connected and ready to be used is online.

opacity (paper) The property of paper which minimizes the show-through of printing from the back side or the next sheet.

opaque 1. (printing) To paint out areas on a negative not wanted on the plate. Usually done with a red-colored "white-out"-like paint. This technique allows items to be easily removed from a page without re-typesetting and re-imaging the negative. The use of direct to plate technology has eliminated this task. 2. (paper) The property that makes paper less transparent. Some grades of paper are called "offset opaque" that exhibit more opacity than the non-opaque grade of the same paper. Such papers are moderately more expensive than non-opaque grades. Often, the same benefit of opacity can be achieved by using a darker (ivory or natural rather than white) or thicker grade (60 lb. instead of 50 lb.).

opaque ink (printing) An ink that conceals all color beneath it. Most inks are translucent and are affected by the color of the underlying substrate. A normal blue ink printed on goldenrod paper will appear to be green, while an opaque blue ink will remain looking blue.

operating system (system software) Software which provides the means by which your computer runs applications, manages drives, prints documents, and so on.

OPI (artwork) Abbreviation for Open Press Interface. An extension to PostScript that automatically substitutes high resolution images for low resolution images used as placeholders. Most of the time book printers want the OPI feature turned off or otherwise disabled. If you have need to use OPI, consult with your printer to properly arrange for it.

Opentype (typography) A recently introduced file format for typefaces. It may contain either a TrueType or PostScript face. It supports unicode and allows as many as 32,000 characters in a single face. File format is the same on both Macintosh and Windows, so Opentype font files work on both platforms. Adobe, the primary promoter of the format, has reissued their complete library in Opentype format. A few Opentype format faces are available from other vendors.

optical disc (storage) A disc (not disk) that is read and written to using light instead of magnetism. Their great advantage is that since the light beam is smaller than the magnetic flux path, they can hold substantially more data. CD and DVD-ROMS are the most popular kind.

option clause (publishing) A clause in a publishing agreement that often requires the author to sell his or her next work to the same publisher on the same terms listed in the current agreement. An option must usually be exercised by the publisher within a short period of time, or the author is then free to seek a new publisher for his or her upcoming book.

orphan (typography) A single word (or syllable) that is left over on the last line at end of the paragraph. Alternatively, a single word or line from the beginning of a paragraph left at the bottom of a column or page, with the remainder on the next column or page. In general, you should try to minimize the number of orphans in your typeset work. See widow.

orthochromatic (photography, prepress) Photographic surfaces, plates, or films that respond to ultraviolet, blue, green and yellow, but not red light. Most photographic materials used in print shops are limited in their response to colors of light—generally being sensitive only to ultraviolet and blue light. Some products have a broader response, yet are not sensitive to red light and can then be handled in a darkroom that is (dimly) lighted with red "safelights." Consumer films are panchromatic—they respond to all colors of light and must be handled in total darkness when outside a camera or protective packaging.

Otabind™ or RepKover™ (binding) A particular method of using a binding tape inside a soft cover to create a perfect bound book that lays open by itself. Usually described as a "lay-flat" binding (not to be confused with "lay flat" laminate). These names refer to a patented process. Usually the patented process is superior to alternative methods. Both names use the same process, but are different licensees of the same patents.

outline font (output) An obsolete term that applies to fonts that use mathematical descriptions of each character rather than bitmapped images. Outline fonts can be viewed and printed at any size and at any resolution. The two types of outline fonts are PostScript and TrueType, although the math is different. When an outline font is printed, the equations are interpreted and a bitmap is created (rasterized) at the resolution of the printer. See RIP.

overhang cover (bindery) A cover larger in size than the pages it covers. This is common with case bound (hard cover) books.

overlay (artwork) A transparent covering over the copy where color break, instructions, or corrections are marked. Also, transparent or translucent prints which, when placed one upon the other, form a composite picture. This was most frequently done to make certain kinds of analog color proofs.

overprinting (printing) Double printing; printing over an area that already has been printed. Sometimes done as a cost savings, for example to insert a new phone number on letterhead when a change was made. Other times, the technique is used with preprinted "shells," for example, a cruise line makes up nice, CMYK brochures, but leaves certain areas open for a travel agent to add specifics about special tours, package deals, or prices.

overrun (printing) Copies printed in excess of the ordered quantity. In book printing, the argument is that there are uncertainties in the number of copies lost (unusable) while setting up various bindery machines. Printing contracts commonly allow a printer to overrun and charge for an additional 10% beyond the basic order. Many printers use this as an excuse to increase their bill by 10%. In reality, most printers can easily estimate the number of unacceptable copies likely to be generated during machine setup, and can then limit the overrun so that the customer receives the exact count.

packet (networking) A series of bytes sent as a group over a network. A packet includes data plus a source and destination address and possibly notes about rejoining the data with other segments sent in separate packets. Networks break data into packets so that lots of computers can share the lines nearly simultaneously.

page description language (output) Abbreviated as PDL. A computer language designed for describing how type and graphic elements should be produced by output devices. The printing and graphic arts industry almost universally uses the PostScript PDL. Many Windows computers and low-cost laser printers use Hewlett-Packard's PCL (printer control language).

page makeup 1. (analog prepress) The assembly (stripping) of all elements to make up a page. This is the work a print shop "stripper" performed. 2. (digital prepress) The electronic assembly of page elements to compose a complete page with all elements in place on a video display terminal and on film or plate. This requires a very different skill-set than a traditional stripper. As time passes, the skills of the traditional strippers is fading away.

pagination (artwork) 1. Describes how books are broken into pages after initial typesetting. Pagination is particularly critical in picture books, where space is usually limited to 24 or 32 pages (including front and back matter) and illustration must be generously accommodated. 2. The process of performing page makeup automatically using computers.

palette (computers) 1. The collection of colors available in a graphic system or program. 2. A selection of options available to a program in a small "floating" window adjacent to or overlaying the main working window. In recent years, many graphics programs have had a significant (indeed, overwhelming) selection of palettes added to their interface. In Photoshop, you can "park' many palettes in "the well" (on the right side of the command bar), turning them into drop down menus. For graphics production, I've added dual monitors to my primary workstation to handle the proliferation of palettes.

pallet (industry) A portable platform for handling cartons in transportation and storage. Usually made of narrow wood boards, but now often made from plastic or even treated wood fiber. The pallet raises the cartons about 4 inches above the floor and allows a fork lift to reach under and lift from all 4 sides. The most common sizes range from 30 x 40 inches to 36 x 48 inches. See skid.

panchromatic (photography) Photographic film that responds to all visible colors.

paper master (printing) A lithographic printing plate made on paper-like substrate. The material is often resin coated paper, but some plastics are also used. These plates are commonly used on small offset duplicators common in quick print shops. Generally, such plates are rated for up to 10,000 impressions. Paper masters tend to stretch during use, making critical registration difficult, particularly on longer runs. Their use on larger presses is rare, since the stretching problem is magnified in larger sheet sizes. Also called paper plates, these are created using an electrostatic method (like a Xerox) or by using a photographic silver-based chemistry. The electrostatic plates are generally of lower quality and resolution than those made with the photographic method.

parallel processing (industry) More than one processor working on the same problem at the same time or working on separate parts of a problem at the same time. A more powerful alternative than a single processor (which most computers use), regardless of how fast that single chip may be. Surprisingly, it uses cheaper hardware. Some schemes, called massively parallel processing, use thousands of processors. Some supercomputers are made of 1000 or more Macintosh G5s.

PARC (industry) (Palo Alto Research Center) A research lab run by Xerox where the basic ground rules of the graphical interface were developed almost twenty years ago. These first rules were then implemented in Xerox's Star computer, developed by their Office Products Division. The Star was one of the inspirations for the original Macintosh when Steve Jobs toured the center and became enamored with the interface. PARC also developed laser printers, Ethernet, PostScript and object-oriented programming.

parity (barcodes) A way of encoding bar/space patterns using an odd or even number of modules bars or spaces. The parity may be calculated by the scanner to provide an error check.

partition (storage) A subdivision in memory or on a hard drive. Partitions are areas that are treated separately even though they are on the same drive physically.

Pascal (programming) A language created by a Swiss computer scientist named Niklaus Wirth and developed at UC San Diego that was popular at the time the Macintosh was being created. The Macintosh's original operating system was written in Pascal, but lately the computer language "C" has become dominant. (C is strongly associated with the UNIX underpinnings of Mac OS X.)

paste (interface) Taking whatever is in the Clipboard and transferring it into the current selection. The contents of the Clipboard can be text, pictures or anything.

paste dryer (printing) An additive for printing ink that uses a combination of compounds to aid in drying.

paste up (prepress) The process of arranging page elements on a master sheet. Use of rubber cement was once popular, but that gave way to the use of hot wax to hold the materials on the backing sheet. Yet another dying art. See mechanical.

patch (programming) A term for a minor modification to existing software.

pathname (programming) A list of names that tells you where an application or document is. The first part of the pathname is the name of the volume (disk or other device), followed by the sequence of nested folders which lead to the file itself. The parts of a pathname are separated by colons (:), slashes (/), or backslashes (\)—depending on the system: Mac, UNIX, or Windows—which is why you can't use these symbols in the name of a folder (directory) or file. As you navigate your way to the file, you move along the path.

PC 1. (hardware) The exciting name IBM selected for their personal computers. Now, it generically refers to almost any brand of desktop computer. 2. (society) Short for politically correct, as in politically correct speech. Oddly, PC speech seems to be a particular enthusiasm of those who claim to be in favor of "free speech"—which under the rules of PC speech, must agree with them. See *Alice in Wonderland*.

PCI (hardware) (Peripheral Component Interconnect) A type of expansion slot used in most desktop computers.

PCL (output) (Printer Control Language) A set of commands by Hewlett-Packard used to control its LaserJet and DeskJet printers. PCL has evolved over time becoming very sophisticated. With the development of PostScript clone software, PCL has become less dominant, as there is little price difference between clone-PostScript and PCL interpreters. See Page Description Language (PDL).

PDA (hardware) Abbreviation for Personal Digital Assistant. A handy, pocket-sized device with address book, calendar, and notepad built in. Some consider it a digital brain. See also PIM.

PDF (industry) Abbreviation for Portable Document File. PDF is a universal electronic file format, with reader software available for Mac, Windows, and UNIX computers. Based on PostScript, PDF can retain very sophisticated and complex formatting in a document, yet allow transfer and output across platforms with little or no significant change to the document. (Printing on some non-PostScript devices is not always accurate.) When created properly, fonts are embedded in the file, so exact typography remains intact. Special standards for exchange of electronic files have been developed, particularly for use by magazine publishers and ad agencies to facilitate error-free file exchange. The standards, PDF-X/1-a and PDF-X/3, will ensure that a file created to the standard will output as expected in a environment that was unknown to the file creator. These standards are also quite useful to book publishers, as file compliance is a relatively sure way to create error free files. I use the PDF-X/1-a standard when I prepare PDFs for the printers I use.

PDL (industry) See Page Description Language.

perfecting press (printing) A sheetfed press that prints both the front and back of a sheet in the same pass. Most book printers use perfecting presses. Web presses are all designed to print both sides of the sheet at the same time.

perfect bound (binding) A method where a paper cover is glued to the book block. Also called paperback or soft cover.

peripheral (hardware) Any hardware that attaches to the computer, including some items that may be inside the computer case. Modems, scanners, disk and CD/DVD-ROM drives, and printers are all peripherals.

permissions (publishing) Describes the acquisition and fee paid in order to use all or part of existing, copyrighted material. Permissions are typically sought, for instance, when a publisher wants to excerpt part of a book to use in an anthology. Authors who wish to incorporate a portion of someone else's work in their own creation are obligated to secure permissions. Contracts often require such authors to obtain and pay for permissions, though the responsibility can often be negotiated. Permissions also apply to photographs or other artwork that may appear in a book.

pH (industry) A number used for expressing the acidity or alkalinity of solutions. A value of 7 is neutral in the most widely accepted scale that ranges from 0 to 14. Solutions with a value below 7 are acid, those above 7 are alkaline. The pH of the fountain solution is one of the ongoing factors that the press operator must monitor. When paper was manufactured with acid residues, the fountain solution would become progressively more acidic during press operation. Now that paper is usually "acid-free," the fountain solution tends to become more alkaline during press operation. Either situation can cause the lithographic process to degrade, causing ink to appear on the sheet in unintended areas.

photoconductor (digital printing) Materials (usually a drum or belt) that become photosensitive after being charged by a corona wire in a copier or laser printer.

photomultiplier tube (prepress) Part of a drum scanner. see PMT.

pica (typography) A typesetting measurement. 1 pica equals 12 points, ⅙ inch or about .167 inch. Type size and line spacing is measured in points, line length in picas. Postscript uses picas of exactly ⅙ inch. The original analog pica was very slightly longer. Be careful that you don't use an old-style pica ruler when measuring computer type. See point.

picking (printing) The lifting of the paper surface during printing. This happens when the tack (pulling force) of the ink is greater than the surface strength of the paper. This is often the source of debris that creates unwanted blobs of ink to appear elsewhere on the sheet.

picture books (publishing) A type of book primarily aimed at children from preschool to age 8. Because of the high cost associated with manufacturing picture books, most picture books are published in 24 or 32 pages, including front and back matter. A high percentage of picture books are printed overseas (usually in Asia) due to the high cost of process color printing in the U.S. and the tight profit margins in publishing.

pigment (printing ink) The solid, microscopic particles that give printing inks their color and opacity. The pigment in black printing ink is carbon black, a pigment that is, essentially, soot. In inkjet printers, the first inks used were derived from chemical dyes rather than pigments. The chemical dyes tend to breakdown in light and to oxidize in the atmosphere, fading rather rapidly. This can be counter acted with specially treated papers and/or coatings to protect the ink. Some more recent inkjet printers use pigment-based inks.

piling (printing) The build up or caking of ink on rollers, plate or blanket that interfere with ink transfer. Also, buildup of paper dust or coating fragments on the blanket of an offset press.

PIM (applications) (Personal Information Manager). Software that converts your computer into a very expensive name-and-address manager, calendar, or the like. Special software is used to coordinate your desktop, laptop, and PDA, but they're probably never all in sync. See also PDA.

pin register (prepress) The use of accurately positioned holes and special pins on copy, negatives, plates and press to ensure proper register of colors. Digital prepress techniques and on-press plate imaging (as done with "DI" presses) eliminates any need for this technique.

pixel (display) Abbreviation for Picture Element. The smallest dot that can be displayed on the screen or a printed page. On earlier computer screens, each pixel was $\frac{1}{72}$ of an inch on a side. More recent computers use higher resolution screens, often with 96 pixels to the inch. On laser printers, one pixel is commonly $\frac{1}{600}$ of an inch on a side. Pixels on color monitors are composed of three spots, red, blue and green.

plastic comb binding (bindery) A type of binding that uses a curved piece of plastic with fingers inserted through small rectangular holes punched along the binding edge of the document. The combs come in a wide variety of colors. In larger volume, are economic to have silk-screened with the title. Not suitable for binding bound galleys. See spiral binding, Wire-O®, and Velobind.

plate cylinder (printing) The cylinder of a press upon which the plate is mounted. Ink is applied to the plate cylinder which, in turn, presses it on to the blanket cylinder which, in turn presses it onto the paper.

platesetter (hardware) A high resolution output device, similar to an imagesetter, that outputs paper, polyester, or aluminum plates instead of photographic paper or film. Some imagesetters can also be platesetters, simply by substituting the output substrate. See imagesetter.

platform (hardware) A buzzword for a computer and its peripheral devices (like a printer or disk drive) which together can run software.

PMA (industry) Publishers Marketing Association. See *www.pma-online.org* for information. Not to be confused with the Photographers Marketing Association, an organization of professional photographers.

PMS (artwork) Abbreviation for Pantone Matching System. A system of color samples used to define standard colors for printing. The PMS establishes standards for color reproduction that can be quantified and compared. The Pantone company offers a wide variety of color charts with definitions of over 1000 colors. The PMS is almost universally used in the United States. There are some other color systems used in Europe and Asia, however, one sign of an overseas printer interested in the U.S. market, is that they are familiar and able to work with the PMS.

PMT (prepress) Abbreviation for photomultiplier tube. A light sensor that can detect very low light levels and amplify the signals for further processing. They are the main sensing element in high quality drum scanners. While horrendously expensive scanners with this technology continue to set the standard for high-quality scans, table top scanners using CCD (charge coupled devices) have been approaching ever closer to the quality of PMT-based scans. For most projects, it's no longer reasonable to justify the extra expense for drum scans.

point (typography) Basic increment of typographical measurement, equal to .013837 inch. Computer programs usually round the point to .0138888 inch making exactly 72 points to the inch. This is a rounding loss of approximately ¼ point per inch. With the ascendancy of desktop computers, the 72 points (and 6 picas) per inch has become a de facto standard measurement. Care must be exercised as point/pica measuring rules can be found both with the traditional measure and the new rationalized measure.

poor trapping (printing) A condition in printing where less ink transfers to previously printed areas than to unprinted paper. Also called under trapping. Yet another aspect of the process requiring vigilance of the press operator.

porosity (paper) The property of paper that allows the permeation of air, an important factor in ink penetration.

portrait (artwork) The vertical orientation of an image or sheet of paper. See also landscape.

port 1. (communications) (noun) A small plug on the back of the computer into which you connect cables. Types include serial, parallel, SCSI, Ethernet, USB, Firewire, MIDI, audio, modem, VGA, DVI and many more. Each port has its particular protocol for information transfer. Sometimes plug adapters (with or without electronic circuits) can be used to convert one kind of port into another. 2. (programming) (verb) Rewriting an application so it runs on a computer system for which it was not originally written. A good port adapts the application to the new computer; a bad port makes your computer look like another brand. 3. (viticulture) A fortified wine that originated on the Iberian peninsula. It takes its name from the city of Oporto, located at the mouth of Rio Douro (River of Gold) in Portugal.

position proof (artwork, printing) A color proof for checking position, layout and/or color break of image elements.

positive (prepress) An image that shares the same characteristics of light and dark as the original. The opposite of a negative. Film positives instead of negatives are used for printing plates throughout much of the world outside of North America. This can be quite a surprise if an overseas printer provides "the film" and a publisher attempts to use it to obtain a reprint in the U.S.

PostScript (output) A popular page description and programming language for defining complex graphics at any resolution, developed by Adobe Systems. This is the page-description language most laser printers use to print high resolution graphics and text. It has become an industry standard. If you are using 37-point outline type and it looks crummy on the screen, it will still print nice and smooth on a PostScript printer. This type of printer has a built-in PostScript processor that does the work of scaling and rasterizing the font. By contrast, with TrueType the computer's processor does all the work of scaling the font for both the printer and screen. PostScript has been updated to PostScript Level 3, but the level 2 version is installed on many machines still currently on the market. Most PostScript print drivers allow you to specify which version to output. PostScript clones (compatible systems) also exist. Most such clones are excellent substitutions, but I've encountered cases where a clone PostScript will fail to print projects that output correctly on devices using true Adobe PostScript.

PostScript Level 3 (output) Introduced in about 1997, PostScript Version 3 is the most up-to-date variant of this venerable PDL. Most of the new features of this version supplement and enhance the high-end output devices used by commercial printers. There are still many printers available with PostScript Level 2 or clones that only incorporate level 2 features. While it may seem "nice" to have the latest version PostScript, I have never encountered any files that failed to print as expected on my Level 2 printer. However the color-related features of PS Ver. 3 are an incentive to seek the most current PostScript version for a color output device.

PostScript printer (output) A printer that can interpret the graphics and fonts that are written in the PostScript page description language. Not all laser printers can read PostScript. Usually if a laser printer is not a PostScript printer, it is a PCL printer. Most inkjet printers use drivers based on simplified processes. They tend to output PostScript graphic images poorly. The easiest workaround is to generate a PDF, then print the PDF to the inkjet printer. The Acrobat software then handles the interpretation of the PostScript code in the file.

posture (typography) The slant of a typeface.

power supply (hardware) The box in the computer (outside, for many laptops) that converts line voltage (AC power) to low-voltage DC needed by the computer's components.

power user (programming) Once you start using keyboard commands, install a few extensions, create a macro or two, and start throwing around words like PCI and PostScript, you can consider yourself a power user. Say words like bandwidth and 64-bit processor in mixed company and you will really get respect.

PowerPC (hardware) Processor chips developed jointly by Apple, IBM and Motorola for Macintosh and other computers. The PowerPC processor is a RISC chip which runs much faster and uses less power (and costs less) than previous chips. The most current are called "G5" for the fifth generation PowerPC chips.

PP&B (publishing) Book publishing industry term that stands for "paper, printing, and binding," which are the biggest costs in manufacturing a book. PP&B for picture books is especially expensive, in large measure because of four-color illustration and the higher grade of paper required.

PRAM (hardware) (Parameter Random Access Memory) pronounced "PEE-ram." A small amount of memory maintained by a battery or two that remembers the date and other important settings. If this information gets messed up, it can really confuse your computer. Consult your computer handbook to learn how to preserve your settings and reset your PRAM in case of corruption.

preferences (software) Files that are created by most software applications to store default settings for the options you may choose as you customize the program's environment. If your program starts acting strangely, one remedy is to delete the preferences file, forcing the program to create a fresh one.

preflighting (prepress) The process used to test and check every component in a project to produce a printing job. Preflight confirms the color mode, color breaks, presence of imported art (TIFF and EPS files), etc. Special software can be used for this process. InDesign has a prepress feature built in that does a respectable job of verifying files. Commercial printers usually have an extensive checklist of items to check as part of their preflight process to avoid any problems when the job reaches the press. The term is borrowed from the aviation industry where aircraft are always checked for obvious problems before attempting flight.

prepress proofs See off-press proofs.

press proofs (printing) A proof of a color project run on a printing press in advance of the full production run. Usually more expensive than off-press proofs. Getting less common as digital prepress workflows are used.

printer's error (typesetting) Marked as "PE" this is the term used to describe a typesetting error. (In the distant past, the printer and the typesetter were often employed in the same shop or were even the same person.)

print quality 1. (printing) A term describing the visual impression of a printed piece. 2. (paper) The properties of paper that affect its appearance and quality of reproduction.

printer (output) Hardware that prints the information in the computer onto paper.

printer's devil (typography) 1. Modern day "apprentice," i.e., in the 1700s and 1800s in order to become a journeyman printer one started as a "printer's devil." An unpaid assistant who was given room and board (typically in the shop itself). His tasks were to keep the shop clean, clean all the presses, clean the hand type with kerosene and perform all duties requested by the journeyman printer. 2. An episode in the fourth season of *The Twilight Zone* featuring Burgess Merideth as the demonic "Mr. Smith" who typesets descriptions of events (for a newspaper) before they occur. Adapted by Charles Beaumont from his own short story "The Devil, You Say?" The program first aired February 28, 1963.

primary colors 1. (artwork) The colors (pigments or of light) used to generate the full range of visible colors. See additive primaries and subtractive primaries. 2. (literature) Name of a book spoofing the Clinton presidency.

process color (printing) The subtractive primaries—cyan, yellow, magenta—with black, used in four color printing.

process printing (printing) Printing with a series of two or more halftone plates to produce intermediate colors and shades.

processor (hardware) A "computer-on-a-chip" that calculates things as opposed to memory chips which store things. Processors are made up of incredibly vast and small electronic circuits containing thousands of transistors and other components. Two types of processor architecture are CISC and RISC.

program (computers) A sequence of instructions for a computer. The same as software.

proof   (typesetting, printing) A sample sheet of the typeset or printed material used to check against the original and/or specifications for the job.

proofreader (publishing) The person who reads the typeset copy against the original to ensure that it is typographically correct.

proofreading (publishing) Checking a manuscript or typeset matter for typographical errors. See also copy editing.

prosumer (industry) A contraction of professional and consumer. Usually used to refer to a class of goods that are at the high-end of consumer needs and at the low end of professional needs.

protocol (industry) Any set of rules for exchanging information.

psychrometer (industry) A wet and dry bulb type hydrometer used to check relative humidity in industrial settings with great accuracy. Maintaining reasonably consistent humidity levels is important in a printing facility. Swings in relative humidity (in either direction) can wreak havoc with paper and presses.

public domain (industry) Any created work, including software, that the public has every right to copy and use in any way they see fit. The authors retain no rights or liabilities, either because they've relinquished them or because they've been dead for a while. Often used incorrectly to describe copyrighted, publicly distributable software for your computer.

publishing agreement (publishing) Contract for sale of literary rights to a publisher. Agreements complex with formal legal writing and typically include 3 to 20 pages, with up to 100 clauses. The contract essentially transfers all or some ownership of a work to a publisher for a sum of money that can vary depending on what rights are sold. It is wise for an author to review a publishing agreement with a suitably qualified intellectual property attorney before signing.

Publishers Weekly (publishing) A weekly trade magazine of the book publishing industry. Issues generally offer book reviews, insider reports, author interviews, bestseller lists, industry analysis and trends, reports on publishing and book selling. Copies are available at larger book stores and many public libraries.

pull quote (editorial) A phrase or short sentence taken from the body text set in larger type in a separate location to draw attention to the material. Pull quotes are usually used only in magazines, but I've seen them used occasionally in books. See call out.

punch (typography) The metal tool which is source for a piece of type. When "punched" against a piece of metal, the convex carving of a letter on the punch leaves the impression of that letter, which is later used to cast the actual pieces of type.

quality control (industry) A program of customer service, process control, and product sampling with the objective of eliminating causes of process variability. Sometimes called statistical process control. Also referred to by various terms like "continual improvement" or whatever the current business fad might be. Mostly involves posting (supposedly) inspirational signs in the workplace. While the analysis aspects of these faddish management techniques are a sometimes helpful tool in industrial operations, an amazing amount of time has been wasted in implementing such programs in business offices. See the related ISO 9000.

query (proofreading, editing) A question about the copy from the proofreader or typesetter directed to the editor or author.

query letter (publishing) A brief letter to an editor that describes a manuscript that an author would like to submit for consideration. If applicable, previous writing credits or experience should be included. If a work of fiction, often a sample chapter is included.

QuickTime (multimedia) Media conversion software from Apple that enables third-party applications to record and playback sounds and movies. It has several image and video compression /decompression methods (codecs) available, including JPEG, MPEG and others.

quiet zone (barcodes) The light space at either side of a bar code which is needed by the scanner to detect the start and end of the code. With ISBN/EAN barcodes, ¼ inch is preferred.

quoin (typography) A small wedge, usually of wood, used for tightening or locking up forms or galleys in pre-electronic printing.

quoin key (typography) A metal key used to tighten wedge shaped wood quoins to brace and secure pages of type in a frame in preparation to be printed.

QWERTY (interface) The standard arrangement of the keys on a keyboard. QWERTY (named after the first six characters on the top row of keys) is awkward and makes typing difficult. Typewriter keys were laid out this way to purposely slow down typing so the keys wouldn't jam in early mechanical typewriters. We live with this inefficiency as proof of the concept of "installed base."

radio button (interface) A button allowing you to select one option from among many. Named for the push buttons on your car radio.

ragged left (typography) Type that is justified on the right margin and ragged on the left. Not particularly common and very difficult to read in long passages. Best reserved for poetry or brief blocks of text.

ragged right (typography) Type that is justified on the left margin and ragged on the right. This does not suggest the complete absence of hyphenation, as well-set type will have a very moderate variation in line lengths.

RAM (hardware) Acronym for Random Access Memory. The temporary memory in which a computer stores information while it's running. Ordinary RAM is volatile—it loses its content when the power goes off. RAM is also extremely fast. Also called DRAM (Dynamic Random Access Memory).

ransom note (design) Use of too many typefaces in a single document. The feeling is one of a ransom note, made up of words torn from magazines or newspapers.

raster (industry) A series of dots or cells arranged in a regular grid.

raster image processor (hardware) The software and supporting hardware in an output device (usually a printer or imagesetter) that calculates the bitmap from the instructions provided by the page description language and directs the marking engine to generate the output image. Most PostScript devices use a built-in RIP. There are software RIPs that run on a general purpose computer linked to an output device, such as an inkjet printer (most common with wide-carriage inkjet printers).

reading fee (publishing) An amount charged by some agents and subsidy publishers to read or evaluate a manuscript, without an obligation to find a publisher or publish the material. Can be a sign of a possible scam. Some agents earn more from reading fees than from commissions generated by sold works.

read-only (storage) A file or a disk is read-only when you can look at and print its contents, but cannot save changes to it. CD-ROMs are read-only. Certain parts of a computer's inner workings are read-only, such as the ROM (read only memory) chip.

ReadMe (applications) A small file on most new software disks which describe what's on the disk, how to install it, and give details of any last minute changes. A variant on the term is Read Me First file.

ream (paper) 500 sheets of paper.

record (applications) One set of related information in a database. In an address book, all the information about each person (name, address, phone and so on) comprises one record. The individual components of a record are held in fields; every record in one database has the same fields (filled or empty). A record is analogous to a row in a spreadsheet.

recto (typesetting) Latin for "right" used to refer to right-hand or odd-numbered pages. See verso.

red, green, blue (color) The additive primary colors. See RGB.

redraw (display) When you move an object across the screen, scroll, or change windows, all or part of the screen has to be redrawn. The computer has to figure out where everything has moved to and redraw them in their new places. This can take a lot of time with complex text or color images (hello Photoshop). Some applications allow you to delay redraw. You can also force a redraw in various ways. Other applications let you work in a lower-resolution mode to speed redraw.

reducers (1. printing) An additive in printing inks used for thinning, if needed, to prepare for use. 2. (photography) Chemicals used to reduce the density of an image or of halftone dots. In the latter case, sometimes called dot etching.

reflectance (industry) The ability of a surface or printed area to reflect light.

reflection copy (photography) Material (mostly flat) that is photographed using light reflected from its surface. In other words, most original artwork used in printing.

refresh rate (display) The speed at which the video screen is redrawn. Faster is better. The original 9″ Macintosh screens refresh at 66 Hz (66 cycles per second), which is much faster than most TVs. Most good monitors refresh at 70-75 Hz. Of course, a fast video redraw still has to wait for software to come up with a new picture for it to draw, which is where processor speed matters most.

register (printing) Fitting of two or more printed images in exact alignment with each other. When the images don't touch, it's called "loose register;" when the images must exactly align, it's called "hairline register." Process color printing is always "hairline register" work.

register marks (artwork, printing) Usually circles with a pair of perpendicular lines through them that are placed on the original artwork to assist in aligning the images during prepress work or on the press. Other symbols, such as simple crosses, are also used. Most page layout programs will automatically generate register marks, if appropriate settings are selected.

register color (layout software) The color designation used in page layout programs for elements that will appear on all printing plates, once an image is separated. Care should be taken to apply this color only to items that appear outside the printing area of the piece.

relational database (applications) A database application where more than one file can be linked together, such as a customer list and an order list linked by a customer number. You can program all kinds of fancy functions; for example, as you type in the customer name, the address and other information automatically pop into the right fields on the order form. They are a must for serious business work. These applications, such as FileMaker Pro and 4th Dimension, are harder to use and cost more than flat-file databases, but they are very powerful.

relative humidity (industry) The amount of water vapor in the atmosphere expressed as a percentage of the maximum that could be present at the same temperature. As the air temperature increases, it is able to hold more water vapor. That's why cold windows or a mirror "steam up" in a bathroom after a shower has been used—the air near the glass is cooled and releases its moisture. High relative humidity during hot weather adds immensely to the discomfort level. Fortunately, I live in California where such is rarely an issue. Note: relative humidity is important in printing because sudden changes can cause considerable problems with ink/water balance stability on press and dimensional changes to printing papers. Covers of finished perfect bound books may curl in high humidity.

remainders (publishing) Leftover books that didn't sell and subsequently are deeply discounted for fast turnover. Publishers strongly resist "remaindering" their titles, but the practice is common for books that failed in the market or were printed in too great a quantity. These are the source of books for bookseller's deep discount tables.

render (display) The creation of a simulated three-dimensional object by taking a model and adding surface textures and lighting effects such as reflections and shadows. Needless to say, this takes a lot of computer muscle and time. The dinosaurs in Jurassic Park were rendered on monster workstations from Silicon Graphics, although they were designed on Macintoshes.

repeatability (prepress) The ability of image/platesetters to keep images in register from one film/plate to the next. Repeatability is measured in micrometers.

repeater (networking) A device used to boost a signal as it passes it along. Most LAN hardware has a relatively limited transmission distance before the signals deteriorate. A repeater will extend the distance of reliable operation.

reprography (printing) Another term for printing and copying.

reset switch (programming) The button with the triangle on it near the power switch (on a Macintosh). Use it as a last resort if the computer freezes. It's better than turning off the computer. But if you're using System 7 or later try a force-quit first by typing Command-Option-Escape to get back to the Finder, where you should then restart the computer. Under OS X, use the "force quit" menu item.

resident font (output) Fonts that come built into the ROMs in a printer. With resident PostScript fonts you also get a disk with the bitmapped screen versions to put on your computer.

resist (prepress) A light-hardened stencil that prevents etching of non-image areas on a printing plate.

resolution (output) A measurement of the clarity and sharpness of a screen or printed document. Resolution is measured in dpi (dots per inch). The greater the dpi the higher the resolution and the smoother the graphics and text will be. A Classic Macintosh screen has a resolution of 72 dpi, which is the standard upon which software is written. Some monitors deviate from 72 dpi, so documents may look larger or smaller than they will print out. Check the application's ruler. Older laser printers had a resolution of 300 dpi while an imagesetter can go up to 2540 dpi. Resolution in halftones is measured in lines per inch, which must be converted to dpi by a sophisticated algorithm such as PostScript.

respi screen (halftone) A contact screen with 110 lpi ruling in highlights and 220 lpi in mid-tones and shadows to produce a longer grayscale and smoother tonal graduation in an image.

restart system (software) A command in the Finder's Special menu that performs the shutdown that you normally do at the end of the day, then immediately reloads the operating system and Finder like it was the next morning. You should restart the computer after you've installed a new extension or Control Panel since those things are activated as the computer boots up. You should also restart if the computer is acting funny or if you've had a freeze or other system crash from which you've been able to get back to the desktop (by pressing Command-Option-Escape). Restarting is much easier on the machine than turning the power off and back on.

retail price (publishing) The cover price (or list price) of a book. Most larger publishers pay royalties based on the cover price, but have contract clauses allowing lower royalties for books sold at "deep discount."

retrofit (industry) Backward integration of a new feature into an older device or program. Many electronic devices now have EEPROMs, that allow "firmware" upgrades that (often) can be downloaded over the Internet.

return key (interface) Just like on a typewriter, the Computer's Return key ends one line and starts another. Weirdly called "carriage return" by some people who seem to not realize that computers don't have paper carriages. In dialog boxes, typing Return will often perform the default action, such as opening the selected file in an Open dialog. In telecommunications, you sometimes must type Return at the end of every command in order to send it, and often insert a return character after every 80 (or fewer) characters. A return character is often used as a delimiter (separator) of whole records in a database or lines in a spreadsheet when exporting.

returns (publishing) Books that are returned to publishers from booksellers because they didn't sell. Returns, a chronic problem in the industry, may run more than 25 percent for any given title originally purchased by the bookseller. In theory such books should be ready for re-sale, but often arrive in the publishers' hands in poor condition, shopworn, or even damaged. Many small publishers offer "hurt" books at a discount or sell less-damaged copies as "used" through the Amazon.com marketplace.

reverse out (typesetting, artwork) A term for setting type as white (paper) against a dark background. Might be used on a back cover. Significantly reduces readability and comprehension. If used, should be limited to short blocks of type. Also, it's better to use a slightly bold typeface to "hold up" (be readable).

revision (publishing) The obligation of an author or illustrator to make changes to an original work in order to make the finished product more saleable. Revision clauses that spell out the degree of revision and time allowed are usually included in publishing agreements. Editors and authors or illustrators typically work closely together during the revision process, with give and take on both sides.

RGB 1. (color) Abbreviation for red, green, and blue, the primary additive colors used in video displays and scanners. Also the color space used in Photoshop for elements that are intended for use on the Internet or for video display only. 2. (display) The primary colors used by video monitors (both LCD and CRT) to generate color. Also used as a reference to that type of display, e.g., it was shown on an RGB display.

rhymer (publishing) An editorial term that describes picture books in which the story is presented in rhyme. Rhymers fall in and out of favor with editors, and are usually more difficult to sell than picture books told in narrative style unless your name is Dr. Suess.

right angle fold (bindery) Term used to describe two or more folds that are at 90° angles to each other. Signatures are given several parallel and right angle folds before binding into the book block and trimming to size.

rights (publishing) The provisions offered for sale by an author or illustrator or photographer to a publishing house for a particular manuscript or work of art.

RIP (output) The acronym for Raster Image Processor. The program inside PostScript laser printers (and other PostScript output devices) that converts the graphics instructions into high-resolution bitmaps before the printing engine places those bitmaps on the page. See raster image processor.

RISC (hardware) Acronym for Reduced Instruction Set Computing. Processor architecture that runs much faster by relying on a reduced set of simplified instructions. The goal is that one CPU instruction can be performed every clock cycle. Although more instructions (therefore longer programs) are necessary to carry out complex processor tasks, RISC still is faster because the bulk of all computing is simple instructions. Apple's PowerPCs are RISC processors, and it is the prospect of greatly increased speed (as well as lower cost) that drove the decision to change from the CISC Motorola 680x0 processor line.

roller stripping (printing) A situation where the ink does not adhere to the metal ink rollers on a press. Yes, this is yet another press operator cause for vigilance.

ROM (system software) Acronym for Read Only Memory. The instructions and data built into the chips of your computer that cannot be changed. Every computer has some kind of ROM, at least to tell it what to do when the power first comes on. Even your microwave oven has a ROM chip. Maybe your washing machine, too.

Roman type style (typography) Letterforms which have vertical stems (as distinct from italic or oblique, which are set at an angle).

root (system) The top-most directory on a disk. Usually the directory you view when you first open a disk. Depending on your operating system, there may be a (relatively) limited number of items allowed in the top level directory. (The reference comes from the idea of the directory structure being like an upsidedown tree, with many branches—folders/directories and sub-folders/directories.)

router (networking) A combination of hardware and a software-referee that divides a busy network into smaller networks (called zones) and keeps the information in those networks separate except when necessary. Routers are used when the amount of information being transferred over the network is too large to be easily carried to all parts of the network or to isolate one portion of a network from another. Also a device that connects both similar and dissimilar network topologies. Sometimes routers have hub, bridge, firewall, and/or gateway functionality at the same time.

royalties (publishing) A percentage of the proceeds from the sale of each copy of a book. Royalties generally range from 3% to 15%, depending on the type of book, amount of experience the author has, its perceived market potential, etc. Authors and illustrators are both paid in royalties unless a flat fee arrangement has been made. Royalties my be calculated on a step-rate based on sales (as more books are sold, the publisher's up front costs are recovered, leaving potential margin for higher royalty rates). See net royalty.

RTF (output) Abbreviation for Rich Text Format. A file format devised by Microsoft that encodes all manner of text formatting information (from text fonts and styles to paragraph indentation to footnotes) among the text to be formatted. An RTF file consists entirely of ASCII characters so it may be transmitted easily between platforms that have software capable of interpreting RTF. Many word processors contain RTF writers to create RTF files, and RTF readers to reformat text from elsewhere. Unfortunately, transfer across platforms of RTF files doesn't always allow proper translation of words with diacritical marks and for some special symbols. (Macintosh and Windows have different arrangements of characters with ASCII values above 128.) The better practice is to move files as a MS-Word document between platforms. When the file is opened in Word, the font translation is usually satisfactory. Usually, I then save the file as RTF before importing it into my page layout program. This step often eliminates some undesirable file fragments that may be in the Word document.

rub proof (printing) Ink that has reached its maximum dryness and does not mar with normal abrasion. In a printing plant, much post press processing must occur before this level of dryness is reached. Most plants prefer use of suction or "air" feed bindery equipment to minimize rubbing and marking that's inherent in friction-feed devices.

rule (typesetting) A line used for a number of typographic effects, including boxes and borders.

run (printing) To print a job.

runaround (typography) Term used to describe type set to fit around an illustration or other page element. See also word wrap.

run in (typesetting) 1. To insert new copy without making a new paragraph. 2. To combine a following paragraph into the previous paragraph.

runnability (paper) Properties that affect the ability of a sheet to run on a press.

running head (typography) A headline or title repeated at the top of each page.

saddle stitch (bindery) A method where a booklet is fastened by a wire through the middle fold of the sheets. Smaller print shops would probably use a saddle staple. The difference between the two devices is that the stitcher works from a spool of wire and the stapler uses pre-formed staples. Depending on the equipment and paper used, saddle stitching is usually limited to binding 20–25 sheets for booklets of 80 to 100 pages. A major failing (for book publishers) is that saddle stitched publications don't have a square spine allowing for a readable spine title. Some printing trade magazines have described "square spine" saddle stitching equipment, but it doesn't seem to be in wide use, as yet.

safe light (photography) Special lamps used to illuminate darkrooms without the risk of fogging the light sensitive materials. While panchromatic materials must be handled in total darkness, most graphic arts materials may be processed in work areas illuminated with safe lights.

sample (digital photography) Basic analog image element taken by an image sensor in a camera or scanner. The sample may be black (gray), or it may relate to a particular color channel. The sample is processed by internal software in the camera or scanner to obtain a pixel.

sans serif (typography) Letterforms without serifs, generally with a straightforward, geometric appearance.

SASE (publishing) Self-addressed, stamped envelope. Most publishers prefer you provide a SASE to enable them to send you a rejection letter at your expense.

save (storage) Storing information so that it is not lost if you turn off your computer (or if there is a power failure or system crash). Normally you save to a disk or some other medium.

scaling (artwork) Determining the proper size of an element or image to be reduced or enlarged to fit an area.

Scan-a-web (press) A device that can be installed on a web press with rotating mirrors that can be speed-matched to the moving web, allowing the viewing of the printed image. Really clever!

scanner (input) A piece of hardware that minutely scans an image, reading dark, light and colored areas as a stream of digital data. You put a photograph, a drawing or even a three-dimensional object on the scanner, close the lid, click your mouse, and the machine sends a copy of it to the computer. It's sort of like a copier. There are also video scanners that can input live stuff. The scanning software usually offers you several file format options. Unless you have a clear idea and a good reason not to, always save scanned images as TIFFs.

score (bindery) To impress or indent the paper in a line to make it easier to fold. Most perfect bound books are scored along the edges of the spine. Sometimes a second score is made parallel to the spine about ¹⁄₁₆ inch from the corner to allow the cover to bend more easily.

screen (prepress) The process used to break up a continuous tone image into dots for print reproduction. See contact screen.

screen angles (process color printing) The screens on each color plate are made at different angles to minimize undesirable moiré patterns when combined into the printed image. The most common screen angles used are: 105° cyan, 90° yellow, 75° magenta, and 45° black,

screened print (artwork) An image that has a halftone screen. These are usually quite difficult to rephotograph or scan without creating undesirable moiré patterns.

screen ruling (artwork) The number of lines per inch in a halftone.

screen saver (software) A utility program that prevents screen burn-in by creating animated effects while you are not working. Effects range from moving patterns or cycling through a series of pictures. If you leave an image on the screen for an extended period of time, it will eventually burn in and leave a permanent shadow—although this is less a problem with newer CRT monitors and doesn't effect LCD panels at all.

script 1. (programming) A user-definable set of instructions for use by an application. 2. (typography) A typeface that resembles handwriting. It may be quite formal, as using a calligraphic pen or more casual as writing with a brush, marker, or even a crayon.

scroll bar (interface) A rectangular bar that may be along the right or bottom of a window. If the scroll bar is empty, the window is showing the entire document. If the scroll bar is active, the window is showing only a part of the document. An active scroll bar is shaded and contains a scroll box in the middle and a scroll arrow at either end. Clicking or dragging in the scroll bar causes the view of the document to change.

SCSI (storage) Acronym for Small Computer Systems Interface, pronounced "scuzzy." A standard for connecting computers and peripheral devices which allows information to be exchanged very quickly. The standard also allows for some communication between devices without the main processor doing anything, although this capability is not used often. The physical SCSI connectors have either 50-pins (the industry standard) or 25-pins (the Apple standard). A revision called SCSI-2 (with 68 pins) now exists. SCSI, in its various forms, has been superseded by Firewire and USB, but some specialty SCSI devices are still on the market.

scum (printing) A film of ink printing in the non-image area of a plate. Usually caused by problems with the ink/fountain solution balance adjustment or fountain solution degradation/contamination. Yet another factor for the press operator to keep under control.

sector (storage) The smallest subdivision of a disk (or other storage medium) ranging in size from 512 bytes on a floppy up to thousands of bytes on a hard disk. The computer reads and writes nothing smaller than one sector, for the sake of speedy operation. A bad sector cannot hold information accurately.

selection (interface) The information affected by the next command. Selected text is usually highlighted, graphic objects sprout handles, icons switch to inverse video. The text insertion point is also a selection.

selection rectangle (interface) An area of the screen shown by a rectangle of dashed lines that is created when you push the mouse button down, hold it and then drag the mouse. A selection rectangle is available in most graphics programs.

A variation is the lasso tool, which snaps tight around the edges of the selected objects when you release the button, as opposed to the rectangle which retains its shape.

self cover (bindery) A printed piece or booklet that has the same paper for the cover as used for the inside pages.

serif (typography) A line crossing the ends of the main strokes making up a letter. There are many varieties of serifs, including slab, square, tapered, and vestigial. Type designed without serifs is called sans serif or gothic. The serif originated with ancient stone masons. When cutting letters into stone, the chisel was turned across the groove at the end of the letter stroke to make a clean, smooth finish to the line. These cross cuts also assisted the eye to line up the letters and see them as positioned evenly along the line of words.

server (networking) When you log onto the Internet, your computer begins a digital conversation with a server. Servers are also used on LANs to provide various services to the client computers, such as storing files (a file server) or handling printing chores (a print server).

set-off (printing) When the ink of a printed sheet transfers to the back of the next sheet above in the delivery section of the press. Also called offset.

SGML (industry) Abbreviation for Standard Generalized Markup Language. A coding scheme for marking up text to be used in a variety of purposes, including typesetting and electronic publishing. With proper planning, SGML coding can make text usable in a variety of situations with a single markup. HTML is a subset of SGML, so if you've ever hand-coded a Web page, you've worked with a simplified form of SGML. While there may be uses for SGML by small publishers, the tools are expensive and the complexities are probably more trouble than they're worth.

shadow (photography) The darkest areas of a photograph which correspond to the largest dots in a halftone.

sharpen 1. (halftones) To decrease dot size, the opposite of dot gain. 2. (Photoshop) A technique where edges of elements within a photograph are enhanced by slightly darkening the darker pixels and lightening the adjacent lighter pixel. Care must be taken to avoid over-sharpening—which causes a halo effect around the darker elements. The feature is accessed through the "unsharp mask" dialog on the filters menu.

shareware (industry) Publicly distributable software that you can use for a time to determine whether you like or need it, and then pay for it. Try before you buy software, usually not marketed through traditional channels but available from users groups, Internet servers, and friends. Shareware is generally pretty cheap ($5 to $25) and you really should pay for it to encourage other people to create nice things for us.

sheetwise (printing) To print one size of a sheet with one plate, then turn the sheet over and print with a second plate using the same gripper and opposite side guide. Compare to work and turn.

shingling (prepress) To vary the gutter (center) margin according to the page position in the signature and the bulk of the paper. See creep.

shoulder (typography) The curved stroke projecting from a stem as with the h, m, and n.

show through (printing) The undesirable condition where printing on the reverse side is visible through the sheet in normal lighting. Show through can be controlled by use of heavier and/or more opaque paper. (The "opaque" paper grades were designed to control this situation.) Also, show through can be affected by the press pressure and characteristics of the ink that reduce its tendency to wick through the paper.

side bar (editorial) A short section of text set off from the main body, usually to one side and sometimes enclosed in a box, that gives a brief exposition on a topic related to the main subject. Side bars are mostly used in magazines, but may be used in books. Chapter 1 of this book has several side bars.

side guide (press) A device in the feed section of a sheetfed press that limits the sideways movement of a sheet as it is fed toward the impression cylinder.

signature (printing, bindery) The name given a printed sheet with multiple pages. Also refers to the folded groups of pages that make up a part of a book. It derives its name from the quality control practice of early printers to sign each sheet as it came off the press.

silhouette halftone (artwork) A halftone of a subject with all the background dots removed.

simultaneous submission (publishing) Submissions that are sent to more than one publisher at a time. Most publishing houses accept simultaneous submissions, but authors who choose this route should note this in their cover letters.

size box (interface) The small area in the lower right corner of most windows which allows you to change the size of the window when you click on it and drag. Some applications will not let you drag with complete freedom; they restrict the window to a maximum or minimum size, or both.

sizing (paper) A treatment that gives paper additional resistance to moisture or vapors.

skid (industry) A platform support for a stack of paper (or other products). Although the term skid is often used interchangeably with pallet—they're generally used for the same purpose—the construction details are different. Skids, usually made from wood, have a solid platform supported by two side rails that raise the platform as much as 10 inches above the floor. The design limits access by a fork lift to two sides. In contrast, pallets generally have a slightly larger platform area made of wood strips that's raised only about 4 inches and usually are designed so that a fork lift can be inserted from any side. See pallet.

slab serif (typography) An abrupt or adnate serif of the same thickness as the main stroke. Slab serifs are a hallmark of the Egyptian and clarendon types. (Examples: Memphis, Rockwell, Serifa)

slitting (bindery) The cutting of printed sheets or a web into two or more pieces with a cutting wheel on a press, folder, or stand alone slitting machine.

slope (typography) The angle of inclination of the stems and extenders of letters, e.g., most italics slope to the right between 2 and 20 degrees.

slot (hardware) The place you plug a card into inside the computer. Usually there's a long thin connector with numerous metal contacts that match traces on the edge of the card. Most computers have several types and sizes of slots.

slug (typography) Character designed to show paragraph breaks in close set type. (¶). Also, a line of type set by a Linotype (hot metal) typesetter.

slush pile (publishing) Stacks of manuscripts that are received by acquisition editors and publishing houses, but not specifically requested. Manuscripts in the slush pile inevitably are read only after material that an editor has sought or received from an agent. Some houses do not accept unsolicited manuscripts; others (typically smaller publishing houses) keep up with slush piles, hoping to find new talent.

small caps (typography) An alphabet of uppercase letters the size of the lowercase available in some typefaces. "Fake" small caps can also be produced in some software applications. If possible, avoid using the software generated version as they are usually too light for the overall text color.

smart quotes (output) Curly "quotation marks" (and apostrophes) that face in opposite directions at either end of the quotation (as used here). Many word processors and other applications have a checkbox in the preferences section to always substitute smart quotes from the plain ones the computer keyboard supplies. Also called "typographer's quotes."

smyth sewn (binding) A method of stitching thread through the signatures of a book before binding. Usually used with case bound, but may be used with perfect bound books. This method allows books to open flat. Is usually considered superior (and more expensive than) adhesive case binding. See notch bound.

SneakerNet (networking) The world's simplest network: none! Just put data on a floppy or other media and carry it between machines. Often the network of last resort.

SNMP (networking) Abbreviation for Simple Network Management Protocol. Invented in the late 1980's to manage the Internet, it has become a de facto standard for network management.

soft cover (book binding) Glue binding with paper cover. See perfect bound.

soft dot (printing) Halftone dot with considerable fringe which causes either dot gain or sharpening during printing. The output depends on the nature of the fringe—if it's too fine, the ink can't adhere, causing sharpening. A thicker fringe carries more ink that its size would suggest, increasing dot gain.

soft hyphen (output) A hyphen that is invisible unless the word falls at the end of the line, in which case the computer will break the word and show the hyphen.

soft proof (prepress) A proof in an electronic form intended for viewing on screen. Such a proof will not show accurate color, but may be adequate for an interior of a book. Generally a PDF file. See hard proof.

soft return (output) A feature of some word processors and page layout programs that forces a line break but doesn't start a new paragraph. This is what the computer does when it word wraps.

software (applications) Hardware is the stuff you can touch. Software is the invisible stuff, the programming, the energy coursing through the chips that makes the computer work. Software includes the applications, the documents and the operating system, in memory or stored on a disk or drive. Some wags refer to the human brain as "wetware."

software piracy (industry) Making copies of commercially-sold software you did not purchase—a form of theft. The term "pirate" typically refers to those who addictively collect software simply for the pleasure of having lots of it. While a serious problem for faddish games (which now use sophisticated security features), it's probably less a problem for business programs than the software industry would like you to believe.

solid type (typesetting) Lines of type set with no leading.

SPC (industry) Abbreviation for statistical process control. See quality control.

special character (output) Generic term for non-alphanumeric characters which have special characteristics, either for enhanced typography or for some purpose special to the computer.

specs (typesetting, printing) The specifications for a job. Includes type specifications and printing specifications.

spectrophotometer (color management) An instrument used for measuring color, often in connection with color calibration and management software. See colormeter, a less expensive, if less accurate, device used for the same purposes.

spectrum (color) A two-dollar word for rainbow—the complete range of colors of visible light from the shortest wave length (violet) to the longest (red). To remember the colors of the spectrum, use the made-up name, Roy G. Biv—which stands for red, orange, yellow, green, blue, indigo, violet.

spell check (typesetting) A technology available in most word processors and page layout programs. It gives you the impression that "you're" work is free "form" errors.

spine 1. (typography) The main curved stroke of a lowercase or uppercase S. 2. (bindery) The bound edge of a book. Sometimes called the backbone.

spiral binding (binderey) A book bound with wire spiraled through holes punched along the binding edge. Similar to Wire-O® and plastic comb bindings.

spooler (output) Software and/or hardware that takes over a task so that the CPU is not tied up. Most often associated with printers where the spooler intercepts the data being sent to the printer and stashes it in RAM or on disk and sends it to the printer at a slow rate the printer can accept. That way you can resume using your computer more quickly. The term comes from an old IBM acronym, "Simultaneous Peripheral Operation On Line."

spot (printing technology) The smallest element in the "addressable grid" of an output device. Similar to a pixel, a spot refers to data, not something which is visible. A mark is what a marking engine (laser) actually places at a spot location. A spot has a spacial aspect (grid location) and a tonal and color aspect. By now you should have spots before your eyes.

spreads and chokes See chokes and spreads.

spreadsheet (applications) An electronic ledger used primarily for calculating numbers. Spreadsheets are made up of cells arranged in rows and columns. Cells can hold data or formulas that take information from other cells, do calculations on it and display the result. Some spreadsheets, such as Microsoft Excel, also have limited database capabilities, do graphs and even some word processing.

spur (typography) A projection smaller than a serif that reinforces the point at the end of a main stroke, as in the uppercase letter G.

spyware (applications) A program, often hidden in another program, that captures information from a target computer and sends it over the Internet to another location. Can be relatively harmless, as a program seeking marketing information about the computer user, or it can be a considerable security risk, capturing credit card and financial data for an evildoer. See malware, virus.

start character (barcodes) The specific bar/space pattern which indicates the start of a bar code.

star target (printing) A pinwheel-like image used to measure resolution of plates during production and their degradation during printing. These are placed in the trim area of the sheet along with registration marks and other control elements.

static neutralizer (printing) A press attachment designed to remove static electricity from the paper to avoid ink set-off and trouble with paper feed. Looks like a silver decorative garland. It's positioned so that the paper lightly brushes it while moving through the press.

step-and-repeat (prepress) A procedure of multiple exposure (film or plates) using the same image by stepping it in position according to a layout template or program. The technique has been adapted in page layout programs to place multiple copies of an element on a page. Particularly useful technique for reproducing business cards or other small items that can be printed several times on one sheet of paper.

stet (proofreading) A proofreader's mark, written in the margin, signifying that a correction mark should be ignored (left as it was).

stochastic screening (prepress) A digital screening process that converts images into very small dots (14–40 microns) of equal size and variable spacing. Color images made with the finest sized dots can look remarkably sharp. Printing with these screens has proved to be more difficult than anticipated, so the technique is not widely used. Also called frequency modulated (FM) screening.

stock (paper) Paper or other material to be printed. For example, "That should be printed on sixty pound stock."

stock photogrpahy (artwork) Ready-made images used to illustrate books, book covers, or other works. May be purchased under royalty free, non-exclusive fee basis, or "controlled rights" contracts. Fees are usually based on usage. Read license agreements carefully, as some "royalty free" art is restricted to non-commercial use. If such art is used on a book, the license is violated and substantial extra charges may apply.

stone (printing) 1. The original printing plate for lithography. The discovery of lithography (which means "stone printing") involved use of limestone treated with chemicals that separated the ink from the non-image areas. As the chemistry became better understood, the process was transferred to more easily used surfaces. 2. The bed used to level and lock metal type in its frame. Stones, ground to a smooth and level surface, were popular as they wore slowly and did not bend under the weight of the type.

stop character (barcodes) The specific bar/space pattern which indicates the end of a bar code.

strike-on composition (typesetting) Use of a typewriter-based typesetting machine. One popular model was built by IBM using a modified Selectric typewriter.

stripping (prepress) The positioning of negatives (or positives) on a flat to compose a page or layout for platemaking. With the adoption of a digital workflow, the skills of a stripper are disappearing.

subsidiary rights (publishing) Sales of your book, either by the publisher or an author's agent, to other outlets such as book clubs, foreign publishers, magazines, or movie studios. If the publisher sells the subsidiary rights, the proceeds are split with the author (usually 50/50). If the agent sells the rights, the author keeps all the proceeds minus the agent's commission. Be sure that the publishing contract addresses how subsidiary rights might be sold and by whom.

subsidy publisher (publishing) A publisher who charges the cost of publishing a book to an author. Sometimes also referred to as "vanity publishers," subsidy publishers typically charge authors for typesetting, printing and promoting their own books. Bookstores often refuse to carry books published by subsidy presses, and such books are rarely reviewed. See *www.aeonix.com* for a full discussion of "the subsidy publishing scam."

substantive editing (publishing) Editing a manuscript for "global" issues, with particular attention paid to overall style, pacing, plot, etc. Senior editors who contract books often deal with substantive editing issues during the revision process with an author. Copy editors later search the revised manuscript for spelling, grammar, and general content mistakes. See chapter 2.

SQL (networking) Abbreviation for Structured Query Language. Database protocol for inter-application communication.

squared serif (typography) a font or type having serifs with a weight equal or greater than that of the main strokes. A feature often associated with "old west" typefaces. See also slab serif.

stand-alone (programming) A program that can be run on its own without other supporting software. A correctly written and compiled program is a stand-alone, as is a demo program with a run-time engine. A program written in Java is *not* a stand-alone program as in requires the presence of the "Java engine" to run.

startup volume (system software) The disk or drive used by the computer to boot up. On a Macintosh the startup volume always puts its icon in the top-right corner of the desktop. On a Macintosh you can change the startup disk by using the Startup Disk Control Panel. Under Windows, you use some weird set of keystrokes during booting to change the boot parameters.

stem (typography) A main stroke that is more or less straight, not part of a bowl. Ex: the letter "o" has no stem, and the letter "l" consists of a stem alone.

stress (typography) The direction of thickening in a curved stroke. This is often a clue used to classify a typeface as old-style, transitional, or modern.

style sheet 1. (output) A list of the text styles and format settings you have applied (or intend to apply) to each paragraph. You can then apply the same style wherever you wish. Best of all, if you make a change in the stylesheet the change is automatically applied to all paragraphs having that style. A wonderful feature of better word processors and all page layout programs. 2. (editorial) A list of treatments, spellings, and style rules for a document. More elaborate style sheets can be devised with styles for a publishing company. In that case, it becomes know as the "house style."

subscript (typesetting) A character, usually in a smaller size, set below the baseline.

substrate (printing) Any material that can be printed on, such as paper, plastic, and fabric.

subtractive primaries (color) Cyan, yellow, and magenta, the hues used for process color printing. (Black is also used to enhance contrast, but is not one of the primaries.)

Sulphate pulp (paper) Paper pulp made from wood chips cooked under pressure in a solution of caustic soda (sodium hydroxide) and sodium sulphide. Known as the kraft process. The resulting paper is most frequently used for packing boxes and "brown" paper products like large envelopes and wrapping paper.

sulphite pulp (paper) Paper pulp made from wood chips cooked under pressure in a solution of calcium bisulphite. It is one of the most common ingredients of printing papers.

supercalendar (paper) A stack of calendering rollers used to make paper with an extra-smooth finish. These rollers may be part of the paper machine or may be a separate device.

superscript (typesetting) A character, usually in a smaller size, set above the baseline.

suprint (prepress) Exposure from a second negative or flat superimposed on an exposed image of a previous negative or flat.

swash (typography) A type design with a flourish replacing a terminal or serif.

SWOP (printing) Acronym for Specifications for Web Offset Publications. SWOP is a set of quality standards for printing magazines on web presses. These highly comprehensive standards can affect all aspects of a project from artwork preparation through to the final printing. While you may encounter references to SWOP, these standards have little to do with book production.

tab (output) On manual typewriters, literally a metal tab that momentarily stopped the travel of the carriage, used to create alignment inside the margins. Computers have expanded the function of the tab character. Now you can specify the alignment the text will take at the tab right, center, left or decimal point or attach leaders for better readability of tables. The tab key is also used to move from field to field in a database or where there are multiple text boxes in a dialog such as Page Setup. When exporting from a spreadsheet or database the tab is often used to delimit (separate) the cells in a row or fields in a record. You can see where the tabs are in a word processor by choosing a menu item such as "Show Invisibles" or "Show Paragraphs." A most useful character.

tack (printing ink) The characteristic of the stickiness of an ink (the cohesion between ink particles). Refers to the separation force needed for proper transfer and trapping on multicolor presses. A tacky ink has higher separation forces and can cause picking or splitting of weak papers leading to hickeys and other undesirable defects.

tagged image file format (artwork) More commonly called TIFF or TIF. A file format for graphics well suited for scanned images and other large bitmaps. TIFF is a neutral format (not proprietary) designed for compatibility with a range of applications. Originally created for storage of grayscale images, it now also accommodates color. Many applications combine TIFF with LZW (non-lossy) compression to reduce file size—but most printers prefer that TIFF images not be compressed. Recent extensions allow preservation of layers at a penalty of large file size as the image is stored both flattened and layered for maximum compatibility.

tail (typography) The part of a Q which makes it look different from an O, or the diagonal stroke of the letter R.

tape drive (storage) Hardware that encodes large amounts of data onto a type of cassette tape. Too slow to replace a hard drive, but relatively cheap backup.

TCP/IP (networking) Acronym for Transmission Control Protocol/Internet Protocol. A set of rules for exchanging information between very different computers on a network. It has become a de facto standard and is the communication protocol of the Internet.

TeachText (system software) A very small word processor that comes with Macintosh system software and allows you to read text documents. In Mac OS X, TeachText has been replaced with TextEdit, a somewhat more sophisticated program that uses RTF as its native file format. TextEdit can read all TeachText files. A similar program for Windows is called Notebook.

tear-off (interface) A menu or palette that you can drag down from the menu bar and leave floating on screen for access. Typically a torn-off menu or palette becomes a mini-window.

teardrop terminal (typography) A swelling which resembles a teardrop at the end of the arm in letters such as a, c, f, g, and y. (Examples of fonts which incorporate teardrop terminals are Caslon, Galliard, and Baskerville.)

telecommunications (communications) The exchange of information in digital format (as opposed to voice) over phone lines.

telecommute (communications) Working at home and sending your work to the office via your modem and fax. Woo hoo!, it's the lifestyle of the '00's !

terabyte (computers) One trillion bytes. Once the sign of a major computer installation, terabyte-sized hard disk storage devices are becoming economically available for desktop computers. What to do with all that space? Fill it with photos of your grandkids taken with your cell phone camera.

terminal (typography) The end of a stroke not terminated with a serif.

text (typography) The body matter on a page of a book as distinguished from headers or footers.

text box (interface) The place(s) in a dialog box where you can type names, etc. Usually identified by a thin black outline and insertion point (flashing vertical bar).

text editor (applications) An application or desk accessory that allows you to type letters and numbers, but which doesn't enable you to specify much formatting. The documents created by these are usually in text-only format.

text-only (storage) A document which contains words but no formatting and which can be read by many different applications, even on other types of computers. A text-only file consists of ASCII characters only.

thermal dye sublimation (hardware) A type of printer that uses heat to vaporize dyes on a carrier material that then floats over and becomes deposited on the specially treated substrate. Similar to Thermal Dye Diffusion Transfer. Generally rather expensive to buy and operate, printers using either of these technologies have become less popular as inkjet printers have improved their color fidelity.

thermal-mechanical pulp (paper) Pulp made by steaming wood chips prior to and during refining. Often produces a better yield and stronger pulp than regular groundwood.

thesaurus (applications) A compilation of words linked to each other. Synonyms and antonyms are linked, as are words similar in meaning. The granddaddy of them all is Peter Mark Roget's Thesaurus, first published in 1852, and unequaled since. There are electronic thesauri, but check out the original to enjoy a stupendous accomplishment. A thesaurus can also be a set of terms used for indexing or classification or a list of keywords.

thick/thin transition (typography) The gradual change in the width of curved strokes.

thixotrophy (printing) A false body in ink. Printing inks may seem quite thick when sitting in the can. When worked (with spatulas) on a mixing table, the ink can become quite free flowing. This is the characteristic of thixotrophy. Printing presses are designed with many ink rollers that work the ink to accommodate this characteristic.

TIFF (output) Acronym for Tagged Image File Format. A graphics format used for saving or creating high resolution bitmaps, grayscale, and color images. TIFFs were invented for scans. TIFFS can have from 1 to 24 bits per pixel and are often compressed. If the image is straight black-and-white, with no gray areas, save it as a line art TIFF. If the image has gray tones, such as a photograph or pencil or charcoal drawing, save it as a grayscale or contone TIFF. Halftones only apply to gray or color areas. A halftone breaks a gray area into dots that a printer can print. If the image is solid black-and-white, you don't need any sort of halftone. If your scanning software can create special effect halftones that you want to use, or if your printer is non-PostScript compatible, save as a halftone TIFF. If there is no special effect halftone that you need, don't bother saving it as a halftone—let the PostScript printer halftone it for you on the way out. Yes, all PostScript printers can take any gray or color and break it into lines per inch. A 300 dpi laser printer creates halftones at 53 lpi; a 600 dpi laser printer creates halftones of 85 lpi; some laser printers with special photo-enhancement software can create halftones of 105 lpi. A high resolution image- or platesetter can create halftones of 120, 133, 150 or more lpi. See also tag image file format.

tilde (output) The little squiggly accent you see over the letter n in some Spanish words. (~ or ñ)

tints (color) Variations of a color made by lightening (adding white). With ink, a special unpigmented "transparent white" is used to mix a tint of a color. In artwork, a tint can often be achieved by "screening back" a color; that is, by printing the color as a screen rather than as a solid area of ink.

tip in (binding) A method where individual sheets are glued into a book after binding. When done well, it's practically invisible.

tissue overlay (artwork) A thin, translucent sheet of paper attached to an art board to protect a mechanical. Often used to indicate color breaks or corrections. Rarely seen in this era of digital art preparation.

title bar (interface) The space at the top of a window containing the name of the window. Clicking on this and dragging moves the whole window.

TK (editorial) Shorthand for "to come." Used to note for the typesetter that some material is missing from the original text, but will be provided soon.

toggle (interface) Change from off to on, selected to unselected, or vice versa.

toner (digital printing) The imaging material used in electrographic printers (e.g. laser printers, copy machines). Usually formulated as a powder, but sometimes as a liquid. Some manufacturers call it "digital ink." This is a marketing affectation, if it uses electrography, it's toner.

toning (printing) Unwanted ink in non-image areas of the plate. See scum.

tooth (paper) The characteristic in paper of the roughness of the surface which allows it to take ink readily.

trackball (interface) A computer mouse replacement consisting of a captured ball that can be spun to move the cursor around the screen. Trackball devices usually have several buttons to initiate functions as the buttons on a mouse.

tracking (typography) Adjusting the letterspacing and wordspacing of a range of characters by the same amount. This is in addition to any kerning adjustments made. Some programs call this "range kerning."

trackpad (interface) A small, flat rectangular pad that Apple developed to replace the trackballs in PowerBooks. It uses capacitance-sensing like the touch-sensitive buttons in elevators, and has 387 points-per-inch sensitivity. Like the mouse, it is a relative-motion device — when you raise your finger and put it back anywhere on the pad, the cursor stays put.

trade paperback (publishing) A book bound with a heavy paper cover, usually with the same typesetting, illustration, and cover design as the hardcover edition. Many books are issued directly as trade paperback rather than being released first in a hardcover edition. The choice between issuing a book as hard cover or trade paper back is primarily a marketing decision.

transposition (industry) The switching of the order of a pair of digits. The most common error in keyed data entry.

transitional (typography) The thick and thin strokes of the letter forms are of greater contrast than in oldstyle faces; but less than modern faces. The characters are usually wider than old style letters. Stress may be vertical or at a small angle. Example is Electra.

transparency (photography) Color positive film. A fancy name for a slide.

transparent copy (artwork) A transparency. Light must pass through it to image or scan.

transparent ink (printing) An ink that does not conceal the color of underlying inks. Process color inks are transparent so that they can blend to display a full range of colors.

transpose (typesetting) To exchange the position of a letter, word, or line with another letter, word or line. Letter transpositions are common in my typing—they often go unnoticed because I know what I thought I typed. Thank goodness for the spell check.

trapping 1. (printing) The ability to print a wet-ink film over previously printed ink. Wet trapping prints over wet ink. Dry trapping prints over dry ink. 2. (prepress) Refers to the slight overprinting of colored inks to eliminate white lines between adjacent colors due to slight mis-register or variations in press or paper. See chokes and spreads. Note: There are a lot of technical aspects to setting up proper trapping. Consult with your printer to get preferred trapping settings for your page layout program, or find a printer who is able to use "in RIP trapping."

trim marks (artwork) Marks placed at the edge of the image to indicate the final size of the page. Requires printing on a sheet larger than the final size. May also include marks indicating the bleed area. See crop marks.

trim size (publishing) The outer dimensions of a finished book. See chapter 3 for a complete discussion.

Trojan horse (applications) A program that appears to be innocuous (such as a screen saver) that harbors a secret function that harms the computer running it. See malware, virus.

TrueType (output) Outline font technology co-developed by Apple and Microsoft. Its primary purpose was to persuade Adobe to release the Type 1 specification (it worked). Unlike PostScript fonts, TrueType fonts are contained within a single file. TrueType, which uses cubic equations to calculate curves instead of the quadratic equations used by PostScript is, in theory, more precise. In practice, this has not been the case with fonts of both types being able to achieve excellent quality. TrueType is the standard in the Windows operating environment. Ultimately, the graphic arts community did not accept TrueType (there were no volunteers who wished to throw away expensive PostScript type libraries to replace them with expensive TrueType libraries), so PostScript fonts continue to dominate graphic arts to this day. Note: Some printers had poor experiences with TrueType fonts when they were first available. Although printers and imagesetters have long been able to properly work with TrueType fonts, many printers still avoid TrueType fonts. Some of their reluctance is reinforced by the presence (on the Internet) of many "free" TrueType fonts. Mostly created by hobbyists, these free fonts often have errors that cause the fonts to fail to output at high resolution. (The errors go unnoticed when output on relatively low resolution laser printers.) Try to avoid using one of these free fonts, or submit a test file to your printer to check the font before using it in your project.

truncation (barcodes) The reduction in bar height below nominal height or that normally specified. Truncation reduces the readability of barcodes and should be avoided if possible. Extreme truncation is often required on some products (with minimum printing areas) such as audio cassettes.

two sheet detector (press) A device that stops the press when more than one sheet attempts to feed into the grippers.

two-sidedness (paper) The characteristic reflecting the difference in appearance and printability between the top (felt) side and the bottom (wire) side of a sheet of paper.

Type 1 font (output) PostScript outline fonts come in two distinct formats for PostScript fonts: Type 1 and Type 3 (Type 2 was a proposed font technology that never made it to market). At first, Adobe Systems, Inc. had a monopoly on Type 1 fonts; they used a secret algorithm to produce them and wouldn't let anybody else have the algorithm, so everybody else had to make Type 3 fonts. Type 1 fonts included technology to help them print well at "low" resolutions (like the 300 dpi common on early laser printers). In 1990, under pressure from the introduction of the TrueType format, Adobe decided to publish the secret algorithm so that everyone (meaning competing font manufacturers) could create Type 1 fonts.

Type 3 font (output) Typefaces that are made without Adobe System's Type 1 technology. They often don't print as cleanly and smoothly, nor as fast, as Type 1 fonts on low resolution printers. Now that the Type 1 font technology is known, Type 3 fonts have practically disappeared. (If you use "free" fonts found online, you may encounter one without ever realizing it.) At high resolution (on an imagesetter) Type 3 fonts were indistinguishable from Type 1.

typeface (typography) The raised surface carrying the image of a type character cast in metal. Also used to refer to a complete set of characters forming a family in a particular design or style.

type gauge (typesetting) A ruler calibrated in picas (and sometimes points) used for measuring type. The most common form has small "wings" at the top that allows it to be held firmly against a page or type form. Sometimes called a "pica pole."

typeset-quality (output) Equal to or greater than 1,000 dots per inch resolution. The resolution at which the eye can no longer distinguish the very small dots that make up the printed letters or graphics. Some laser printers output at 1,200 dpi, however, their output is usually short of "typeset-quality" due to splatter and uneven toner disposition. Still, such output looks rather good.

typo (typography) An error in typesetting.

UCA (prepress) Acronym for Under Color Addition. Used with GCR, ink is added to shadow areas to increase color saturation.

UCC (industry) Abbreviation for Uniform Code Council. The U.S. organization which administers the UPC and other retail specifications.

UCR (prepress) Acronym for Under Color Removal. When making a color separation, black ink is used to replace cyan, magenta, and yellow ink in neutral areas only (that is, areas with equal amounts of cyan, magenta, and yellow). This results in less ink and greater depth in shadows. Because it uses less ink, UCR is used to make separations for printing on newsprint and uncoated stock, which generally have greater dot gain than coated stock. See GCR.

uncial (typography) A bookhand used from the fourth to the eighth centuries in Latin and Greek manuscripts.

unit (press) Refers to the combination of inking, plate, and impression mechanisms for each color on a multi-color press. A true two-color press would be said to have two units. Typical presses running process color jobs will have four to six units. (Four units run the CMYK inks, additional units might be available for spot color inks or varnish coatings.) Some multi-unit presses have a feature that flips the sheet to print the other side. A typical configuration might have 4 units for process color and a fifth unit that can either print an additional color or varnish or—using the sheet flipping feature—print black (or colored) ink on the back of the sheet. Some presses are made with eight to ten units so elaborate printing can be done on both sides of the sheet in one pass. Web presses, which are designed to print on both sides of the paper at the same time, may be single unit, or may have four or five units for process color jobs.

UNIX system (system software) A powerful operating system developed at Bell Labs in the 1960's. UNIX is a multi-user, multi-tasking system, since it was designed in the days when one mainframe or minicomputer had many users. Basic UNIX commands are the same on all computers, which is to its credit, but they are quite user-unfriendly. (I would say "user-hostile.") UNIX was written in the language C, so it can be linked to lots of different computers. Hidden deep under the friendly surface of Mac OS X is a UNIX heart. When I was a UNIX system administrator, twenty-plus years ago, I don't think I would have considered it credible if someone told me that one day UNIX would be the basis of a consumer-level operating system.

-up (printing) Printing an item more than once on a sheet to take advantage of the full press size. Business cards, for example, are often printed two- or four-up.

UPC (industry) Abbreviation for Universal Product Code. The standard bar code type for retail products in the U.S. With the changes to the ISBN/EAN now being progressed, the requirement for books sold outside bookstores to have UPCs will be eliminated. A welcome savings to small publishers.

update (industry) A minor improvement in software or hardware, often just bug fixes. These ought to be free, since they are akin to warranty repair service on your car. Updates are indicated by the decimal component of the version number, as in 2.5 replacing 2.3. Hardware updates usually involve driver software, although use of EEPROMs have allowed use of software to make "firmware" updates.

upgrade (industry) A major improvement in software or hardware. Software upgrades occur when the publisher issues a new version of the product and makes it available at a reduced cost to existing owners. An upgrade should include whole new features and possibly a new file structure. Minor fixes are called updates, and ought to be free. Upgrades are indicated by the integer component of the version number, as in 3.0 replacing 2.5. Hardware upgrades involve adding or exchanging existing equipment for newer, usually at full cost. Donald Trump regularly does this with his wives.

upload (communications) Transferring a file from your computer to a remote computer, using a terminal program and a transfer protocol (for example, FTP).

uppercase (typography) Capital letters such as A, B, C, etc. Derived from the practice of placing these letters in the top (upper) case of a pair of typecases by printers when laying out text.

utility (applications) An application that does only a few things and which probably doesn't create a document. Most often utilities are applications which can perform maintenance functions on software and hardware (Norton Speed Disk), enhance workability (Font Reserve), or increase your computing pleasure (screensaver).

UV ink (printing) Ink made without solvent that requires exposure to ultraviolet light to cure (dry). They are used extensively in screen printing and are especially useful for printing on plastics.

UV coating (printing) Similar to a varnish or aqueous coating, but somewhat thicker. UV coating adds a shiny, protective layer over the ink. Might be used on a book cover, but plastic laminate is more durable—and probably a better choice.

vacuum frame (prepress) A device designed to hold flats or other artwork in tight contact with plate material during exposure through use of a vacuum.

vaporware (industry) Any computer product that is announced, and possibly even demonstrated at trade shows, but which is not released on time. Some programs continue for years at this stage. See Microsoft.

varnish (printing) A thin clear coating applied over printing for protection or for appearance. Varnish may be shiny (gloss) or dull (matte). For appearance or special effects, a varnish may even be applied over a plastic laminate. See aqueous coating, UV coating.

VDT (industry) Acronym for Video Display Terminal. See CRT.

vehicle (ink) The fluid component that acts as a carrier for the pigment. Usually a heavy petroleum grease or a vegetable oil. Not a significant source of volatile organic compounds.

vellum finish (paper) A toothy finish (slightly rough) that is receptive to ink. Papers with vellum finish are usually somewhat thicker and exhibit less glare. Characteristics usually desirable for books.

Velobind (bindery) A type of mechanical binding where the "teeth" on a strip of plastic is placed through small holes punched along the binding edge of the sheet. In some machines, the teeth are melted by an electrically heated blade to permanently attach to a matching plastic strip. Other models use a "clincher strip" where the teeth have a rough edge that locks by friction into the matching strip. The excess length is cut off by the binding device that requires no electricity. Finally, in a third Velobind system (for light duty use) the teeth are simply 'bent over' in the matching strip that has a flap that covers a groove where the excess tooth length is hidden. Velobind is not usually appropriate for trade books.

verifier (barcodes) A device used to measure the bars and spaces of a bar code and report on their accuracy in relation to the specified tolerances. Usually actual barcode readers are more forgiving of slightly out-of-spec barcodes than verifiers.

version number (applications) The numbers after a program name which indicate its seniority. The higher the number, the more recent (and supposedly more capable) the program. Version numbers are typically given in ones and tenths, and occasionally in hundredths. It goes like this: "ProGram 1.4.3". The 1 means it's the first version to be released to the public. The 4 means they've added features and upgraded that first version four times, but not overhauled it. The

3 means they've updated this, fourth revision of the first version, three times because of bugs or incompatibilities. Before software is released to the public, its version numbers are prefixed by a letter. Many software developers follow this pattern: D for development (the earliest versions), A for alpha testing (still working on major bugs or features), and B for beta testing (most features working, still smoothing out rough edges), and sometimes FC for "final candidate" stage. When the program is finally prepared for release, it's said to be the "golden master."

video card (display) The card that controls the video display on your screen. You can get different kinds of video cards for different kinds of monitors that display different levels of grayscale or color. Accelerated video cards are helpful if your software requires frequent redraws of huge image files—like Photoshop.

vignette 1. (artwork) An illustration or halftone in which the background fades gradually away until it blends into the unprinted paper. 2. (photography) A lens that does not image the film or digital sensor completely to the edges. Usually observed with low quality or inexpensive lenses. Sometimes a filter is used to create this effect on purpose.

virgule (typography) A diagonal slash or solidus (/). Sometimes called fraction slash.

virtual memory (computers) Virtual memory is the use of space on the hard drive as though it were real RAM memory. Available on most computers, virtual memory creates a memory space greater than the amount of installed real RAM memory and uses the hard drive for the "extra" memory. Virtual memory swaps information between the main RAM and the hard drive as different pieces of information are needed. While hard drive space is less expensive than physical RAM, the memory used on the hard drive is many, many times slower than the internal computer memory. Used properly, virtual memory can allow additional programs to run, but it can also bring a significant reduction in computer performance. If you find that you are frequently using virtual memory, you should consider expanding the RAM in your computer, if possible.

virus (applications) A bit of code written to do such things as: corrupt your operating system, lock you out of your own machine, eat your applications and documents. They can even wipe out the contents of an entire hard drive. Viruses have sub-species named for the way they operate, such as worms and Trojan horses. Viruses travel from computer to computer via disks, email, and the Internet. You don't always know you have a virus; they often have delayed reaction time, so that you continue using the sick application until one day it eats itself. You'll know you have a virus when things start acting funny on your computer. Windows may not function properly, printing might not work right, files may be changed, programs may be "damaged." If you suspect a virus, get virus-protection software and disinfect your hard drive and every single disk in your house and office. Then never open a file on your computer without checking it first. See malware, Trojan horse, worm.

VOC (industry) Acronym for Volatile Organic Compounds. These are chemical compounds that easily evaporate and contribute to smog. Printing can be a significant source of VOCs, primarily due to alcohol used in fountain solutions and solvents used to clean presses. In recent years, printers have converted to products that have reduced or eliminated production of VOCs.

volume (storage) All or part of a hard disk, floppy disk, tape drive or other storage device.

WAN (networking) Acronym for Wide Area Network. Computers that are connected together across long distances and can exchange information. The computers need not be connected constantly to form a WAN. Nobody ever says this word, but people do write it.

warm color (artwork) A color with a yellow or red cast.

washup (printing) The process of cleaning the press (rollers, ink fountain, etc.) either to prepare for using a different color ink or at the end of the day. Some inks "stay open" in the fountain and may be left overnight without drying out or forming a dry skin on the surface.

waterless plate (printing) A printing plate with silicon rubber in the non-image areas that may be run without dampening solution. Most waterless printing must be done on presses designed with special cooling features, as the fountain solution normally provides cooling through evaporation that keeps the rollers from overheating.

web (paper) A roll of paper used in web printing.

web press (press) A press that prints on a roll of paper.

web tension (printing) The amount of pull applied in the direction of travel of the web of paper by the action of the web press.

whiteletter (typography) The generally light Roman and italic letterforms favored by humanist scribes and typographers in Italy in the 15th and 16th centuries. In architecture, comparable to the Romanesque style.

wide/narrow ratio (barcodes) The ratio between the narrow bar and the wide bar widths in certain bar code symbologies

widow (typography) The final line of a paragraph that runs over onto the top of the next column or page. Typographers aren't cruel even though they seem to always want to get rid of widows and orphans. See orphan.

window (interface) The rectangular area that displays information on the desktop and through which you view documents. Every application (and every document) has its own window. You can open or close a window, move it around on the desktop, scroll through it if the document is larger than the window, and change its size and edit its contents. When windows are stacked slightly offset below and to the right of each other, they are said to be tiled. Some programs let you tile the windows automatically.

Wire-O (binding) A binding method using a wire comb (similar to a plastic comb) that fastens the spine through rectangular holes punched in the binding edge of the sheets. The advantage over wire spiral is that the sheets remain parallel. The bindings are durable and are compatible with the punches made by a comb binding punch, although a different device is required to close the Wire-O binding.

wire side (paper) The side of the sheet that was next to the wire as it went through the paper making machine. Opposite to the felt side.

with the grain (paper) Feeding paper through a press with the grain parallel to the axis of the impression cylinder.

wood cut (artwork) An illustration made to look as if it was carved in relief as on a block of wood. Wood cuts were the earliest means of including illustrations with printed material and were commonly used with a letter press. A more modern variant is the cut linoleum block, which produces a smoother print. There is a wide variety of wood/linoleum block styled illustrations in EPS format. Also, there are several decorative typefaces fashioned to have a wood or linoleum block look.

word break (editorial) Breaking a word on a syllable at the end of one line and continuing the word on the following line. In most programs, controlled by the "hyphenation" routine. The best automatic hyphenation uses a dictionary, but most programs also allow an algorithm to break words. The former usually fails to break words (that should be broken) if they are missing from the dictionary. The second method often does a poor job of breaking words, often making inappropriate selections of the breaking point.

word processor (applications) An program that lets you type on a computer and which has additional features including a choice of multiple fonts and styles, paragraph formatting and tab stops. A more powerful word processor has other helpful features such as automatic pagination, footnoting, indexing, style sheets, spell checking, etc. While word processors have added more page layout features, they lack the sophisticated typographic control available in programs designed specifically for page layout. If you want to get the most out of your word processor, read *The Macintosh is Not a Typewriter* by Robin Williams.

word spacing (typography) The space between words. Usually word spacing is adjusted to create justified type.

word wrap (output) The automatic moving of words from one line to the next when the line gets too long for the margins. This is one of the nice features which distinguishes a computer from a typewriter.

work-and-tumble (printing) To print one side of a sheet, then to turn it gripper-edge to back (using the same side guide) to print the other side with the same plate. Artwork on the plate is arranged with both the front and back art such that the resulting sheet will have at least two images of the final piece.

work-and-turn (printing) To print one side of a sheet, then to turn it over, using the same gripper-edge (but opposite side against the side-guide) with the same plate. You will yield two or more copies of the printed piece from each sheet. Both work-and-turn and work-and-tumble are strategies to gain the advantage of using the full press size to produce a job more efficiently. For example, a run of 10,000 printed both sides can be run with 5,000 sheets and a single plate.

work for hire (publishing) An arrangement where an author or illustrator is paid for work performed, but the "authorship" is designated to the individual or company paying for the work. See ghost writer.

workstation (hardware) At one time a desktop computer that was bigger/faster/more expensive than a microcomputer, but the distinction has blurred in recent years.

WORM (hardware) Acronym for Write Once Read Many. A type of optical drive designed to accept a single write, but be reread many times. CD and DVD drives are a successor to the original WORM drives. Optical drives using the original WORM concept are no longer being sold.

worm (applications) A type of malware designed to infect a computer. Similar to a virus, but is operationally different. The user experience is, however, similarly unpleasant. Common with Windows computers. Uncommon on computers using the Mac OS.

wove (paper) Paper with a uniform plain surface with a soft, smooth finish. Contrast to vellum-finish, which has more tooth, and to smooth-finish, which has a harder surface.

wrinkles 1. (paper) Creases occurring during printing. 2. (ink) An uneven surface during drying, most obvious in large, uniformly colored areas.

wrong font (typography) An error during typesetting and marked with "WF" during proofreading. While such errors are usually quite obvious, it can take a sharp eye to spot it if the erroneous font is similar to the correct font. With modern page layout programs, the error most likely occurs when multiple style sheets were created during the design phase and a final font choice was not completely implemented in the style sheet structure.

wristwatch (interface) The shape of the cursor on a Macintosh when the computer wants you to wait while it does something. On recent versions of the Macintosh OS, the wristwatch has been replaced with a spinning, colorful beach ball. Windows computers use a sand clock symbol, an odd choice for up-to-date technology.

Writer's Market (publishing) An annual guide by Writer's Digest Books that details which publishers are looking for what manuscripts, and prices they're willing to pay. The book is generally available in bookstores' reference sections. Writer's Digest Books also offers variations of the guide, specifically targeting the children's publishers, mystery publishers, agents, etc. See Literary Marketplace.

WWW (industry) Acronym for World Wide Web. This is the interconnected set of hypertext (http) servers accessible through the Internet. A common assumption is that the WWW is the Internet. In reality, the Internet existed long before the WWW, but used (and still uses) more primitive tools for access.

WYSIWYG (interface) Acronym for What You See Is What You Get. The state when the display on your computer screen matches the image on the printer page—more or less. Pronounced "wizzy-wig."

x dimension (barcodes) The nominal width of a narrow bar.

X Files (television) A popular science fiction television show. "The Truth is Out There."

x-height (typography) The height of lowercase characters excluding ascenders and descenders. The x-height is extremely important in determining the relative readability of different typefaces. A larger x-height generally makes a face more readable.

xerography (printing) A process for reproduction by the action of light on a photoconductive surface that has been given a static charge with a corona wire. The latent image is developed with a resinous, pigmented liquid or powder (toner) that is then transferred to a substrate (also given a static charge with a corona wire). The substrate then passes through a set of heated rollers that melts and presses the toner, affixing it permanently to the sheet.

XML (industry) Acronym for eXtensible Markup Language. A successor to SGML Shares many of the advantages of SGML, but is easier to work with. XML allows designers and/or programmers to create tags with new meanings, hence the term "extensible." XML was created to allow richly formatted documents to be distributed via the Web.

YA books (publishing) Refers to Young Adult books, which are most often targeted at readers ages 12–18.

yellow (color) Hue of a subtractive primary and one of the four process color inks. It reflects red and green light and absorbs blue light.

zero suppression (barcodes) The method of shortening a bar code number by eliminating 0's.

Resources

THE AUTHOR OF THIS BOOK HAS DEVELOPED a list of over 100 printers who specialize in books. There is also an article discussing the vanity press scam. Additional information and updates to subjects covered in this book will also be posted there. Visit the Æonix Publishing Group website at *www.aeonix.com*. The links page on the website includes links for the following resources:

www.bookexpoamerica.com – **BOOK EXPO AMERICA.** The national trade show for booksellers. It features over 2,000 exhibitors, books in all formats and covering all the categories, plus gifts, toys, stationery, music, and related equipment for booksellers; an educational forum that looks at the business of books from many viewpoints; and a center of rights activity where sales of book rights for other markets take place.

http://hometown.aol.com/catspawpress/ToolShed.html – **THE TOOL SHED IN THE CAT'S BACKYARD** by Patricia Bell, publishing educator and publisher of Cat's Paw Press. You'll find a rather large assortment of the various "tools" publishers need in the course of getting their enterprise started or in finding new avenues. Here you'll find items that were deemed necessary, valuable, or simply useful.

Pat Bell wrote *The Prepublishing Handbook: What You Should Know Before You Publish Your First Book.* Unfortunately, the book has gone out of print. If you're thinking of self-publishing, see if you can find a copy in a library.

www.copyright.gov – THE U.S. COPYRIGHT OFFICE. Here's where you can get basic information and forms to register a copyright.

www.bowkerlink.com or *www.isbn.org/standards/home/index.asp* – **BOWKERLINK** is the **R.R. BOWKER** main page or use the second link to go directly to the ISBN/SAN agency. This is where you can order your ISBNs online with a credit card. Bowker will offer to register you for a SAN (Standard Address Number). Small publishers (with only one office location) don't need to spend the money for a SAN. Bowker will also offer to "store" your ISBN ledger for free for the first year (which implies it won't be free after that). How tough can it be to track ten numbers? Tell 'em you'll track them for yourself.

www.bookmarket.com – The website for **John Kremer**, author of *1001 Ways To Market Your Book*. His web site also has a ton of information for self-publishers.

www.frugalmarketing.com – **Shel Horowitz**, author of *Marketing Without Megabucks: How to Sell Anything on a Shoestring*. Offers assistance in developing publicity releases and marketing programs for authors. Many tips on his website.

www.midwestbookreview.com – **The Midwest Book Review,** Editor-in-Chief Jim Cox is very supportive to self-publishers and self-published works (when they meet all trade book standards). Lots of useful information on the Midwest Book Review website.

www.parapublishing.com – **Dan Poynter's website**. Dan Poynter is the "guru" of self-publishing. His book, *The Self Publishing Manual* is in its 13th edition. His website has lots of information. Some of it is available for free. (He also has a marvelous book about older cats. Follow the link from his main web page.)

www.pma-online.org – **Publishers Marketing Association** (PMA) is a trade association of independent publishers. Founded in 1983, it serves book, audio, and video publishers located in the United States and around the world. Its mission is to advance the professional interests of independent publishers. To this end, PMA provides cooperative marketing programs, education and advocacy within the publishing industry. PMA's membership of more than 3,400 publishers continues to grow. PMA offers worthwhile programs and information to all member publishers, regardless of their size or experience. It also has an excellent newsletter.

www.spannet.org – SPAN, **Small Publishers Association of North America,** is another national organization for small publishers. Similar benefits as PMA. Also has an excellent newsletter. Their website states: The Small Publishers Association of North America (SPAN) is the premier voice of independent publishing and the second largest such association in the world. We concentrate on your need to sell more books. Founded by Tom and Marilyn Ross, who have retired from the organization.

The Ross' recently released their updated version of *The Complete Guide to Self Publishing*, fourth edition. The Ross' book is an excellent resource to the beginning or even more experienced small publisher. The Ross' book and the Poynter book are the backbone of the self-publishing industry.

www.baipa.net – The (San Francisco) **Bay Area Independent Publishers Association** (formerly the Marin Small Publishers Association) was founded in 1979 and incorporated as a nonprofit public benefit organization in 1981. BAIPA is an educational institution and business association dedicated to elevating the art of the independent author-publisher. Membership consists of creative people who, under their press or business names, are involved in various aspects of publishing. Meetings are held on the second Saturday of each month. See their website for details. (If you're not in the SF Bay Area, look for affiliated organizations at the PMA website or for partner organizations at the SPAN website for a publishers organization near you.)

http://publishingcourses.stanford.edu – **Stanford Publishing Courses for Professionals.** Stanford University offers highly tailored publishing courses for working publishing professionals who want to expand their expertise in the business of publishing. Participants are drawn from publishing companies, associations, corporations, and non-profits throughout the U.S. and around the globe. While excellent, these are very expensive programs. I attended the course in 1995 and the "Executive Refresher" course in 1996.

www.quality-books.com/qb_pcip.html – **Quality Books, Inc**. for information on obtaining a PCIP from their service. Quality Books is a distributor who specializes in library sales. As a distributor, they are "friendly" to independent publishers. Also see **Cassidy Cataloguing** at *www.cassidycataloguing.com/PublisherCIP.htm*

Annotated Bibliography

Book Design

Adler Elizabeth W., *Everyone's Guide to Successful Publications,* Berkeley, Calif., Peachpit Press, 1993.

Non-specific to book design, but a good beginners guide to all types of publications with many ideas and good, practical advice. Used copies at Amazon are as little as $4.38.

Adobe Creative Team, *The Official Adobe Print Publishing Guide,* San Jose, Calif., Adobe Press, 1998.

It has good information on the printing process and working with color. Since it's a few years old, it isn't quite up-to-date on color management issues, but it's a start. Although nicely produced, the off-yellow color used throughout the book reminds me of colors I encountered near a chicken coop—a bit of a turn-off.

Bringhurst, Robert, *The Elements of Typographic Style,* Second Edition, Point Roberts, Washington, Hartley & Marks, 1996.

Bringhurst's book is one of the standards for aspiring book designers. While I can't say I agree with everything he suggests, this is a book for more in-depth study of book design and typography. He has an especially interesting section that relates book proportions to the musical scale. I can only recommend this book for those wanting a more advanced study of the topic.

Holleley, Douglas, *Digital Book Design and Publishing,* Elmira Heights and Rochester, New York, Co-published by Clarellen and Cary Graphic Arts Press, 2001.

Richly illustrated with color images, this is an excellent book on book design. Written for professional designers, it focuses exclusively on Quark Xpress. Includes a good discussion of working with color. Useful for someone who wishes to get past the basics, although Adobe InDesign is probably better for laying out books.

Kinross, Robin and Jost Hochuli, *Designing Books: Practice and Theory,* London, Hyphen Press, 2003 (First published in hardback in 1996).

   Could be used as a text book. Written for designers, but even a beginning non-designer can benefit from the discussion of sample layouts. Moves from all-text layouts through to rather complex layouts with multiple images quite rapidly.

Lee, Marshall, *Bookmaking: Editing/Design/Production,* Third Ed., New York and London, W.W. Norton & Company, 2004.

   A widely used text book. Highly detailed, particularly with background aspects of the book business (such as paper-making). Probably a little much for a beginner, but a more advanced publisher may find the book useful.

Lewis, John N., *The Twentieth Century Book, Its Illustration and Design,* New York, Reinhold Publishing Corporation, 1967.

   Nice history of book design, particularly for the evolution of covers.

McLean, Ruari, *Modern Book Design; from William Morris to the Present Day,* Fair Lawn, New Jersey, Essential Books, 1959.

McLean, Ruari, *The Thames and Hudson Manual of Typography,* Reprinted 1997, London, Thames and Hudson, Ltd., 1988. (First hardcover edition, 1980).

   One of the classics of book design and typography. An excellent choice for someone seeking a deeper understanding of typography and book design.

Powers, Alan, *Front Cover: Great Book Jacket and Cover Design,* London, Mitchel Beazley, an Imprint of Octopus Publishing Group, 2001.

   A review of cover design throughout the 20th century, this full color book is filled with examples and cogent discussion of the designs and their evolution throughout the era. Probably more suited to intermediate to advanced designers.

Rice, Stanley, *Book Design: Text Format Models,* New York, R.R. Bowker, 1978.

   This book is an attempt to standardize book designs. Probably of greatest use in a large publishing operation that is producing many books each year. If you want to have your book look like a trade book, this is one place to see fragments and pieces of the more common book designs—but this is no "how to" book. It can be confusing and obscure to a beginner. Book may be a bit hard to find, as it appears to be out of print.

Samara, Timothy, *Making and Breaking the Grid: A Graphic Design Layout Workshop,* Gloucester, MA, Rockport Publishers, Inc., 2002.

   The concept of a grid is very important in layout, particularly in large format books, newspapers, and newsletters. This book explores grid-use quite thoroughly.

Strizver, Ilene, *Type Rules!: The Designer's Guide to Professional Typography,* Cincinnati, Ohio, North Light Books, 2001.

This book is more intermediate than the Robin Williams titles, but a beginner with a strong interest in typography will benefit from it. A good reference work for basic typographic techniques, but there's little information on putting the elements all together to create a comprehensive design.

Sutton, James and Alan Bartram, *Typefaces for Books,* New York, New Amsterdam Books, 1990.

I am often asked, "What's a good typeface for books?" Well, here's the book that can tell you. (The answer is: it depends on your book.) With an extensive section of samples, this will allow you see what typefaces appeal to you and how readable each one might be. I guess it really doesn't give you any easy answers, though.

University of Chicago Press, *The Chicago Manual of Style: The Essential Guide for Writers, Editors, and Publishers,* 15th Ed., Chicago, University of Chicago Press, 2003.

The main reference work used in creating books and other long documents. The subtitle says it all. I can't imagine designing books without my copy of *Chicago* at hand.

Wheildon, Colin, *Type & Layout: How Typography and Design Can Get Your Message Across—or Get In the Way,* Berkeley, California, Strathmore Press, 1997.

This book has been republished by The Worsley Press (Australia) under the title of *Type & Layout: Are You Communicating or Just Making Pretty Shapes.* A must-have book if you're serious about design. Available from Amazon.com.

Williams, Robin, *Beyond The Mac is Not a Typewriter, Berkeley, California,* Peachpit Press, 1996.

An excellent book for beginners to learn about typography. A sequel to Williams' excellent *The Mac Is Not a Typewriter.* Covers the topic in a very approachable manner.

Williams, Robin, *The Mac is Not a Typewriter,* Berkeley, California, Peachpit Press, 1995.

Now available in its second edition. An excellent book for beginners to learn about typography. A very approachable book. Plenty of examples and illustrations. There was also a version for PC/Windows computers, but, for the most part, the book isn't particularly platform specific.

Williams, Robin, *The Non-Designer's Design Book: Design and Typographic Principles for the Visual Novice,* Berkeley, California, Peachpit Press, 1994.

Now available in its second edition. An excellent book for beginners to learn about the general principles of design. As with the other books by Robin Williams, this is a very approachable book.

Williams, Robin, *The Non-Designer's Type Book: Insights and Techniques for Creating Professional-Level Type,* Berkeley, California, Peachpit Press, 1998.

A follow-on book to her earlier books, *The Non-Designer's Type Book* takes a deeper look at good typography at a beginner to intermediate level.

Wilson, Adrian, *The Design of Books,* New York, Reinhold Publishing Corporation, 1967.

Wilson's book is one of the classics of book design. Now available in an inexpensive reprint edition from Chronicle Books with foreword by Sumner Stone (an excellent contemporary type designer). Most suitable for the more advanced student of book design.

Yelland, Jill, *Type Survival Kit,* 3rd Revised Edition, Maplewood, New Jersey, In U.S. by fivedegreesbelowzero press, LLC.; Originally by Press for Success, Perth, Australia, 2003.

Probably of greatest interest to professional designers, this "survival kit" has a lot of information in a small package. Includes E-guage, depth scale, and copy fitting guide for measuring type. Also a good example of a wrapped Wire-O binding with spine text.

General Books on Typesetting, Design, and Production

Anderson, Laura Killen, *Handbook for Proofreading,* Chicago, NTC Business Books, 1990.

Now published by McGraw-Hill. A nice guide to proofreading, including some discussion of typefaces and readability. A good overview for a beginning proofreader.

Andersson, Mattias, et al., *PDF Printing and Publishing: The Next Revolution After Gutenberg,* Torrance, California, Micro Publishing Press, 1997.

Written in support of Adobe Acrobat, Version 3.0, this book gives a good overview of the advantages and disadvantages of digital publishing. Acrobat has moved on (version 7 is now shipping), so the book is somewhat dated now. Appears to be out of print, but used copies are available at Amazon.

Balkin, Richard, *A Writer's Guide to Book Publishing,* (two chapters by Jared Carter), New York, Hawthorn Books, Inc., 1977.

Barker, Malcolm E., *Book Design & Production for the Small Publisher,* San Francisco, Londonborn Publications, 1990.

A complete guide to book production for self publishers using pre-computer methods. Out of print. This was the primary inspiration for the book you are reading.

Basiliere, Pete, *Successful Print Buying: A Guide to Cost-Effective Procurement of Quality Printing,* Paramus, New Jersey, NAPL (National Association of Printing Leadership), 2003.

Written for a printing trade group, this is a fairly comprehensive overview of the factors involved in buying printing. Has a step-by-step approach that walks you through the prepress and printing processes.

Beach, Mark and Eric Kenly, *Getting It Printed,* 3rd Ed., Cincinnati, Ohio, North Light Books, 1999.
This is a classic for the print buyer. Takes you through the printing process (from a buyer's standpoint) and discusses the nuances of dealing with printers.

Beaumont, Michael, *Type: Design, Color, Character and Use,* Cincinnati, Oh., North Light Books, 1987.
This is a designer's view of typography. There's not much content on books, but as an idea generator it's got a lot of examples.

Bove, Tony, Cheryl Rhodes, and Wes Thomas, *The Art of Desktop Publishing: Using Personal Computers To Publish It Yourself,* New York, Bantam Books, 1986.

Brownstone, David M. and Irene M. Franck, *The Self Publishing Handbook,* New York, A Plume Book, New American Library, 1985.

Bryan, Marvin, *Digital Typography Sourcebook,* New York, John Wiley & Sons, 1997.
An in-depth reference to typefaces available in digital versions. Includes a CD with some useful typefaces.

Burke, Clifford, *Type from the Desktop: Designing With Type and Your Computer,* Chapel Hill, NC, Ventana Press, Inc., 1990.

Campbell, Alastair, *The New Graphic Designer's Handbook,* 2nd Ed., Philadelphia, PA, Running Press Book Publishers, 1993.

Collier, David, *Collier's Rules of Desktop Design and Typography,* Reading, Mass, Addison-Wesley Publishing Company, 1991.

Cost, Frank, *Pocket Guide to Digital Printing,* New York, Delmar Publishers division of Thompson Learning, 1997.
Overview of on-demand production and publishing with digital technology. Plenty of illustrations to make this subject somewhat less opaque.

Craig, James, *Basic Typography: A Design Manual,* New York, Watson-Guptill, 1990.
This book is a good basic manual of typography with lot of examples. Previously published as *Phototypesetting: A Design Manual* but was given an updated title when it was updated to reflect the desktop publishing revolution as it was occurring.

Ede, Charles, *The art of the book,* New York, The Studio Publications, 1951.
Interesting discussion of book design as it stood in the post-World War II period.

Fawcett-Tang, Compiled and Edited by Roger, *New Book Design,* U.S. edition, New York, Harper Design International an imprint of HarperCollinsPublishers, 2004.
A typical design book with many photos emphasizing artsy stuff. Interesting to professional designers, but not particularly helpful to a beginner.

Felici, James, *The Desktop Style Guide,* New York, Bantam Books, 1991.

Felici, James, *The Complete Manual of Typography: A Guide to Setting Perfect Type,* Berkeley, California, Adobe Press by Peachpit Press, 2003.
This book is a very detailed, in-depth look at typography. It makes an excellent reference.

Firmage, Richard A., *The Alphabet Abecedarium: Some Notes on Letters,* Boston, Mass., David R. Godine, Publisher, Inc., 1993.
For the hard core typographer and type junkie, this book examines the history of each letter of our alphabet. It is truly remarkable how few changes have been made to the overall structure of the alphabet, though the forms of the individual letters have certainly evolved. Has far more detail on the topic than you could ever imagine.

Gill, Bob, *Forget All the Rules You Ever Learned About Graphic Design, Including the Ones In This Book,* New York, Watson-Guptill Publications, 1981.
An inspirational work for designers, but of no use in helping design a book.

Greenfield, Howard, *Books: From Writer to Reader,* New York, Crown Publishers, Inc., 1989.
A good overview of the publishing process, but suffers from being out of date with respect to current production techniques.

Heller, Steven and Philip B. Meggs, editors, *Texts on Type: Critical Writings on Typography,* New York, Allworth Press, 2001.
This book is an anthology of philosophical discussion of typography. Oddly, the book itself is (in my humble opinion) rather poorly designed with too-small type and overly long lines. If you can get past the reading comprehension problems, it's an interesting book for the type junkie.

Henderson, Bill—Editor:, *The Publish It Yourself Handbook,* 3rd Ed., Wainscott, NY, Pushcart Press, 1987.

Kipphan, Helmut, *Handbook of Print Media: Technologies and Production Methods,* Berlin and New York, Springer, 2001.

Everything you wanted to know about printing in excruciating detail. This 1,227 page book printed on coated stock has 1,227 illustrations (most in color) and 92 tables to help explain every concept. The book was published in association with Heidelberg, one of the world's largest press manufacturers. Nothing was left out. Mortgage your house to buy a copy or sell your car to get a used edition.

Lawson, Alexander, *Anatomy of a Typeface,* Boston, David R. Godine, Publisher, 1990.

This book is perfect for the hard-core type junkie. Others may find it a bit arcane. Discusses the specifics and background of about 30 different typefaces.

McLean, Edited by Ruari, *Typographers on Type,* New York, W.W. Norton & Company, 1995.

An anthology of articles and interviews of book designers, publishers, and other shining lights of the publishing industry from 1895 to 1995. Not much help in designing a book, but does let you know that the same issues continue to confound the industry.

McMurtrie, Douglas C., *The Golden Book; The Story of Fine Books and Bookmaking—Past and Present,* 3rd Ed., New York, Covici and Friede, 1934.

Primarily useful for historical background only.

McNaughton, Harry H., *Proofreading and Copyediting; A Practical Guide To Style For The 1970s,* New York, Hastings House, 1973.

Although somewhat dated, not much changes in proofreading.

Moye, Stephen, *Fontographer: Type by Design,* New York, MIS:Press (subsidiary of Henry Holt and Company, Inc.), 1995.

Essentially, a manual for working with the font creation program, Fontographer. Useful background on typography. Now out of print.

Parker, Roger C., *Looking Good In Print,* 3rd Ed., Chapel Hill, North Carolina, Ventana Press, 1993.

Now in its fifth edition, this book is one of the basic books for learning general design with desktop publishing.

Potter, Clarkson N., *Who Does What and Why In Book Publishing,* New York, A Birch Lane Press Book, 1990.

Sorts out all those editors and stuff.

RJ Communications, *Publishing Basics for Children's Books: A Guide for the Small Press and Independent Self-Publisher,* New York, RJ Communications, 2001.

Written by and for a printing broker, this book is an excellent introduction to publishing children's books. Nonetheless, it *is* a promotional product (available free from the BooksJustBooks.com Web site) and it is tailored to support their marketing efforts.

Shaffer, Joseph Marin and Julie, *The PDF Print Production Guide,* Pittsburgh, Pennsylvania, GATF Press, 2003.

In depth exploration for getting the most out of the Adobe Acrobat PDF file format. Has a good troubleshooting guide. Probably beyond the needs of most readers of this book.

Shepard, Aaron, *Books, Typography, and Microsoft Word,* Los Angeles, Shepard Publications, 2003.

Sold online as an ebook. If you really, *really* must use Word to layout a book, this publication tells you how to do it. It's a good example of the best typography you can get out of Word.

Updike, Daniel B., *Printing Types, Their History, Forms, and Use; A Study in Survivals,* 3rd Ed., Volume 1, Cambridge, Massachusetts, The Belknap Press of Harvard University Press, 1962.

Gives an extensive history of type up to 1800; useful for background only—too esoteric otherwise.

Updike, Daniel B., *Printing Types, Their History, Forms, and Use; A Study in Survivals,* 3rd Ed., Volume 2, Cambridge, Massachusetts, The Belknap Press of Harvard University Press, 1962.

Woolf, Gordon, *Publication Production Using PageMaker,* Hastings, Victoria, Australia, The Worsley Press, 2002.

Although the book is aimed at newspaper publishers, it is a well written guide to using PageMaker at a high level. Gordon Woolf has written (or co-written) several other books on self publishing.

Self Publishing

Bell, Patricia, *The Prepublishing handbook,* Cat's-paw Press, Eden Prairie, Minnesota, 1992.

Spells out some of the hard lessons you need to learn before jumping into the deep end of the self-publishing pool. Has excellent advice from someone who's been there.

Bruno, Michael, editor, *Pocket Pal: The Handy Little Book of Graphic Arts Production,* 19th Edition, Memphis, Tennessee, International Paper Corporation, 2003.

This is *the* reference for those working in graphic arts or printing. It's filled with succinct descriptions of most processes. Actually fits in a pocket. It isn't as impressive as the *Handbook of Print Media,* but it covers much of the same material and you won't need a cart to carry it around.

Hemmerly, Sylvia, *Unlocking the Secrets of Publishing: Simplified Guide to Independent Publishing,* Port Richey, Florida, Inkling Press, 2002.

A basic overview of the publishing process for self publishers. Unfortunately, it screams, "self-published."

Huenefeld, John, *The Huenefeld Guide To Book Publishing,* Revised 6th Ed., Bedford, MA, Mills & Sanderson, Publishers, 2001.

A text book for managing a publishing company. It's hard not to say that a beginner should have this book, but it probably is not reasonable to expect a beginner to want to get this deep in the business aspects of publishing. It's a good reference, though.

Kremer, John, *1001 Ways To Market Your Book,* 5th Ed., Open Horizons, Fairfield, Iowa, 1998.

If this book doesn't give you some ideas on how to sell your book, then I guess your book simply won't sell.

Poynter, Dan, *The Self-Publishing Manual,* 13th Ed., Santa Barbara, Calif., Para Publishing, 2002.

Dan Poynter's classic book is "the user manual" for self publishers. Nobody should be without it.

Reiss, Fern, *The Publishing Game: Publish a Book in 30 Days* (self publishing), *The Publishing Game: Find an Agent in 30 Days* (traditional publishing), and *The Publishing Game: Bestseller in 30 Days* (book promotion), all, Boston, Peanut Butter and Jelly Press.

A newer voice in self publishing, Fern Reiss has written several helpful books for self publishers. Focus is mostly on the business and procedures of self publishing. Worth a look.

Ross, Tom and Marilyn, *The Complete Guide to Self-Publishing,* 4th Ed., Cincinnati, Ohio, Writer's Digest Books, 2002.

This book, along with Dan Poynter's *The Self Publishing Manual,* are the most influential books of the self-publishing community. I suggest you own both books as comparing them will often make concepts and procedures more understandable. Otherwise choose one or the other—that way, you won't have to ask basic questions like, "How do I get an ISBN?"

Ross, Marilyn, *Jump Start Your Book Sales,* Communication Creativity, Buena Vista, Colo., 1999.

　　Contains lots of useful information and ideas to try for selling your books.

Children's Books

Lee, Betsy B., *A Basic Guide to Writing, Selling, and Promoting Children's Books: Plus Information about Self-Publishing,* Brunswick, Georgia, Learning Abilities Books, 2000.

　　Short, only 40 pages, with a definite self-published look, this is more of a primer on the subject rather than a comprehensive discussion. It's only $4.95.

Litowinsky, Olga, *It's a Bunny-Eat-Bunny World: A Writer's Guide to Surviving and Thriving in Today's Competitive Children's Book Market,* New York, Walker & Company, 2001.

　　Written by a former major publisher children's book editor, this book explores the ins and outs of the children's book publishing world. Besides, I love the title.

Mogilner, Alijandra, *Children's Writer's Word Book,* Cincinnati, Ohio, Writer's Digest Books, 1999.

　　One common error for beginning children's book writers is to ignore the concept of "graded" words. (A concept that I find abhorrent, but it is the current educational fashion and can only be ignored at your own peril.) This book will help you keep your manuscript on track at the right age profile for your story.

Pope, Rebecca Chrysler and Alice, *Children's Writer's & Illustrator's Market,* Cincinnati, Ohio, Writers Digest Books, 2004.

　　The children's book version of the annual publishing directory.

Shepard, Aaron, *The Business of Writing for Children,* Shepard Publications, 2000.

　　Apparently self-published, this book is by a successful children's book author.

Suben, Berthe Amoss and Eric, *Writing and Illustrating Children's Books,* Cincinnati, Ohio, Writer's Digest Books, 1995.

　　As the title indicates, focus is on writing and illustration, not on production.

Underdown, Harold D., *The Complete Idiot's Guide to Publishing Children's Books,* 2nd Ed., New York, NY, Cahners Business Information, Inc. 2001.

　　As with most of the "Complete Idiot's Guides," this book is comprehensive with considerable useful information. It looks at children's books both from a self-publishing approach and from the "sell it to a publisher" approach.

Sources for Buying Type

I STRONGLY ENCOURAGE YOU TO BUY TYPE FROM RELIABLE SOURCES, such as Adobe, Agfa, International Typeface Corporation, Bitstream or vendors offering libraries from these type foundries. Watch for good deals—at one time a modest collection of Bitstream fonts was sold with Font Reserve and with some Corel products. Bitstream offers "The Cambridge Collection" for about $200. It has several excellent typefaces for setting books. Individual fonts can be purchased and downloaded over the Internet from *www.myfonts.com* and other online vendors. MyFonts offers good quality fonts from over 200 font foundries including some of the smaller design houses offering some very eclectic selections (mostly suitable for display purposes).

A typeface library created by URW, a European font foundry that went bankrupt some time ago, is available at modest prices from a variety of sources. At one time the complete library (1,000+ typefaces) was given away with Macromedia Freehand. (Adobe offered a collection of 200 fonts from their library with both Illustrator and PageMaker in competition with the Macromedia offering.) So, similar offers may appear again sometime. A 500 typeface sub-set of the URW library is offered at FontSite.com for only $39.95. While some of the URW fonts aren't properly kerned, most are okay—so this represents a good way to get acceptable typefaces for typesetting a book.

Beware of most other "cheap" typefaces. Many are poorly drawn knock-offs of the large font foundry offerings, but are without kerning tables and often have many technical flaws. The Internet is also filled with free and shareware fonts. While some are of excellent quality, most are drawn by hobbyists and beginners learning to use font drawing software. Often these free fonts have serious technical errors—though they may work on your home printer, they often will fail on the high resolution imagesetters used by printers. Not an experience you'll want to have!

Index

About the Author

IN THE EARLY 1980's PETE MASTERSON WAS ASKED to "look into" alternatives for producing Southern Pacific's tariffs (price lists) as the system being used was slow and expensive. The deregulation of transportation was at hand, and the railroad would need to produce revised tariffs more quickly to remain competitive. After some research, Pete identified a computerized system, using proprietary software, that could be used. Made up of a combination of microcomputers, a minicomputer and the company's mainframe, the system was used to convert over 10,000 tariff pages (split into about 100 separate volumes) into digital files. After only 8 months, the system had paid for itself in savings over the previous methods of typesetting and printing. After many years doing railroad things, Pete had discovered a new vocation.

Leaving the railroad in 1987, he then opened a retail printing center featuring desktop publishing, photocopying, and printing. The shop grew rapidly and soon became one of the most successful in the franchise in Northern California. He sold the shop in 1991. (To grow further, the shop needed additional investment beyond his resources.)

Soon after, Pete was hired as general manager of a book-oriented typesetting service in San Francisco. There, he worked with some of the most distinguished book designers while producing books for HarperSan Francisco (New Age and religious imprint of HarperCollins), Addison-Wesley, McGraw-Hill, University of California Press, University of Washington Press, and several other large- and medium-sized publishers.

Next, he was hired by a contractor to NASA at Ames Research Center to supervise the publications and graphics function. The staff of twelve produced over 200 publications each year ranging from high-end full color projects through ultra-short run books produced on a networked Xerox Docutech.

In 1996, the contract was reduced and Pete opened Æonix Pubishing Group to help independent and self-publishers with their projects. He operates his business from El Sobrante California, where he resides with his charming wife of 30+ years and his cat.

Book Design and Production
A Guide for Authors and Publishers
Cover and interior designed by Pete Masterson, composed by Æonix Publishing Group in Adobe Minion with subheads in Myriad and display lines in Rubino Sans using Adobe InDesign CS. Minor revisions and corrections for the second printing were done with Adobe InDesign CS3.

I f you enjoyed this book and would like to order additional copies for yourself or for friends, please check with your local bookstore, favorite online bookseller or visit *www.beaglebay.com* and place your order directly with the distributor.

Please visit *www.aeonix.com* for our list of printers who specialize in books, our extensive list of publishing links, and other publishing information not directly covered in this book.

Feedback to the author may be sent by email in care of the publisher at *info@aeonix.com*.